MW01625528

Process Architecture in Biomanufacturing Facility Design

Process Architecture in Biomanufacturing Facility Design

Edited by
Jeffery Odum
NNE, Durham, North Carolina, USA

and

Michael C. Flickinger
Biomanufacturing Training and Education Center (BTEC),
North Carolina State University, Raleigh, North Carolina, USA

This edition first published 2018

Registered Office
John Wiley & Sons, Inc., 111 River Street, Hoboken, NJ 07030, USA

Editorial Office
111 River Street, Hoboken, NJ 07030, USA

For details of our global editorial offices, customer services, and more information about Wiley products visit us at www.wiley.com.

Wiley also publishes its books in a variety of electronic formats and by print-on-demand. Some content that appears in standard print versions of this book may not be available in other formats.

Library of Congress Cataloging-in-Publication Data is available for this title.

978-1-118-83367-4 (hardback)

A catalogue record for this book is available from the British Library.

Cover Design: Wiley
Cover Image: Courtesy of NNE

Set in 10/12 pt Warnock Pro by SPi Global, Chennai, India.

10 9 8 7 6 5 4 3 2 1

This book is dedicated to our families, who tolerate and endure the consistently inconsistent nature of our schedules and timeline when developing this type of knowledge tool. Your patience is both appreciated and amazing.

Jeffery Odum and Michael C. Flickinger

Contents

Contributors

Josh Capparella
Precis Engineering, Inc.
Ambler
Pennsylvania
USA

Samuel Colucci
Precis Engineering, Inc.
Ambler
Pennsylvania
USA

Daniel Conner
Precis Engineering, Inc.
Ambler
Pennsylvania
USA

Jonathan Crane
HDR, Inc.
Atlanta
Georgia
USA

Robert Dick
Precis Engineering, Inc.
Ambler
Pennsylvania
USA

Beth H. Junker
Bioprocess R&D
Merck Research Laboratories
Rahway
New Jersey
USA

Flemming K. Nielsen
NNE A/S
Gentofte
Denmark

Jeffery Odum
NNE
Durham
North Carolina
USA

Larry Pressley
IPS
Morrisville
North Carolina
USA

Kip Priesmeyer
Kip Priesmeyer & Associates, LLC
St. Louis
Missouri
USA

Hartmut Schaz
NNE
Frankfurt
Germany

Henriette Schubert
NNE A/S
Gentofte
Denmark

Amanda Weko
AGW Communications
Haddonfield
New Jersey
USA

Mark F. Witcher
NNE
Durham
North Carolina
USA

and

IPS
Morrisville
North Carolina
USA

Foreword

Architecture evokes an interaction between fabricated space, form, and human activity. It can be a modest human-scale design or an awe-inspiring monumental work of art. Every architecture has both synergy and limitations. The synergy and limitations can be site, materials, artisanship, climate, government regulations, and certainly resources and budget.

Process Architecture in Biomanufacturing Facility Design is the first volume specifically written for architects, designers, and facility engineers to introduce the unique synergy between bioprocessing–biological medicines–people–facility and the limitations imposed by government regulatory requirements for the safe production of complex human biologics or vaccines by living cells.

Jeffery Odum has carefully assembled contributions into a unique and useful volume that explains the synergy of biopharmaceutical product, a carefully controlled process, equipment, GMPs, and facility design. This work explains how bioprocess architecture—the actual steps in the manufacturing process—affects the design of the manufacturing space and how people operating biological processes within this space produce, purify, formulate, and minimize contamination of potent human medicines.

Training a new generation of architects, designers, and engineers is particularly important now as the types and scale of biological medicines derived from living cells are rapidly expanding. Facilities to manufacture biological medicines for long-term care of chronic or degenerative human diseases are being built today. These new facilities will be capable of producing very large volumes of biologics (10,000 kg of active pharmaceutical ingredient or biologic per year) to meet the anticipated demand for treating hundreds of millions of patients over a long period. Examples of these large-scale biomanufacturing needs are facilities to produce biological medicines for treating diabetes, or Alzheimer's.

These facilities also need to be flexible. Many new facilities will be significantly smaller (reduced footprint, reduced construction costs, reduced utility, water, and solid waste requirements per kg product) to speed construction, reduce financial risk, and to meet requirements of new sites. Examples are new facilities to rapidly manufacture millions of doses of vaccines and adjuvants to meet global pandemics that have recently been built to improve regional vaccine quality. Vaccines can be complex multicomponent biologics manufactured by multistep processes. However, they must be manufactured as new products each year (often combining new antigens) seasonally. These facilities need to be operated efficiently with standardized automation for biologics to be compliant and cost effective.

In contrast, the emerging field of patient-specific biologics and cell therapy requires a completely different process architecture approach to facility design as each product may be derived from cells from a single patient destined to be immunologically or genetically altered and returned to that same patient. Product isolation and facility design to minimize product bioburden contamination are critical for these types of processes.

Process Architecture in Biomanufacturing Facility Design is an important new work to accelerate the design of a new generation of efficient, flexible, cost-effective, and compliant biomanufacturing facilities to deliver life-changing biologics and vaccines to patients worldwide.

Michael C. Flickinger, Professor, PhD

Preface: Why a Book on Process Architecture?

Process Architecture may not be a term that many people are familiar with in the context of this field. Most people relate this term to computer hardware and software design or business processes such as logistics or enterprise systems. The structure of a process system, or its architecture, is viewed as the hierarchy and relationship of a process system's components. But within the global biopharmaceutical industry, the term has a completely different meaning and holds a key strategic place in the list of activities necessary to develop a drug or biologic manufacturing facility design that meets regulatory compliance requirements.

For biopharmaceutical drug and vaccine manufacturing facilities, the relationship between product, process, and facility requires synergy across architecture and engineering disciplines, which is driven by a set of legally binding guidelines known as current good manufacturing practices, or the cGMPs.[1] These strict regulatory guidelines are the foundation of current drug safety practice related to how these specialized facilities are licensed and operated. The critical first steps in developing concepts for proper operational effectiveness and regulatory compliance involve the execution of *process architecture* as presented for the first time in this book.

Biopharmaceutical process architecture involves the integration of process understanding and facility architecture concepts to create a design that meets the regulatory compliance guidelines for the manufacture of quality, safe drug, and biologic products. Product attributes, process parameters, and operational philosophy are defined and integrated via architectural programming activities that will define the solution(s) to an architectural problem, in this case, how best to provide a regulatory compliant manufacturing facility design.

This book focuses not only on the regulatory considerations driving facility design decisions but also on many of the different aspects of design specific to biomanufacturing pilot and production facilities for both biological therapeutics and vaccines. As the global biotech industry continues to grow

1 Code of Federal Regulations, 21CFR, Part 211.

into its fourth decade, more facility design companies will find themselves in a position to both understand and define facilities in a manner where the architectural program not only addresses function, form, economics, and time but also does so in a manner that will withstand regulatory scrutiny from different global agencies. Operational efficiency, flexibility, high-utilization rates, and a stringent focus on quality will take precedent over image. The practitioner leading the programming effort must not only establish the considerations, limits, and possibilities of the design but must also understand the product–process–facility relationship and be able to apply cGMP in all elements of the conceptual design.

The global biotechnology industry focused on drug and vaccine manufacturing will continue to grow; the execution of sound process architectural programming to define compliant facilities will be an essential part of this growth process. The authors contributing to this book are thought leaders within this industry and bring, along with their skills, a broad depth of experience in process engineering, architectural programming, regulatory compliance, facility operation, and aseptic manufacturing. Their willingness to contribute their time and energy to this project is not only greatly appreciated but is also a valuable and unique contribution to the expansion of the industry's body of knowledge.

Jeffery Odum, CPIP

Chapter 1

Introduction to Biomanufacturing

Mark F. Witcher

NNE, Durham, North Carolina, USA

1.1 Introduction

While the book covers designing biopharmaceutical and vaccine manufacturing facilities, this chapter is a brief introduction to biopharmaceutical manufacturing covering the essential elements of the overall biomanufacturing enterprise. Biomanufacturing is very complex and challenging. To be successful, the facility must be designed with a basic understanding of the overall manufacturing enterprise and how it functions. To simplify the discussion, vaccines will be combined with protein products for the purposes of this chapter.

The objective of this introduction is to:

- Identify and describe a biopharmaceutical manufacturing enterprise's three basic elements (process, facility, and infrastructure);
- Briefly describe the constituents used in biopharmaceutical manufacturing processes to appreciate the complexity and fundamental issues of operating the process's unit operations (UOs) within the facility;
- Identify the basic process UOs typically used in biopharmaceutical manufacturing;
- Provide a framework for describing, evaluating, and controlling the facility and process UOs, and;
- Provide an understanding of how the overall manufacturing enterprise is operated and controlled.

In order to understand the challenges of biomanufacturing, we begin by looking at the overall manufacturing enterprise that surrounds and operates the manufacturing facility. As shown in Fig. 1.1, the manufacturing enterprise's myriad of components can be divided into three elements. The facility, process, and infrastructure are integrated and operated in concert to produce the product. As will be discussed, the three elements contain a wide variety of components

Process Architecture in Biomanufacturing Facility Design, edited by Jeffery Odum and Michael C. Flickinger.

required to achieve the overall objectives of the enterprise. Although the distribution used here provides structure to this introductory chapter, many variants or alternative distributions are possible.

In older enterprises, the facility and process are interdependent because the two were designed and constructed at the same time, usually for a specific product. In future enterprises, the process and facility will much less dependent on each other, with the multiproduct facility capable of running a wide variety of different processes. Because the process is not directly integrated with the facility, the processes can be moved in and out of a facility or moved to a different facility depending on manufacturing capacity or logistical requirements [1].

Manufacturing facilities are harder to run than they are to build. For the enterprise to be successful, the facility must be designed to be operated. All three elements must be carefully developed, so they can be integrated to assure the overall enterprise's success. They must work together in order to support and facilitate adequate control of all production activities to assure the efficient production of high quality product over the entire lifecycle of every product produced by the facility. This concept is emphasized by understanding that conformance lots or PPQ (process performance qualification) batches are more than just a test of the process [2]. Conformance lots are a test of the enterprise's ability to operate the process. Many conformance lot issues are encountered because the staff, part of the infrastructure, are not adequately trained and do not have the experience to execute the many tasks and activities required to successfully run the process and facility.

Starting from the beginning, the basic constituents of biopharmaceutical processes are cells, nucleic acids, and proteins.

1.2 The Basic Constituents of Biopharmaceuticals

Biopharmaceutical manufacturing uses relatively simple processes composed of UOs employing relatively simple pieces of equipment as compared to other industries, particularly those in the chemical and petrochemical industries. However, the constituents within the UOs are many orders of magnitude more complex. The purpose of this section is to provide a very brief introduction to the formidable challenges of managing and manipulating these constituents.

A defining element of biopharmaceuticals is the use of cells to produce the product. The product is most frequently a protein, but the product can be the cells themselves used for cellular therapies or in some cases the product can be genetic material manufactured in large quantities to modify or control genetic constructs and control mechanisms within the patient.

The cell is the largest and most complex element. The growth and characteristics of the cells are defined primarily by the genetic information stored in

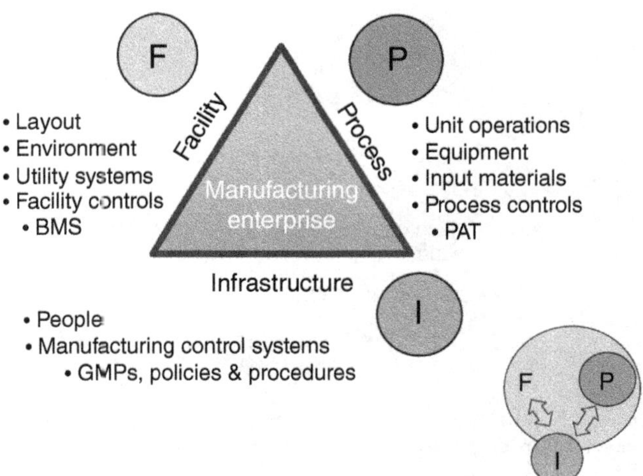

Figure 1.1 The manufacturing enterprise is composed of three elements. Basically, the enterprise operates the process within the facility under the control of the infrastructure. The process and the facility are separate entities that can be run independently with the infrastructure providing the interface between the two. As shown in the side figure, the process is contained within the facility. The infrastructure element resides partially within the facility because the infrastructure is composed of both facility-specific and companywide, multifacility components.

the cell's DNA. The machinery of the cells that use the genetic information is operated by proteins. The discussion will begin by describing the simplest element, proteins. Most pharmaceutical products are proteins manufacturing by the cell under the control of genetic constructs inserted into the cells using recombinant technology.

1.2.1 Proteins

Biopharmaceuticals are proteins. The product is created by cells in a complex environment of many complex biological molecules (carbohydrates, nucleic acids, lipids, and proteins) and cellular processes required for cell growth and product manufacturing. The discussion begins with an overview of proteins and their structure and behavior.

The fundamental structure of proteins is shown in Fig. 1.2. Proteins are polymers of 20 specific amino acids shown in Fig. 1.3 assembled by the cell according to the genetic code contained within the cell.

The sequence of amino acids forms the primary structure of the protein. Depending on many different structural and environmental factors, the amino acid chain is folded into higher order structures. These higher order structures can be defined as follows:

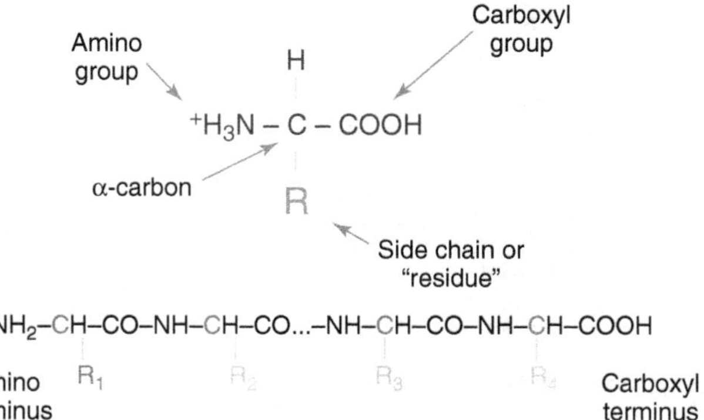

Figure 1.2 Proteins are a polymer of the 20 amino acids listed in Fig. 1.3. They are typically between 100 and 1500 amino acids in length. The term *peptide* refers to amino acid polymers of <25 amino acids.

Amino acid	Abbrev.	Amino acid	Abbrev.
Alanine	Ala (A)	Methionine	Met (M)
Cysteine	Cys (C)	Asparagine	Asn (N)
Aspartic acid	Asp (D)	Proline	Pro (P)
Glutamic acid	Glu (E)	Glutamine	Gln (Q)
Phenylalanine	Phe (F)	Arginine	Arg (R)
Glycine	Gly (G)	Serine	Ser (S)
Histidine	His (H)	Threonine	Thr (T)
Isoleucine	Ile (I)	Valine	Val (V)
Lysine	Lys (K)	Tryptophan	Trp (W)
Leucine	Leu (L)	Tyrosine	Tyr (Y)

Figure 1.3 While the Earth's natural environment contains hundreds of different amino acids, all life forms use only the 20 amino acids shown.

- *Primary*: the linear amino acid sequence of the polypeptide;
- *Secondary*: the localized folding of the primary structure into substructures sometimes called *helices*, *strands*, and *sheets*;
- *Tertiary*: the combining of the secondary spiral and flat elements to form larger three-dimensional structures;
- *Quaternary*: combining of tertiary structures to form a much larger complex three-dimensional structures.

The structures determine the protein's properties, including its safety and efficacy as a therapeutic product.

Additional protein complexity comes from post-translational modifications (PTMs) added to the protein during and after the four structural features described earlier are formed by the cell. PTMs change the behavior of the protein in solution and affect their therapeutic impact, safety, and efficacy in a wide variety of ways. The four structural features and PTMs along with the various altered product forms resulting from degradation, alternations, misincorporation, and aggregation combine with various impurities and contaminates from the manufacturing processes to determine the complex set of critical quality attributes (CQAs) that define the product's overall quality target product profile (QTPP) as defined in ICH Q8 (R2) [3]. The objective of the manufacturing process is to consistently produce a product described by the product's QTPP defined during development, tested in preclinical testing; and demonstrated to be safe and effective by clinical trials. For more details on the definition of biological products, ICH Q6B and ICH Q5E should be consulted [4,5]. Biochemistry text books can be consulted for more information related to the structure, composition, and behavior of proteins [6–8].

All cells use thousands of proteins, called *enzymes*, to operate the machinery required to perform the myriad of internal maintenance functions, reproduce, and manufacture the product protein [9]. The production of proteins within the cell is carried out by complex metabolic pathways based on the genetic information contained within the cell. The genetic information is stored in the cell's *DNA* (deoxyribonucleic acid) formed by polymers of nucleic acids described later.

1.2.2 Nucleic Acids (DNA and RNA)

DNA provides a stable and efficient mechanism for storing and maintaining genetic information critical to the cell's ability to consistently and efficiently replicate without losing the ability to produce the target protein over many generations. DNA has a complex double helix spiral structure formed by two complimentary sequences of nucleic acids. The shape of the double helix can be found in many text books on the subject.

The management of DNA within the cell is a very complex and poorly understood sequence of internal processes. DNA has six different structural features much like the four structural features of proteins, but much more complex involving other supporting protein molecules to assist with storage, protein expression, and duplication. Further, DNA is subject to additional complex modifications involved in many cellular regulatory functions controlling how the DNA is used for cellular functions [10].

RNA or ribonucleic acid is a second nucleic acid very similar to DNA used by the cell for a variety of functions. The difference between RNA and DNA is the substitution of the nucleotide uracil for thymine in the sequence. While DNA is stable for information storage, RNA is more biologically active, less stable, and used in a number of important biological functions to transfer information and regulate protein production. RNA is also the basis for some biopharmaceuticals

Central dogma:

- Replication - DNA ⇒ DNA
  - Provides for cell duplication, but retains genetic information
- Transcription - DNA ⇒ RNA
  - Transfer of information from genetics to protein manufacturing mechanism
- Translation - RNA ⇒ Protein
  - Proteins from genetic information

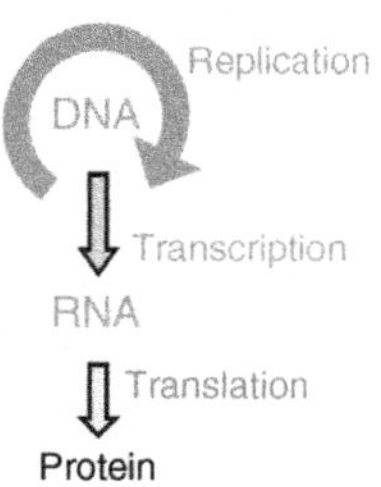

Figure 1.4 The central dogma describes the *replication* of DNA that allows the genetic information to be passed to future generations; *transcription* converts the DNA sequence code for the protein into RNA for transmission to the *translation* mechanism that converts the RNA sequence code into the amino acid sequence that determines the protein's primary structure.

not covered in this introduction. However, the same basic concepts apply to manufacturing biologically active RNA therapeutics.

Nucleic acids and protein expression are connected by the central dogma that describes the three steps required to support cell reproduction and protein production. The central dogma is shown in Fig. 1.4 [11].

Protein expression within the cell is controlled by a wide variety of feedback and feed forward control mechanisms based on the central dogma. Many of these control mechanisms have not yet been identified and much work remains by the scientific community to better understand the myriad of mechanisms used by the cell to control protein expression.

The use of living cells to manufacture the product is the primary distinguishing feature of biopharmaceuticals. While peptides can be manufactured by chemical synthesis, long amino acid sequences and the resulting structural forms and PTMs required for therapeutic safety and efficacy can only be achieved using living cells.

The history of cells goes back more than 3.5 billion years. The evolutionary history of cell development has resulted in extraordinary capabilities and complexity. The following is very brief introduction of the basic features required for the cells to make therapeutic proteins and vaccines.

1.2.3 Cells

In nature, there are virtually an unlimited number of different types of cells that possess a wide variety of capabilities, behaviors, and properties. The field of microbiology is very diverse and covered by many good books [12–14]. Biopharmaceutical manufacturing typically uses and manages one of two types of cells. For the purposes of understanding the fundamental principles associated with biomanufacturing, only two of the myriad of possible cells will be covered here.

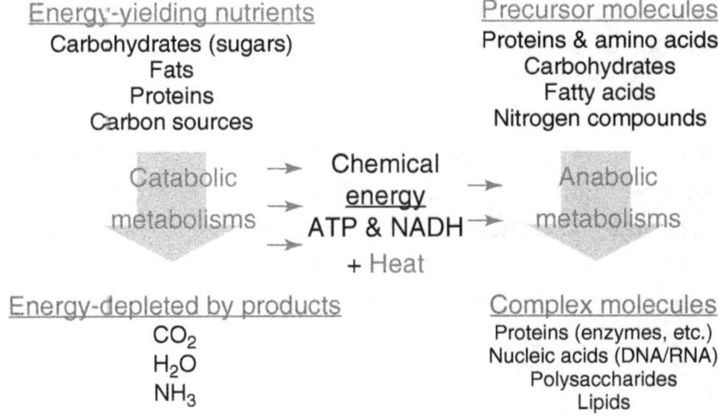

Figure 1.5 All cells operate using catabolic metabolisms that generate energy for the cell by converting energy-containing raw materials into energy-depleted by-products. Anabolic metabolisms use the energy to convert other raw materials into complex molecules required to operate the cells metabolisms and support cell reproduction. The product protein is one of the complex molecules.

The first cell type is prokaryotes, which are relatively simple cells and grow very rapidly. Many prokaryote cells double in number every 30–60 min [15,16]. *Escherichia coli* is one of the types frequently used to make recombinant proteins. Prokaryotes are grown in fermenters designed to meet the requirements for supporting rapid growth.

The second type is eukaryote cells derived from plants, animals, and insects. The most frequently used eukaryote cells are mammalian cells [17]. Of the mammalian cells, Chinese hamster ovary (CHO) cells are frequently used for manufacturing therapeutic proteins. Eukaryotes have roughly a thousand times the internal volume of a prokaryote cell, although the sizes of both types of cells can vary over a wide range. Eukaryote cells are grown in cell culture bioreactors.

Both types of cells operate using metabolisms driven by a large number of specialized protein called *enzymes* necessary to make the myriad of required reactions work efficiently [9]. The metabolisms have evolved over billions of years and are extremely diverse and highly efficient in performing their functions. Taking a simplistic view, the metabolisms can be divided into two categories. Catabolic metabolisms shown in Fig. 1.5 on the left take energy containing raw materials and break them down to provide energy to the anabolic metabolisms on the right. Anabolic metabolisms take complex raw materials and produce the complex molecules required for cell functions and reproduction.

As will be explained, growing the cells and producing the product are determined by the manufacturing process's ability to manage the inputs and outputs of the two metabolisms. Adequate raw materials must be supplied to the cells along with the removal of by-product waste materials. An environment that supports the cell's metabolic enzymes with the correct operating temperature by removing or adding heat to the cell's environment is required. Thus, the necessary energy and mass transfer conditions and capabilities must be provided by the fermenter or bioreactor.

With the primary constituents identified, the discussion turns to the processes and equipment used to manage the cell's internal processes to grow the cells, produce the target protein, and then purify the product to provide a therapeutic product with the required purity, potency, and activity. The processes and equipment will be described in the context of the manufacturing enterprise are shown in Fig. 1.1.

1.3 Enterprise Element #1—Manufacturing Processes

Expanding Fig. 1.1, the components of the enterprise's process element are shown in Fig. 1.6.

1.3.1 Process—Unit Operations

The distinguishing feature of biomanufacturing is the growing of cells to produce the biopharmaceutical or vaccine. Biopharmaceutical processes or UOs use relatively simple equipment to manage and protect the complex constituents described previously. Some specific requirements of the cells and proteins in the process are poorly understood. The complexity of the cells,

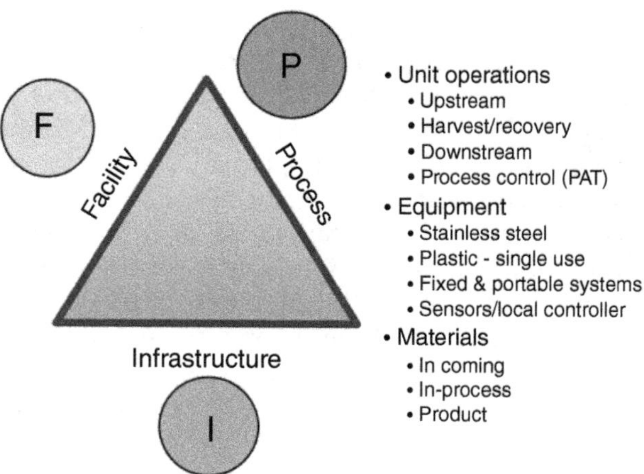

Figure 1.6 The process element of the manufacturing enterprise includes the unit operations required to make the product along with the equipment implementation of the process and the raw and in-process materials required to make the product.

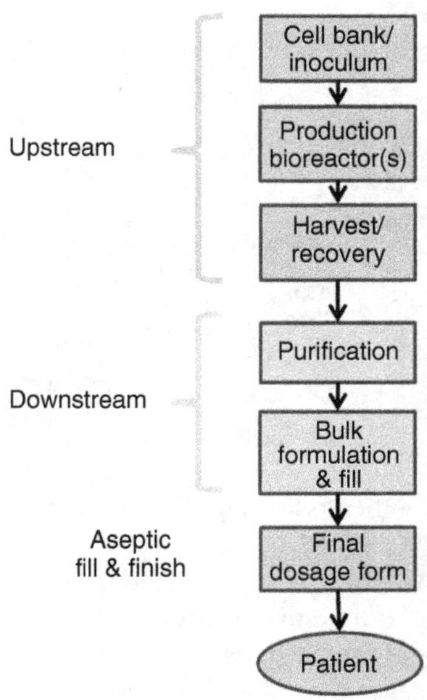

Figure 1.7 A typical biopharmaceutical process is initiated from a cell bank containing cells whose DNA have been modified using recombinant technology into an inoculum train composed of a sequence of small bioreactors. After a sufficient number of cells are grown and the product produced in the main production bioreactor, the cells and product are separated in a harvest or recovery step. The product is then purified to an appropriate level for eventual formulation into a final dosage form for administration to the patient.

proteins, and other constituents contributes to the uncertainty and risk for managing the processes for manufacturing therapeutic products.

Biopharmaceutical processes generally follow a relatively straightforward sequence shown in Fig. 1.7.

Good manufacturing practices (GMPs) require that all process steps be separated from each other using an effective validated segregation strategy [1,18,19]. The segregation strategy must also keep all products and processes within the facility separated from each other.

As shown in Fig. 1.7, the process begins with breaking a vial from the cell bank to begin growing a sufficient number of cells required to manufacture the required amount of product. These initial processes are called the *upstream processes*.

1.3.2 Upstream Processes—Inoculum through Production Bioreactor

Most upstream processes fall into two categories: fermentation and cell culture. Rapidly growing, very hardy organisms such as prokaryotes and yeast require bioreactors that can support large energy and mass transfer requirements for growing the rapidly dividing organisms to high final cell densities. The second category is cell culture processes necessary to grow fragile, slow growing eukaryote organisms such as CHO cells. These two categories share many basic principles and characteristics, although the major differences between the two will be identified.

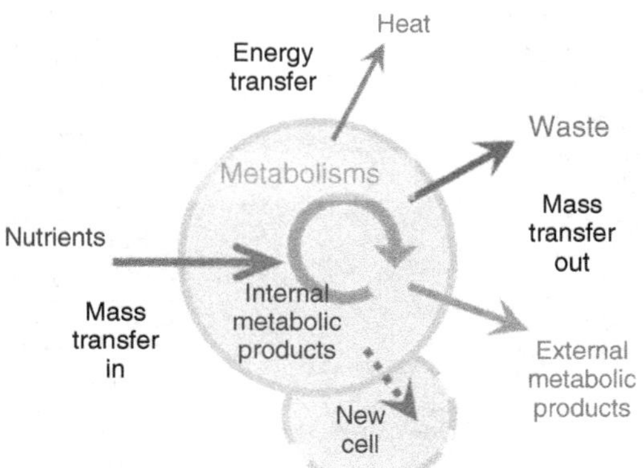

Figure 1.8 An overview of the mass and energy transfer required for the cell metabolisms to function correctly.

The following discussion will focus primarily on cell culture processes. Recent advances over the past 20 years have advanced cell culture to the point where it is used for many commercial biopharmaceutical produces because of inherent advantages over fermentation processes. The primary product difference between the two processes is that fermentation of prokaryotes cannot produce products with PTMs required for therapeutic activity. As will be explained later, PTMs may be critical to the safety and efficacy of the product. Cell culture processes not only provide the necessary PTMs to the product's structure but also excrete the product into the media greatly simplifying the separation of the cells from the therapeutic protein. Fermentation processes will remain important for manufacturing nonprotein products, large-scale commercial manufacturing of nontherapeutic proteins such as commercial enzymes, and making the few therapeutic proteins not requiring PTMs.

The overall internal cellular processes required to grow the cells are shown in Fig. 1.8. The metabolisms within the cells were summarized in Fig. 1.5. The upstream bioreactors must provide the correct operating environment to support adequate mass transfer of nutrients including oxygen into the cell and removal of waste products shown in Fig. 1.5 without damaging the cell's outside surface membrane owing to shear forces. In addition, the bioreactor must provide for the removal of energy generated by the cells to maintain the correct internal cell temperature for the metabolic enzymes to function properly. Energy removal for cell culture bioreactors is relatively small; however, in the case of fermentation, energy removal of heat from the cells and work added from agitation of the viscous, high cell density media is a significant issue.

The upstream processes are typically initiated by opening a 1-mL cryogenically preserved vial or ampule and aseptically placing roughly one million cells in a T-flask bioreactor for cell culture or a 1-L shake-flask for fermentation. The shake-flask is grown for approximately 12 h and then used to inoculate the production fermenter. Smaller intermediate seed fermenters are

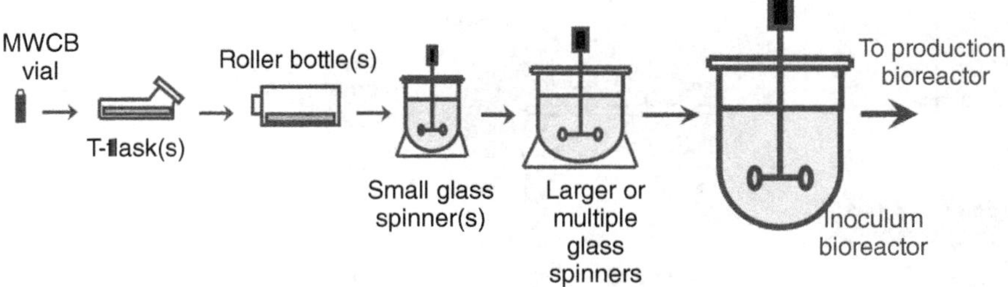

Figure 1.9 Upstream cell culture unit operation sequence, including initiation from a cell bank through to the final inoculum bioreactor. The inoculum train can vary widely depending on the sequence of small-scale bioreactors used to expand the culture. A key element of the inoculum train is to maintain the cell density above the minimum required for growth and below the maximum cell density supported by the bioreactor's mass transfer capabilities. Fermentation seed sequences are much simpler and faster because of the very rapid growth rate. No minimal cell density is required for propagating prokaryote cells.

sometimes used to provide more efficient use of equipment by increasing the turnover rate of the high asset value production fermenter. In cell culture, the T-flask is grown a few days, and then the cells removed and placed into a slightly larger bioreactor. For cell culture, the cells must be maintained at a cell density between approximately 10^5 and 10^7 cells per milliliter for the culture to remain viable. Maintaining the necessary cell density range requires splitting the culture into successively larger bioreactors every few days to keep the cell density within the operating range while increasing the total number of cells. One possible inoculum sequence is shown in Fig. 1.9. Typical inoculum development sequences can take 3–5 weeks depending on the total volume of cells required to inoculate the production bioreactor.

Fermenters and production bioreactors support the growth of cells in a stirred tank reactor that provides agitation while adding raw materials and oxygen. The bioreactor must also provide various sensors, inoculum and media addition, and removal ports depending on the operating requirements and sequence of the culture. Specifics on fermenter design can be found in Stanbury [20] and cell culture bioreactor design can be found in Lubiniecki [21] and Ozturk and Hu [22]. The basic elements of a fermenter and bioreactor are shown in Fig. 1.10.

Cell culture bioreactors range from very small (<1 L) to up to 15,000 L for large stainless steel (SS) production bioreactors. Disposable plastic single use bioreactors (SUBs) range up to 2000 L, although functionally most SUBs are 500 L or less because of mass transfer and agitation limitations. SUBs are frequently used in inoculum trains for large SS bioreactors. Fermenters can be up to, and in some cases larger than, 100,000 L for large-scale production of enzymes and antibiotics.

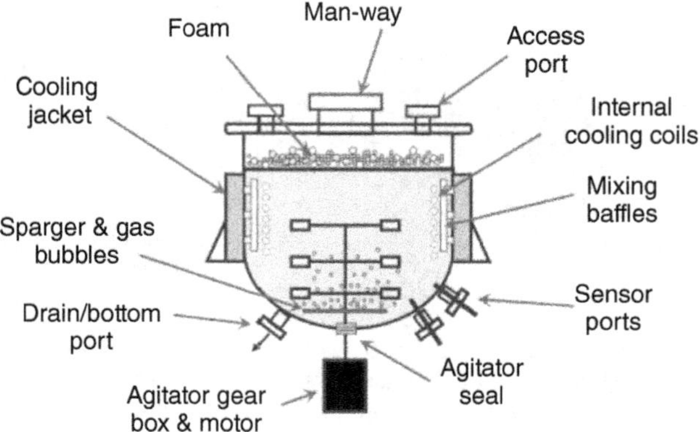

Figure 1.10 The key features of a production bioreactor required for operating and managing the mass and energy balances of the cell metabolisms shown in Figs. 1.5 and 1.8.

After the final cell density is reached and the product protein produced to the final titer, the next step is to remove the product from the cells. For cell culture, the cells excrete the product protein into the media. The media also contains a wide variety of leftover media components, cell waste products, and some cell debris and metabolic enzymes from cells that have died and broken open. For fermentation, the product remains within the cell as insoluble inclusion bodies.

1.3.3 Upstream Processes—Harvest and Recovery

For cell culture processes, the cells and media containing the product protein are usually separated by some combination of normal filtration, tangential flow filtration (TFF), or centrifugation. Cell culture process can use all three types of UOs; however, fermentations typically are limited to centrifuges because of high cell densities and high solution viscosities from the fermenter. TFF is also widely used in protein purification for a number of different operations and will be covered in the purification section. The most important objective of harvest and recovery operations is to preserve as much active product as possible by minimizing degradation of the product.

1.3.3.1 Normal Filtration

Normal filtration consists of passing the solution containing suspended cells, proteins, or particles through a mesh that retains the larger particles or cells while allowing the solution and smaller particles or proteins to pass through. A wide assortment of filters with different pore sizes and filter material are available for the diverse filtration applications. Large surface area filter presses are available to remove the cells from large production bioreactors. Normal filtration is used in a wide variety of applications ranging from removing unknown and undetected viral threats by viral filtration to protect UOs from large particles that could interfere with the process's performance.

1.3.3.2 Centrifuge

Centrifugation is the accelerated settling of cells, proteins, or particles suspended in solution by high gravitational forces induced by rapid rotation of the solution. Centrifuges are used for separating cells from media and after protein precipitation in some large-scale processes. Disk stack and tubular centrifuges are the most common type used in manufacturing [23].

The primary advantage of centrifuges is that they do not add anything to the solutions and recently developed designs can be cleaned and sanitized effectively. Centrifuges come in a variety of sizes. Some disposable centrifuges have recently been developed and can be used for small- and medium-scale applications.

For cell culture processes, centrifuges can be used as cell separators to remove the cells from the media. The cells may be returned to the perfusion bioreactor, so they can continue to produce product or discarded in the case of cell harvesting in an operation similar to the normal filtration operation described earlier. After the cells are removed, the clarified solution containing the product is sent directly to purification or frozen for later purification.

In the case of fermentation, centrifuges are typically used to remove water from the media to produce a thick cell paste to assist with breaking the cells open to release the product (cell disruption).

1.3.3.3 Cell Disruption

After prokaryote fermentations, the target product protein is retained within the cells as insoluble inclusion bodies. The cells must be broken open to remove the product from the cells. A variety of cell disruption devices such a colloid and ball mills borrowed from other industries are used to grind up and/or dissolve the cell walls to make the target protein available for solubilizing. The milling process may be supplemented by freezing, detergent additives, or other UOs to enhance the cell destruction and increase the efficiency of the disruption processes. Usually, the solution is centrifuged a second time, then filtered to remove the insoluble cell debris from the solubilized protein. Once the target protein is solubilized, it can be renatured in the appropriate buffer solution into an active form for further purification in the downstream processes.

1.3.4 Downstream Processes

A simplified downstream process for purifying a monoclonal antibody (mAb) is shown in Fig. 1.11 [24]. Most downstream processes contain similar UOs

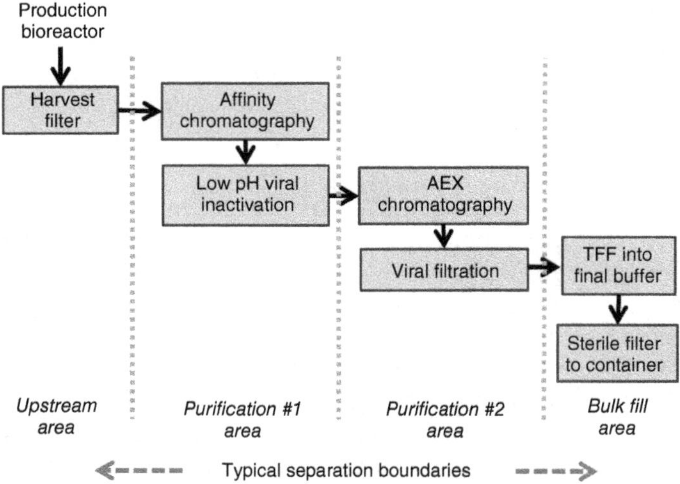

Figure 1.11 Typical monoclonal antibody (mAb) purification process. The mAb is selectively removed by the AC step. Protein A is usually the active ligand that captures the product mAb. The mAb is eluted in a low pH buffer that is then held under controlled conditions to inactivate viral contamination. The solution is then passed through an anion exchange (AEX) chromatography column to remove negatively charged contaminates, including any contaminating protein A from the affinity column. After viral filtration, the mAb is placed into the correct buffer for cryogenic storage in a sterile container. Additional steps may be added depending on the nature of the impurities and contaminates and the purity requirements of the final product.

depending on the extent of purification required and the nature and type of the upstream processes. After filtration, usually by normal filtration, the remaining media solution containing the target protein along with contaminating impurities pass through the sequences of UOs shown.

Typical downstream processes are composed of two to four chromatography columns, one or more TFF processes, and two orthogonal viral clearance steps. *Orthogonal* refers to operating by different physical principles.

1.3.4.1 Viral Clearance

Viral contamination can occur from a variety of sources. The possibility of viral contamination should be minimized using nonanimal sourced raw materials, particularly the cell culture media, and using GMPs to minimize the likelihood of operator or environmental contamination during processing. Viral clearance through inactivation or removal is possible with a variety of UOs. Figure 1.11 shows a low pH inactivation step from the elution buffer from the first chromatography step. The second viral removal step by normal filtration is shown just before final formulation. Validation of the viral removal steps using a portfolio of three to five model viruses is a key process validation requirement [25,26]. Very high concentrations of the model viruses are used to challenge the viral removal UO in a process validation exercise to show that it is capable of removing unknown or undetected viral threats to assure product safety.

1.3.4.2 Tangential Flow Filtration

The basic components of TFF are shown in Fig. 1.12. A solution containing cells, particles, proteins, and/or ions is separated by forcing the smaller components through a membrane using high pressure generated by a pump. TFF can be used to provide separation as an ultrafiltration process, or used to concentrate solutions as a diafiltration step. For ultrafiltration, the contents of the feed tank are pumped under high pressure through the TFF unit and the two output solutions sent to separate vessels. Either the retentate or permeate may contain the product. For diafiltration, the retentate containing the product is returned to the feed tank as the liquid is removed through the permeate side. One buffer can be exchanged for a second buffer using diafiltration by first removing most of the first buffer, then adding the second buffer to flush out the first buffer in a sequence designed to remove the first buffer to the required levels.

The particle separation process is controlled by a membrane selected to provide the desired separation. Many different membrane porosities and material compositions are available. Figure 1.13 shows the different membrane pore sizes for the broad classifications of TFF UOs.

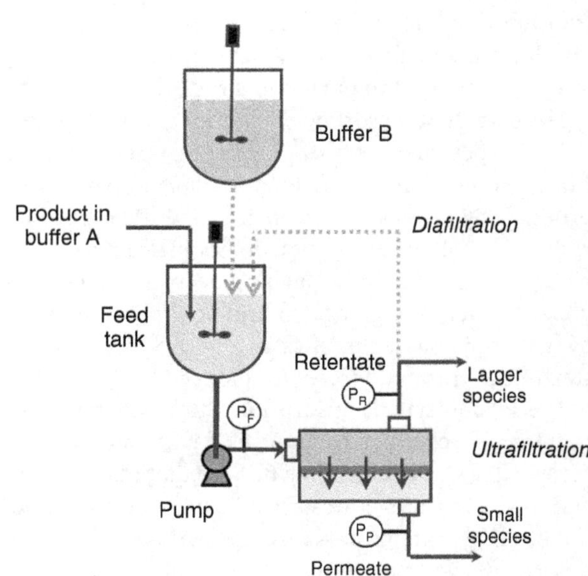

Figure 1.12 Tangential flow filtration (TFF) process is a pressure-driven process. The TFF unit is supplied by a pump fed from the feed tank. The permeate contains the liquid and smaller particles or ion species that have passed through the membrane. The retentate, which can be removed or returned to the feed tank, contains the particles or ion species retained by the membrane.

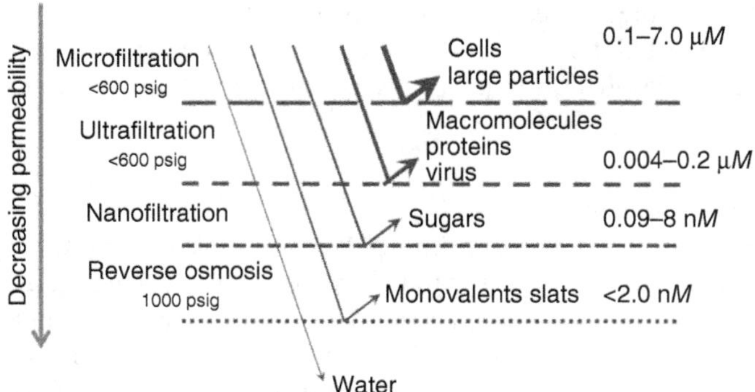

Figure 1.13 Different types of membranes used in TFF provide a powerful tool for selecting particles or ion species in solution and can be used for a wide range of separations from harvesting cells from media to purifying sea water through reverse osmosis membranes capable of removing monovalent salts.

TFF is a versatile tool for separating and purifying proteins and other species. Most frequent TFF applications in biomanufacturing include exchanging buffers and concentrating product solutions between chromatography steps or for storage [27].

1.3.4.3 Chromatography

Chromatography uses the diverse properties of proteins described earlier to separate the therapeutic protein from other contaminating proteins and molecular species. Its processes are very "protein friendly" because they use solution conditions that preserve the important protein properties required for the product to remain a safe and effective therapeutic. It can be highly selective and used to purify the product to very high purities.

The overall process flow of column chromatography is shown in Fig. 1.14.

Two types of chromatography are used. Size exclusion chromatography (SEC), sometimes called *gel filtration*, separates the species, including the product, by size as shown in the simplified chromatogram in Fig. 1.15. The resin and proteins in solution do not interact other than the physical hindrance owing to size. SEC is valuable because it can separate undesirable product variants such as aggregates and fragments that have the same charge and hydrophobic characteristics as the product. It is frequently used as the last step in the chromatography sequence.

The second type is adsorption chromatography where the proteins in solution interact chemically because of active sites on the proteins in solution with active sites on the fixed resin. By manipulating the properties of the protein and the resin's active sites, specific proteins can be adsorbed on the resin while nonbinding proteins pass through the bed. Adsorption chromatography can be described using the simplified chromatogram shown in Fig. 1.16.

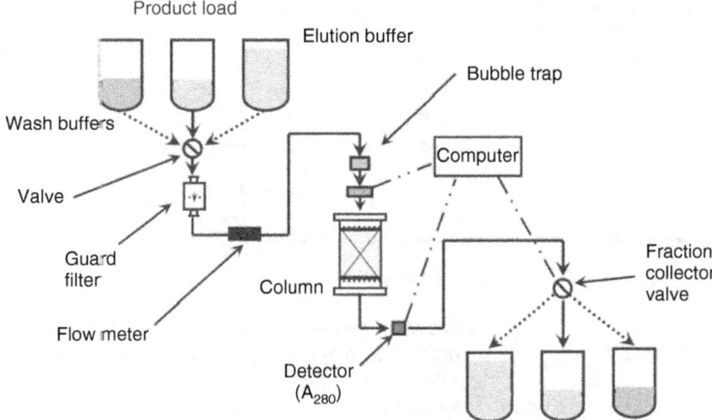

Figure 1.14 Chromatography process flow diagram—the column is fed through the required sensors from a bank of buffer feed tanks for preparing the column resin, loading the feed material, eluting the product from the column, and cleaning and washing the column after use. After passing through a detector that measures protein concentration, the outflow of the column is sent to a variety of vessels appropriate to receive the material. One of the vessels receives the eluted product depending on the time the product is scheduled and detected to elute from the resin bed.

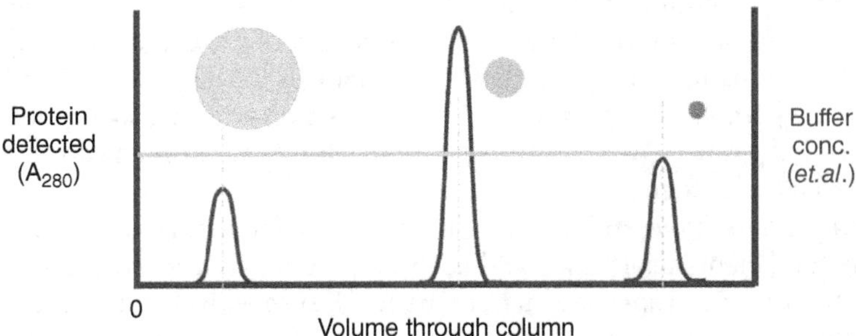

Figure 1.15 Basic principle behind size exclusion chromatography (SEC). The larger species come out first because their large size limits them to a much shorter pathway through the resin. The small species have a longer effective path length because they can go through the numerous smaller channels in the resin and thus take longer to travel the bed's length. In the example shown, the product is the center peak while the first peak might be aggregates of the product and the last peak a product fragment caused by degradation of the product during upstream processing.

The following three types of adsorption chromatography are most frequently used:

- *Ion-Exchange (IEX).* The charge of the protein interacts with charged sites on the resin bed. Depending on the type of resin selected, the proteins can be bound to anion exchange (AEX) resins if the protein has a negative charge

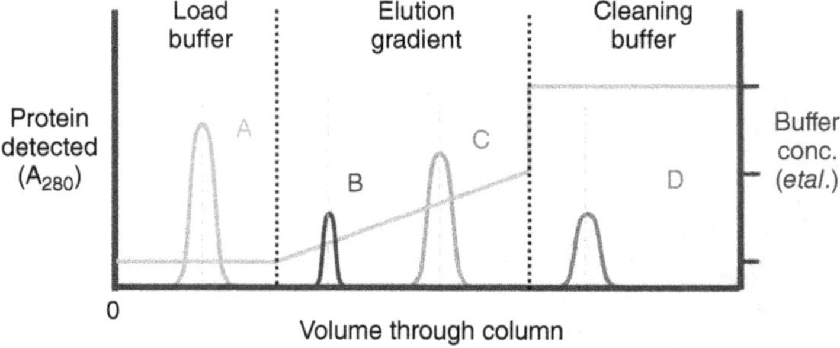

Figure 1.16 Illustrative example of the options for separating protein species in complex solutions using differences in reversible adsorption on a fixed resin chromatography bed. In this example, the column is loaded at time = 0 with four different protein species (A, B, C, and D). Protein species A passes through the column, whereas B, C, and D are bound. The buffer pH and/or counter-ion concentration flowing through the column is changed to form a gradient as shown. Weakly bound B is removed first, followed by C as the buffer concentration of one or more components increases. In this example, C is most likely the product. After C is eluted, the tightly bound D is removed using a high concentration cleaning or stripping buffer.

and cation exchange (CEX) resins if the protein has a positive charge under the ionic and pH conditions of the load buffer. A wide variety of IEX resins are available. The sign of the charge and the strength of the charge of many proteins can be manipulated considerably by changes in the buffer pH and counter-ion type, concentration, and charge. IEX is a very flexible adsorption chromatography process that can be used to efficiently separate many diverse proteins from each other.

- *Hydrophobic Interaction (HIC).* As discussed earlier, depending on the protein's structural elements, history, and environment, it can have a variety of hydrophobic surface properties that can be used to selectively bind it to a hydrophobic resin. The hydrophobic properties of both the resin and the proteins in solution can be manipulated by changes in the buffer pH, ion concentrations, and the concentration of protein compatible organic modifiers such as ethanol (EtOH).
- *Affinity Chromatography (AC).* Some therapeutic proteins have sites on the protein surface that can be used to selectively remove the protein from a complex solution of many other proteins and chemical species. mAbs have a highly specific, controllable reactivity with Protein A shown in Fig. 1.11. Affinity chromatograph is sometimes called *capture chromatography* or a capture step. Some ion exchange interactions can be very specific and selective and can be used as a capture step.

A detailed discussion of process chromatography can be found in Hagel *et al.* [28] and Carta and Jungbauer [29].

With the various UOs for both upstream and downstream processes reviewed, the discussion switches to the very important topic of describing, characterizing, and controlling the process UOs of the manufacturing enterprise shown in Fig. 1.6. The following section provides a lexicon for describing and understanding the process inputs and outputs.

1.3.5 Process Performance and Control

One of the difficult challenges of working with biopharmaceutical processes is describing the process's performance and control strategies in terms of inputs and outputs. The process's performance can be developed, evaluated, and described by either treating each UO individually or combining UOs into larger logical groups such as upstream or downstream.

The current regulatory literature describes the process's behavior in terms of CQAs and critical process parameters (CPPs). The regulatory definitions for CQA and CPP are:

- *CQA*. ICH Q8 (R2) definition:

  > A physical, chemical, biological, or microbiological property or characteristic that should be within an appropriated limit, range, or distribution to ensure the desired product quality [3].

- *CPP*. ICH Q8 (R2) definition:

  > The parameter whose variability has an impact on the critical quality attribute and therefore should be monitored and controlled to ensure the process produces the desired quality [3].

However, in practice, CQA and CPP do not provide an adequate description of the process's inputs and outputs to properly define the behavior of the processes. CQAs that describe the product are outputs. However, CPPs can be either inputs or outputs. In order to properly understand the characterization and control of processes, additional definitions are required. The first term is critical process response or CPR.

- *CPR*. A measureable process output that describes or represents the process's performance. CPRs are frequently measured on-line and can be used as measured inputs to automatic control loops.

The term *CPR* is *not* included in the current ICH Q8 (R2) framework, but it is a critical concept to understand and control the performance of the process. A *CPR may or may not be the same as a CQA*; however, almost all CPRs will have some impact on the CQAs of the product. Many CPRs can be measured directly in real time and are thus available to control the process using the principles expressed in FDA's (Food and Drug Administration) PAT (process analytical technology) initiative [21]. An example of a CPR is the temperature of a fermenter.

Having defined the outputs, we turn to a better description of the complex set of inputs to a biopharmaceutical process. Inputs or CPPs can be subdivided into the following three categories:

- *Critical Materials Parameter (CMP).* Input parameters are the raw materials that are fed into the process. Note that the output of one process may be the input to the following process, so that a CQA of the first process is the CMP to the second. Examples of CMPs would be a media component to the bioreactor and an impurity from the upstream process into the downstream process.
- *Critical Design Parameter (CDP).* Parameters are fixed by the selection of equipment types and features. Design parameters do not, or at least should not, change during operation. Design parameters are managed through change control. Examples of a CDP would be the type of impeller used in a bioreactor and the amount of surface area used to cool a fermenter. (Note that the fouling of the surface area changes the effective cooling capability of the cooling coil would be a change in a CDP.)
- *Critical Operating Parameter (COP).* Operating parameters that can be manipulated to achieve control over the process. Some operating parameters are held constant, whereas others may be actively controlled via open loop or feedback control systems. Some operating parameters are critical control parameters or CCPs as described later. Examples of a COP would be the speed of the agitator in the fermentation vessel and the coolant flow rate through the fermenter cooling coils to control the temperature of the fermenter.

The reason for breaking the input CPPs into three different groups is because each group must be treated very differently. Failure to understand and deal with each group in the context of their impact on the process's behavior leads to considerable confusion.

Further discussion why the temperature of a fermenter can be viewed as a COP is important. Typically, temperature is viewed as an input because it is set on a controller and the control system maintains the temperature of the fermenter at the prescribed temperature. If the fermenter heat exchanger/control system is designed properly, the fermenter temperature can be tightly controlled. However, in some fermenters, the temperature may vary because of the extreme energy generation from agitation and metabolic heat. Temperature excursions away from the set point are important and can impact product quality. The true COP for such a case is the coolant flow rate to the fermenter controlled by the feedback control loop measuring the fermenter's temperature, a CPR.

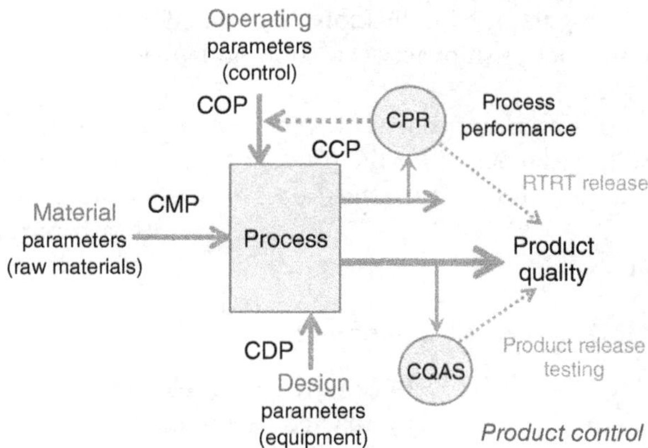

Figure 1.17 Summary of process inputs (parameters) and outputs (attributes), and how they are used to describe the process's performance in terms of CPRs and CPPs (COP, CMP, and CDP) to control the quality of the product as defined by the product's CQAs.

For developing effective process control systems, an additional term must be defined. A COP used to control a CPR can be defined as a CCP.

- *CCP.* CCPs are COPs used to actively control CPRs. ASTM E2476-09 defines CCP as: "…that subset of Critical Process Parameters which can be changed under automated control to modify the trajectory of the process, and hence to modify the CQAs of the intermediate process product or the final drug product" [35].

A key process development objective is to identify the impact of each COP on each CPR. For each CPR, the COP that provides the best control should be identified and used in the process control strategy. An example of a CCP would be the coolant flow rate to a fermenter used to control the fermenter's temperature CPR by a feedback control loop. ASTM E2476-09 provides some guidance in selecting CCP/CPR combinations using QRM methods [35].

A generalized process flow diagram that shows the elements of the process control strategies are shown in Fig. 1.17.

The combination of all the attributes and parameters forms the design space described earlier. ICH Q8 (R2) defines the design space as:

> **Design Space** – The multidimensional combination and interaction of input variables (e.g., material attributes) and process parameters that have been demonstrated to provide assurance of quality. Working within the design space is not considered a change. Movement out of the design space is considered to be a change and would normally initiate a regulatory post-approval change process. Design space is proposed by the applicant and is subject to regulatory assessment and approval.

The ICH Q8 (R2) definition has two parts. The first sentence is the definition, whereas the remainder attempts to clarify the definition by establishing the regulatory implications associated with the definition and the important relationship between the regulatory agency and the applicant's design space.

An important role of process development is to build or define the ICH Q8 (R2) design space for the product's process by identifying and establishing the relationships between the process's inputs as defined by the three types of input CPPs (COP, CMP, and CDPs) and outputs (attributes) as defined by the CQA and CPRs.

1.3.6 Process—Equipment

The relationship between the facility element and the process element shown in Fig. 1.1 is greatly impacted by the equipment type used to implement the UOs. Process equipment is the second set of elements shown in Fig. 1.6. This relationship is depicted in Fig. 1.18. In older enterprises designed around fixed SS equipment, the three elements were developed simultaneously and are thus, out of necessity, very interdependent on each other for operation and control. More recently developed single use systems (SUSs) allows decoupling of facility and process elements. The two are integrated not by connected piping, but solely by the procedural control provided by the enterprise's infrastructure element. If the SUSs are skid mounted, then further decoupling is possible because the processes are no longer required to be operated in a specific location [1]. They can be located in any operating area that provides the required environmental and support capabilities.

SUSs have a number of significant advantages for manufacturing biopharmaceuticals. SS systems rely on clean-in-place (CIP) systems for cleaning and sanitization. Presterilized SUSs do not require extensive cleaning and cleaning validation studies because they are used once and then discarded. SUSs may also provide simple, relatively standard platform equipment that can be readily configured and used to support a wide variety of manufacturing UOs. "Off the shelf" skid-based SUSs are also easily cloned to increase production rates or move a process to a different location.

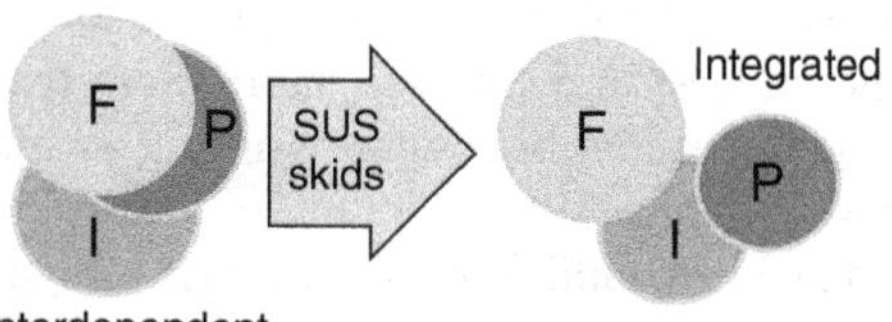

Figure 1.18 Cartoon showing the impact of skid-mounted SUSs on facility interdependencies. The enterprise elements shown in Figs. 1.1 and 1.6 are integrated using the infrastructure element that coordinates and controls the performance of the process element within the facility element to assure a high quality product.

1.3.7 Process—Materials

Control of raw materials is critical to the successful operation of a manufacturing enterprise. The industry is replete with operational problems likely introduced by unknown or undetected variability or changes in raw materials. The raw materials attributes (outputs of the vendor) are described by the CMP-inputs identified in the process control and performance section. Incoming identification, acceptance, QC testing, and release of all raw materials are required by regulatory GMPs. All vendors supplying raw materials should be controlled by appropriate QA auditing and review procedures [18,19].

1.4 Enterprise Element #2—Manufacturing Facility

With the manufacturing processes briefly summarized, the discussion turns to the basic components of the facility element shown in Fig. 1.19. The facility has four basic components: layout configuration, environmental controls, utility systems, and facility controls required for supplying the process with raw materials, services, and a well-controlled surrounding environment.

When the facility is designed, each facility component must be considered with respect to supporting all the process's UOs. We will look at each component and how they impact the process's operation. The layout determines how the process elements are arranged in the facility to satisfy adjacency and material, personnel, and waste flow requirements. Although not addressed here, the facility's layout will be significantly impacted if products with intermediate or final steps involving materials with biological safety level hazard classifications (BSL-2 or 3) are produced [31,32].

1.4.1 Facility—Layout

In facilities where the process is hard piped together to transfer in-process materials, adjacencies that simplify and aid in the flow of material are critical to minimize the amount of expensive piping. In Fig. 1.20, the relationship between various support (warehouse, media, and buffer) preparation areas and process UOs are shown. Personnel flow between areas is through airlocks. In skid-mounted smaller SUS-based processes, adjacencies are not so important because material can be transferred between processes in portable totes or vessels.

In developing a facility design, the flows of personnel, raw, intermediate, and in-process materials must be defined and controlled. Unidirectional flows that keep all the flows separated are desirable. However, balancing all the different flows can be an over-constrained problem that results in compromises and trade-offs between competing priorities.

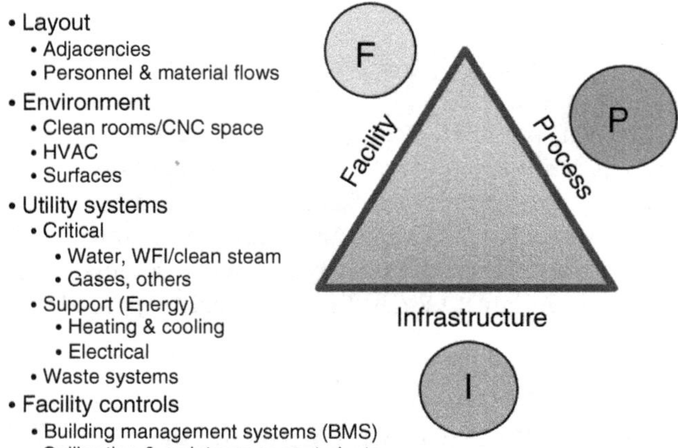

Figure 1.19 Expanding on Fig. 1.1, the facility element is the facility's layout, environmental control features, utility systems, and facility controls required for providing the process with the necessary support and protection.

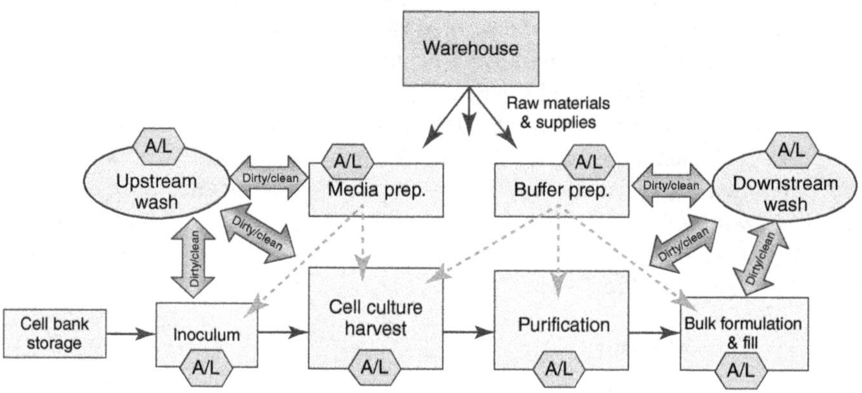

Figure 1.20 Basic in-process, personnel, and waste flows are required between unit operations, including supporting preparation areas.

To achieve the required flows and adjacencies, several layout options are possible. The layout shown in Fig. 1.21 is large common spaces that may contain several UOs for a single product or may contain operations associated with several different products if appropriate segregation strategies are employed to maintain separation between the products. More segregated layouts are possible and should be considered depending on the process's requirements, equipment used, and segregation strategies employed to achieve the required process separation imperatives. Any segregations strategy must be fully validated using an appropriate lifecycle approach [1].

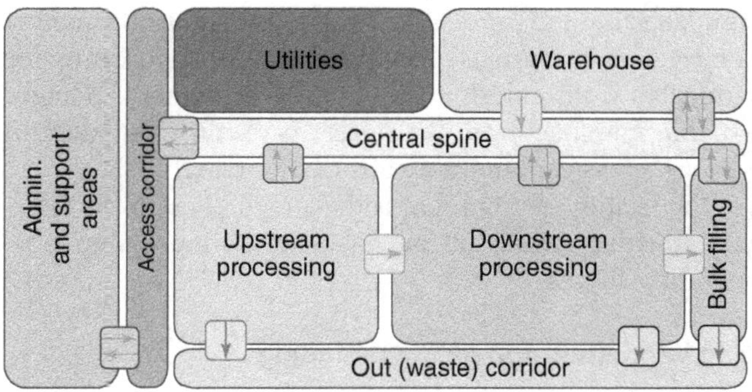

Figure 1.21 Basic facility layout with flow control between areas. Processes located in the common areas would be commingled and separated by a segregation strategy based on closed equipment of the type provided by single use systems or highly automated SS systems. Note that media and buffer preparation operations in this layout would be carried out within the operating space.

Depending on the segregation strategy and open process requirements, more physically separated operating areas may be desirable. The most frequent facility segregation strategy used for mAb processes, as shown in Fig. 1.11, are four separate areas divided such that the output of each viral clearance step is transported directly to a separate area. As each viral clearance step must be validated to be separate, a segregation strategy that does not use physical separation by separate facility spaces must have a fully validated segregation strategy based on the process equipment, such as closed SUSs, to provide the required separation.

1.4.2 Facility—Environment

Providing the process's UOs with the environment required to support and protect the process from contamination is required. The process requirements can vary widely from completely closed processes that are sealed against the environment by hard piping to open processes that rely completely on a sterile or aseptic environment to prevent contamination. The selection of the appropriate space depends on a wide variety of factors. In all cases, the operating areas should be physically secure with access limited to appropriately qualified personnel.

1.4.3 Clean Rooms/CNC Spaces

Facility spaces can be assembled using a variety of different space classifications depending on the segregation strategy used to keep material and operations separated. Controlled, nonclassified (CNC) areas are used for processes that incorporate segregations strategy utilizing closed process equipment, either automated SS or closed plastic systems (SUSs) designed to prevent leakage of process material to the environment or intrusion of environmental contaminates into the process. If the processes must be opened during operation, then clean rooms are used to minimize the opportunity for environmental contaminates from entering the process equipment. If more than one product is being manufactured, then great care must be taken to prevent actual or perceived cross contamination from occurring.

A key component of the facility environment is the air supplied to the operating rooms. The HVAC (heating ventilation and air-conditioning) system provides the air to the operating spaces.

1.4.4 HVAC—Heating Ventilation and Air-Conditioning

A facility's airflow and room pressure distribution within the facility are controlled by the HVAC system. The AHUs' (air handling units) assignments to the different rooms depend on the segregation strategy used to separate the processing areas. A biotech manufacturing facility may have between 6 and 20 or more AHUs depending on the size of the facility and the segregation strategy. The HVAC system is also used to control, minimize, or prevent the migration of contaminates from one room to other rooms. One contaminate that must be carefully controlled are biocontaminates from biological safety (BSL-2, 3, 4) containment areas into outside areas. For details on BSL categories and requirements, the reader is referred to appropriate references [31,32].

The basic elements of an HVAC system are shown in Fig. 1.22. Clean room conditions are achieved using HEPA (high efficiency particulate air) filters that remove between 99.97% and 99.9999+% of the particles from the air. The distribution of the particles removed and the specifics of the different types of HEPA filters are beyond the scope of this chapter.

The HVAC air classifications of clean rooms are summarized in Fig. 1.23. The particle limits, or specifications, for the various areas are complex and depend on the particle size being measured. Various reference sources should be used to define specifics on particle size specifications and methods of measuring room performance under operation and nonoperational periods [30,33,34]. The measured particle load in a room is highly dependent on the amount of particles generated during operation, the cleaning procedures used to clean and sanitize the areas, internal airflow patterns around equipment and people, and the amount of filtered air supplied by the HVAC systems.

Five different types of areas are used to support biopharmaceutical manufacturing. These classifications are:

- *CNC*. CNC (controlled non-classified) spaces are not HEPA filtered and thus not considered clean rooms. CNC rooms would be used for closed processes that can be assembled, or cleaned and sanitized without being opened to the environment. CNC spaces are relatively less expensive to build and operate than clean rooms.

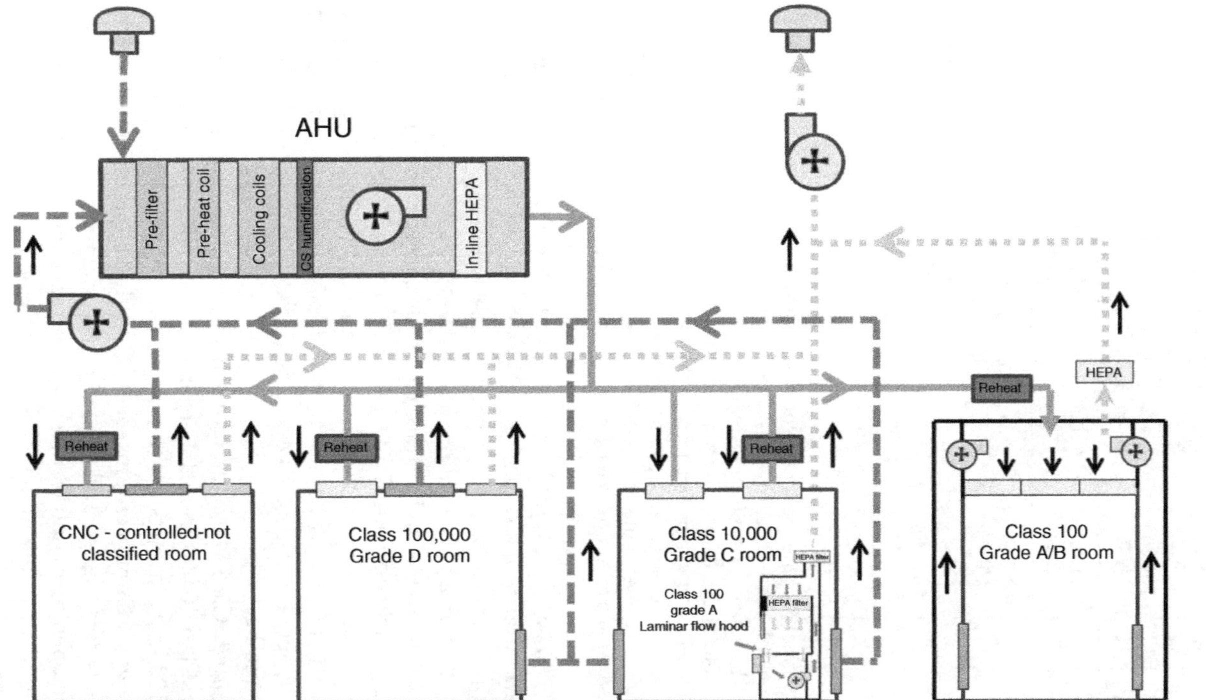

Figure 1.22 The basic HVAC elements are shown. The air handling unit (AHU) pushes air into the rooms through appropriate filters. Clean rooms are supplied through in-line HEPA (high efficiency particulate air) filters in the ducts or terminal HEPAs in the duct outlets into the operating spaces. The air is removed from the room via air return duct to the AHU or vented outside the facility after appropriate filtration to remove possible process contaminates. For discussion purposes, the AHU shown supplies all four areas. In practice, the AHUs would be limited to supplying areas with common process separation requirements based on a well-defined segregation strategy.

	Standards			Airflow velocity M/S (ft./min.)	Air changes per hour	HEPA ceiling coverage	Notes
	Fed. Std. 209E	EU-cGMP	ISO 14644-1				
Common classification	100,000	D	8	0.005–0.041 (1–8)	5–48	5–15%	• Upstream operations, cell culture • General access area; airlocks
	10,000	C	7	0.051–0.076 (10–15)	60–90	15–20%	• Purification areas
	100	B	5	0.127–0.203 (25–40)	150–240	25–40%	• Aseptic operations, tight bioburden control • Low air returns
	100	A	5	0.203–0.406 (40–80)	240–480	35–70%	• Used for aseptic formulation and fill • Biosafety cabinets • Requires laminar flow field

Figure 1.23 Summary table of the most frequent clean room classifications according to the three most commonly used standards. The archaic, out of date Federal Standard 209E is still frequently used by many to identify the categories formally defined in the ISO 14644 and EUA-cGMP standards [33]. Although not an exact correlation, the three standards generally agree. Table also shows typical ranges for airflow velocities in the room and/or in laminar flow areas, room air change rates, and percent ceiling coverage for the classifications. Airflow velocities at the surface of HEPA filters is usually between 70 and 110 ft/min. Airflow rates above approximately 100 ft/min are turbulent.

- *Class 100,000, Group D.* Group D rooms have HEPA filtration and may or may not have low air returns. Class 100,000 spaces are frequently used for corridors, support, and upstream processing areas where closed systems are used for fermentation and cell culture operations.
- *Class 10,000, Group C.* Group C rooms have a considerable amount of HEPA filtration. The particle load limits for Class 10,000 areas is about an order of magnitude less than for Class 100,000 areas. In Fig. 1.26, the Group C Room shows a Class 100, Group A Laminar Flow Hood used for operations that require aseptic operations such as open transfers of cell cultures from one bioreactor to another.
- *Class 100, Group B.* Class 100 areas have two order of magnitude lower particle limits than Class 10,000 areas. In Group B, the overall room environment meets Class 100 specifications.
- *Class 100, Group A.* Group A rooms meets Class 100 requirements; in addition, the airflow is designed and validated to be laminar with the airflow protecting the open operation from environmental contamination from air turbulence circulating contamination from the operator and the environment to the product.

As shown in Fig. 1.22, the AHU provides heating, cooling, and if necessary humidification, usually from clean steam addition. Temperature control of the

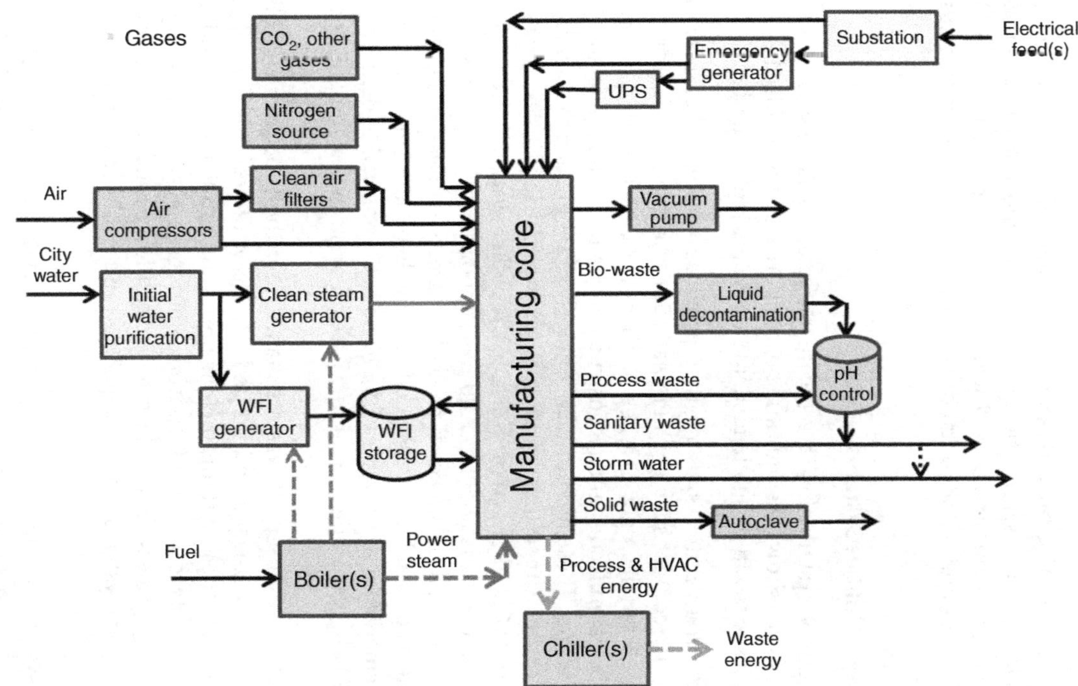

Figure 1.24 Schematic of the utility systems relationship to the manufacturing core containing the process unit operations.

rooms is accomplished by reheating the air supplied to the rooms to adjust the room temperature. Heating and cooling of the HVAC/AHU systems are provided by the facility's utility systems.

1.4.5 Surfaces

Clean rooms are constructed with smooth cleanable surfaces that maximize the ease of cleaning and sanitization of all vertical and horizontal surfaces. Piping in the room is minimized and windows have sloping sills or are flush with the walls to minimize horizontal surfaces that could collect settled dust or particles.

1.4.6 Facility—Utilities Systems

The overall utility system flow for the manufacturing facility is shown in Fig. 1.24. The figure shows the multitude of support systems required.

The utility systems can be subdivided into the following categories:

- *Critical Utilities.* Utility systems that provide materials that may have direct product contact such as WFI (water for injection), clean steam, and in-process gases (clean air, CO_2, etc.). These systems are utility processes that must be carefully designed, constructed, qualified, and operated.
- *Support Utilities.* These nonproduct contact utilities include the steam boilers that provide primary energy sources for the critical utilities, waste systems, and HVAC. Other systems include the chiller systems to remove heat for HVAC and process cooling. Another support utility is the instrument air system to provide air power, if needed, and feed material to the clean air system. The electrical feed is designed to provide reliable power for all systems with emergency and uninterruptable power supply (UPS) distributed to process and utility systems as required base on a risk analysis and mitigation strategy.
- *Waste Systems.* A variety of waste streams from the processes and utility systems must be managed and properly disposed of according to all local and federal laws. These systems include both solid and liquid wastes. If biological safety waste is generated, appropriately designed waste handling systems must be in place to assure the safety of employees and the environment [31,32]. All waste systems should be qualified and validated to an appropriate level.

All utility systems should be appropriately validated for their use [19]. All systems can be process validated over their lifecycle according to FDA's 2011 process validation guidelines. The level of validation effort should be appropriate for their application. The critical systems described earlier should be fully validated using each of the validation stages defined in the FDA's process validation guidance [2].

1.4.7 Facility—Control Systems

The final component of the facility elements is the control systems required to run the facility element. These systems include the building management systems (BMSs) that control the HVAC and utility systems. They are systems very similar to the automated process control systems discussed earlier. In addition, maintenance control, including instrument and control tracking systems, should be included.

Having discussed, the enterprise's facility and process elements, the third element is the manufacturing infrastructure required to operate and control the enterprise.

1.5 Enterprise Element #3—Manufacturing Infrastructure

The final enterprise element, the infrastructure, is used to manage and control the process and facility elements. The components of the infrastructure are shown in Fig. 1.25. The people provide the labor and expertise to perform a wide variety of nonautomated operating and oversee automated functions. The people are directed and guided by the enterprise's written policies, practices, and procedures.

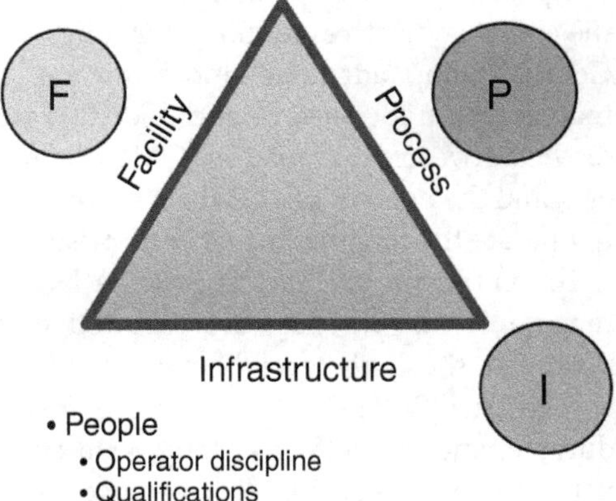

Figure 1.25 The infrastructure element of the manufacturing enterprise provides command and control of the entire manufacturing operation composed of the process and facility elements.

1.5.1 Infrastructure—People (Operating Staff)

All manufacturing operations are as good as the performance of the staff operating the facility. All personnel must use operating discipline to execute all tasks and perform all the functions required to operate the facility. The staff must be qualified to understand and perform all the tasks and activities. The staff becomes qualified through proper training and experience. A comprehensive training program that guides and documents all general and specific training and qualifications of every staff member is essential.

The most important factor to a manufacturing enterprise's success is the personal discipline of all management and operating personnel to identify, develop, write, and use the enterprise's procedures and practices for all manufacturing operations. Appropriate enterprise policies and procedures, including standard operating procedures (SOPs) and manufacturing procedures (MPs), should be in place to guide the personnel in the performance of all activities.

1.5.2 Infrastructure—Enterprise Practices and Procedures

Procedural control systems, better known as *GMPs*, are a key element to successfully control a manufacturing enterprise. GMPs are frequently regarded as those practices defined by regulatory guidelines and statues. However, regulatory GMPs are not sufficient to adequately control biopharmaceutical manufacturing. Regulatory GMPs are based on terms such as *adequate* and *sufficient*. Regulatory GMPs are designed to be used by regulatory agencies to measure compliance and are not sufficiently detailed or comprehensive to achieve all the control objectives required for robust and reliable long-term performance. Compounding "adequate" and "sufficient" features and systems often leads to marginal or poor performance. For this reason, a clear distinction should be drawn between the regulatory GMPs defined in 21CFR 211, etc. and ICH Q7A, etc., and the enterprise's GMPs developed or adapted specifically for, and used to operate the manufacturing enterprise [18,19].

The enterprise's GMPs should be based on excellent practices. Excellent manufacturing practices are required to control and maintain the organization at the high level of performance required to consistently produce a high quality product and assure manufacturing that provides efficient, durable, and cost-effective long-term supply of high quality product to the patient community. Excellence must be based on continuous improvement of best, better, and in some cases optimal practices for all activities and tasks required to manage the entire lifecycle of the process and facility's operations.

The final topic for the successful operation of a manufacturing enterprise is the control of the process and facility elements using the control systems contained within the process, facility, and infrastructure elements.

1.6 Controlling the Manufacturing Enterprise

A manufacturing enterprise is controlled by three complementary control systems. Each control system performs two functions. A control system's first function is to provide the guidance and direction to *establish and maintain control* of the product, process, and facility to predefined specifications and limits. The second function is to provide sufficient documentation to *prove that control was maintained* during the operation of the manufacturing and facility processes. Thus, in developing or evaluating any control system, two tests must be met. First the system must control and it must prove control.

The three control systems are:

Product Controls—Historically, product control has been achieved by end product testing to release the final product. Such approaches are insufficient for biopharmaceuticals. Product release, or control, is best achieved through a comprehensive evaluation of QC product testing and real-time evaluation of the manufacturing process's performance against the process's expected or benchmarked performance. Such methods are called *real-time release testing* (*RTRT*) that combines end product testing of CQAs with the performance of CPAs and CPPs controlled by the Process's control systems according to predefined performance limits, including defined target and variability specifications. RTRT is defined in ICH Q8 (R2) as:

> The ability to evaluate and ensure the quality of in-process and/or final product based on process data, which typically include the valid combination of measured material attributes and process controls.

Sequential in-process release testing and RTRT, when combined with QC release testing against regulatory approved specifications, provide an effective tool for releasing safe and effective biopharmaceutical products for subsequent manufacturing processes or to the patient.

Process Controls—Process control are part of the process element, but is a critical element of controlling the operation of the process. The process UOs must be controlled by a process control strategy that measures CQAs and outputs (CPRs) and manipulates appropriate defined CPPs as shown in Fig. 1.17. The process control system uses a combination of open-loop control defined during process development, feedback control of CPRs by appropriate COPs (CCPs), and possibly feed forward control manipulating COPs based on measurement and estimated impact of variations in CMPs. Process controls and the associated data gathering are typically implemented by direct digital control (DDC) computer systems. Because utility systems are processes, the same process control principles apply.

Procedural Controls—The infrastructure, including the facility and process operation, is controlled by the enterprise's GMPs defined in the enterprise's policies and procedures. The enterprise's GMPs are carefully constructed to control the execution of all enterprise elements to assure a high quality product. The enterprise's GMPs should provide appropriate levels of control and proof of control required by the process and utility UOs and their relationship to the quality of the product. All the enterprise's GMPs should meet and exceed all applicable regulatory guidelines and regulatory GMPs governing the product being produced [18,19].

The required enterprise GMPs are shown in Fig. 1.26. As shown, the enterprise's policies and procedures (GMPs) are used to control all three elements of the manufacturing enterprise shown in Fig. 1.1 along with the incoming raw materials and the released product. The enterprise GMPs must be designed to provide adequate control and proof of control to a level appropriate for the nature and application of the product. The definition of appropriate control is based on various risk assessments and the associated risk control and mitigation strategies defined during the creation and operation of the enterprise.

The second responsibility of a control system is proof of control. Proof of control is shown in Fig. 1.27. The enterprise's GMPs should require the flow of all of the required information into the enterprise's many documentation databases to assure both control and proof of control for all three product, process, and enterprise GMPs control systems. The documentation should meet two requirements. It should provide assurance to outside agents (regulatory or otherwise) that complete control was maintained over the entire product

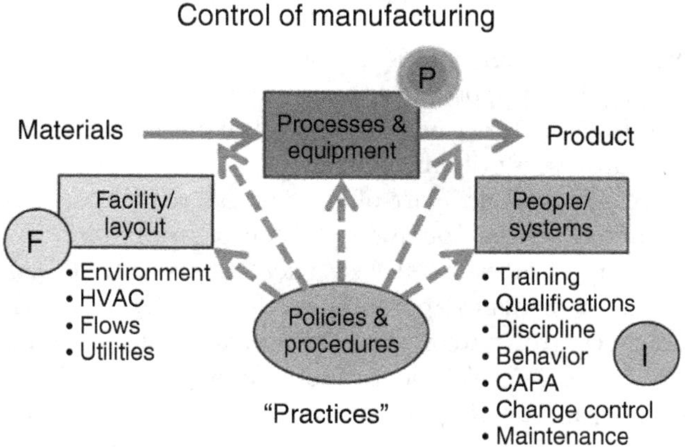

Figure 1.26 The enterprise elements are controlled by the practices defined in the infrastructure's practices defined in the policies and procedures. These practices are referred to as the *enterprise good manufacturing practices* (*GMPs*).

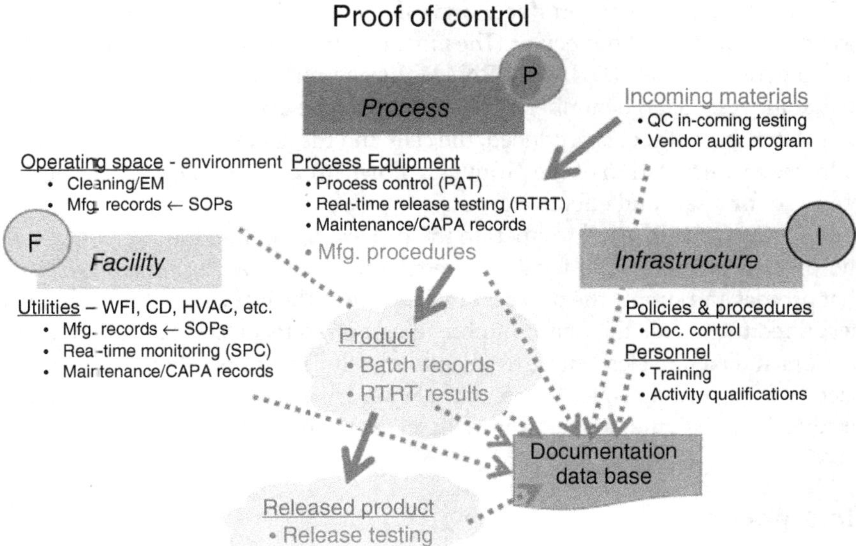

Figure 1.27 The second responsibility of the infrastructure is to provide "proof of control."

lifecycle, and secondly provide sufficient information to allow for investigating internal process and product problems and issues. The second requirement for complex processes and products can be very demanding and require that considerable information is gathered.

The acquisition and flow of information from the enterprise elements is shown in Fig. 1.27. The process flow is shown by the red arrows from the incoming materials through the process element to the product and ultimately for release of the product. Information required by the enterprise's GMPs is gathered from all the elements and placed in the documentation databases.

1.7 Summary

This chapter contains a very brief introduction to the biomanufacturing enterprise required to manufacture biopharmaceuticals. The enterprise can be broken down into three elements. These elements are:

- The *process* is the sequence of UOs and the equipment and systems required to manufacture the product.
- The *facility* provides the surrounding environment and utility systems required to protect and supply the process.
- The *infrastructure* is the personnel and practices used to operate and control the process and facility.

The biopharmaceutical product is manufactured using cells composed of and uses basic constituent molecules. The most important constituents are proteins and genetic materials (DNA and RNA). The cells are grown in upstream UOs to manufacture a large number of cells that produce the product, usually a protein. After the product is produced, the cells and the protein must be separated, followed by purification of the product in downstream operations to the purity required for a safe and effective product.

The UOs typically used to operate the process are defined in the context of the process element. The facility element utility and environmental systems that protect and supply the process are described. The infrastructure element is described to provide the final components required to successfully operate the process and facility elements. The chapter concludes by describing the systems used to control and prove control of the enterprise, so that a safe and effective product can be manufactured and released to the patient population.

References

1 Witcher MF. Impact of facility layout on developing and validating segregation strategies in the next generation of multi-product, multi-phase biopharmaceutical manufacturing facilities. *Pharm Eng* 2013 Supplement to; **33**(1):74–82. November/December.
2 FDA. Guidance for industry—process validation: general principles and practices; Revision 1; January, 2011.
3 FDA. Guidance for industry—Q8 (R2)—pharmaceutical development; FDA Website; Revision 2; November, 2009.
4 ICH Q6B. Specifications: test procedures and acceptance criteria for biotechnology/biological products; March, 1999.
5 ICH Q5E. Comparability of biotechnological/biological products; EMEA CPMP/ICH/5721/03; June, 2005.
6 Harvey RA, Champe PC, Ferrier DR, editors. *Lippincott's illustrated reviews, biochemistry*. 3rd ed. Baltimore: Lippincott Williams & Wilkins; 2005.
7 Creighton TE. *Proteins—structure and molecular properties*. 2nd ed. New York: W. H. Freeman & Co; 1993.
8 Branden C, Tooze J. *Introduction to protein structure*. 2nd ed. New York: Garland Publishing, a Taylor Francis Group; 1999.
9 Dressler D, Potter H. *Discovering enzymes*. New York: Scientific American Library; 1991.
10 Watson JD. *DNA, the secret of life*. New York: Alfred A. Knopf; 2003.
11 Hodge R. *The molecules of life—DNA, RNA, and proteins*. New York: Facts-on-File, Inc.; 2009.
12 Salway JG. *Metabolism at a glance*. 3rd ed. New York: Blackwell Publishing Co; 2004.

13 Wheelis ML. *Principles of modern microbiology*. New York: Jones and Bartlett Publishers; 2008.
14 Glick BR, Pasernak JJ. *Molecular biotechnology; principles and applications of recombinant DNA*. 2nd ed. Washington: ASM Press; 1998.
15 Gottschalk G. *Bacterial metabolism*. 2nd ed. New York: Springer-Verlag; 1986.
16 White D. *The physiology and biochemistry of prokaryotes*. 2nd ed. Oxford: Oxford University Press; 2000.
17 Stacey G, Davis J. *Medicines from animal cell culture*. Chichester: John Wiley & Sons, Ltd; 2007.
18 FDA. Guidance for industry—Q7A good manufacturing practices guidance for active pharmaceutical ingredients; August, 2001.
19 CFR 21. Part 211—current good manufacturing practice for finished pharmaceuticals.
20 Stanbury PF, Whitaker A, Hall SJ. *Principles of fermentation technology*. 2nd ed. New York: Butterworth Heinemann; 1995.
21 Lubiniecki AS, editor. *Large-scale mammalian cell culture technology*. New York: Marcel Dekker, Inc; 1990.
22 Ozturk SS, Hu W, editors. *Cell culture technology for pharmaceutical and cell-based therapies*. Boca Raton: Taylor & Francis; 2006.
23 Leung WW. *Centrifugal separations in biotechnology*. New York: Elsevier; 2007
24 Rathore AS, Kumar J, Tugcu N, *et al.* Evolution of the monoclonal antibody purification platform. *BioPharm Int* http://www.biopharminternational.com/biopharm/Downstream+Processing/Evolution-of-the-Monoclonal-Antibody-Purification-/ArticleStandard/Article/detail/827131?contextCategoryId=43048; November, 2013.
25 Schmidt S, Mora J, Dolan S, *et al.* An integrated concept for robust and efficient virus clearance and contaminant removal in biotech processes. *BioProcess Int* 2005;**3**:1–7.
26 Immelman A, Kellings K, Stamm O, *et al.* Validation and quality procedures for virus and prion removal in biopharmaceuticals. *BioProcess Int* 2005;**3**:38–45.
27 Cheryan M. *Ultrafiltration and microfiltration handbook*. 2nd ed. New York: CRC Press; 1998.
28 Hagel L, Jagschies G, Sofer G. *Handbook of process chromatography; development, manufacturing, validation, and economics*. 2nd ed. New York: Academic Press; 2008.
29 Carta C, Jungbauer A. *Protein chromatography*. Wiley-VCH; 2010. –Weinheim.
30 FED-STD-209E. Federal standard airborne particulate cleanliness classes in cleanrooms and clean zones; September 11, 1992.

31 U.S. Department of Health and Human Services. *Biosafety in microbiological and biomedical laboratories.* 5thDecember ed. 2009.
32 AIChE—Center for Chemical Process Safety. *Guidelines for process safety in bioprocess manufacturing facilities.* Hoboken: John Wiley & Sons, Inc; 2011.
33 ISO. International Standard ISO 14644-1: cleanrooms and associated controlled environments—part 1: classification of air cleanliness; Ref. # ISO 14633-1: 1999(E).
34 Lydersen BK, D'Elia NA, Nelson KL. *Bioprocess engineering; systems, equipment, and facilities.* New York: John Wiley & Sons, Inc; 1994.
35 ASTM E2476-09. Standard guide for risk assessment and risk control as it impacts the design, development, and operation of PAT processes for pharmaceutical manufacture; July, 2009.

Chapter 2

Product–Process–Facility Relationship

Jeffery Odum

NNE, Durham, North Carolina, USA

2.1 Introduction

The relationship between the product being manufactured and the process by which it is being manufactured will drive the facility design. While this is common to many industries, for the manufacture of biological products, the relationship is much more complex. There is a simple essence to biomanufacturing that sets it apart from traditional chemical synthesis drug manufacturing. This is due to the fact that products are developed around living organisms instead of chemical raw materials. For chemical synthesis products, we usually refer to them as *active pharmaceutical ingredients* (*APIs*) or small-molecule drugs. APIs are usually not proteins and, therefore, not living organisms, they may be organic or inorganic, and they can usually be replicated easily with an understanding of molecular weights and raw materials.

Biologics are quite often obtained from proteins where they may be extracted from the cells or developed through recombinant cell technology. Protein extraction is common in blood fractionation and enzyme production but can also be accomplished from natural products, plants, or nutraceuticals. In the current biotech manufacturing paradigm, recombinant DNA technology uses genetically modified single-cell organisms to produce the protein of focus (Fig. 2.1).

The current paradigm of the biotech industry centers on the analogy that the process is the product. The entire manufacturing process determines the quality of the therapeutic being produced. This translates to the attention to detail in the following:

- Raw/starting materials
- Fermentation
- Purification

Process Architecture in Biomanufacturing Facility Design, edited by Jeffery Odum and Michael C. Flickinger.

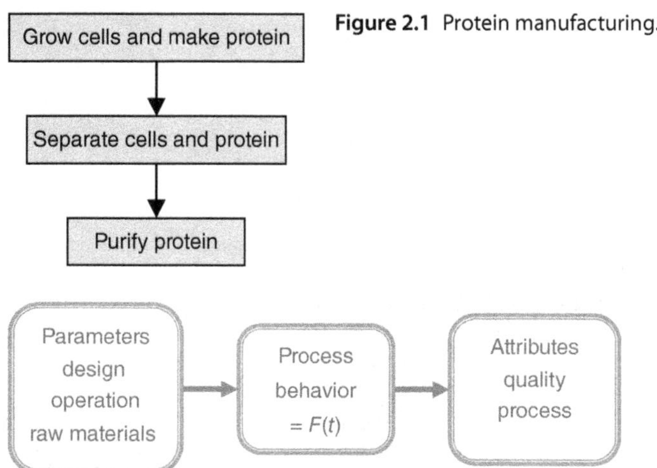

Figure 2.1 Protein manufacturing.

Figure 2.2 The product-process-facility continuum.

- Formulation
- Filling.

The entire manufacturing process must be defined and controlled with tremendous detail so as not to impact the quality, safety, or efficacy of the product.

Therefore, for biomanufacturing, we must focus on traditional biological- and chemical-based unit operations.

Proteins, made of sequences of amino acids, have complex three-dimensional structures, implication for the equipment and facility design in terms of methodology of the manufacturing unit operations, and the relationship between the design of the manufacturing system and the facility. This relationship is best defined by the continuum shown in Fig. 2.2.

The number of amino acids may illustrate the complexity of some common therapeutic proteins (Table 2.1).

2.2 The Characteristics of Biological Therapeutic Products

To understand biomanufacturing facility design, it is important to first understand the unique characteristics of biological products and how those characteristics influence so many of the facility design elements. While the regulatory guidelines may be similar between pharmaceutical and biological products, the differences are striking and result in more complex and challenging facility designs.

Table 2.1 The Number of Amino Acids in Common Therapeutic Proteins

Protein	# Amino Acids	Molecular Weight (kDa)
EPO—glycosylated	166	36.0
Mab (IgG)	~1100	150
Interleukin-2 (human)	133	15.4
GCSF	175	18.8
Insulin (Humalog®)	51	5.8
FVIII	1438	170
Hemoglobin	574	65
Tumor necrosis factor (TNF-α)	471	52
Lysozyme (egg white)	129	13.9

Abbreviations: EPO, erythropoietin; GCSF, granulocyte-colony stimulating factor.

As previously mentioned, the manufacture of a biologic product involves the use of living organisms to manufacture the required amounts of materials needed to produce the final drug product. It should be defined here the difference in terminology between drug substance and drug product.

Drug substance is referred to as the bulk form of the targeted biologic. The FDA (Food and Drug Administration) defines drug substance as follows [1].

- A substance recognized by an official pharmacopeia or formulary.
- A substance intended for use in the diagnosis, cure, mitigation, treatment, or prevention of disease.
- A substance (other than food) intended to affect the structure or any function of the body.
- A substance intended for use as a component of a medicine but not a device or a component, part, or accessory of a device.
- Biological products are included within this definition and are generally covered by the same laws and regulations, but differences exist regarding their manufacturing processes (chemical process vs biological process.)

Drug product is the finished dosage form that contains the drug substance that is delivered to the patient. The majority of biological drug products are in the form of sterile injectables. There are times when these terms are interchanged incorrectly based on these definitions.

Biologics are very fragile molecules that easily degrade in the digestive system. That is why they are generally manufactured as sterile injectable drugs in product form. They are designed to interact with molecules outside the cell and are typically more difficult and costly to manufacture. They require aseptic processing applications in order to preserve the viability of the cell line and

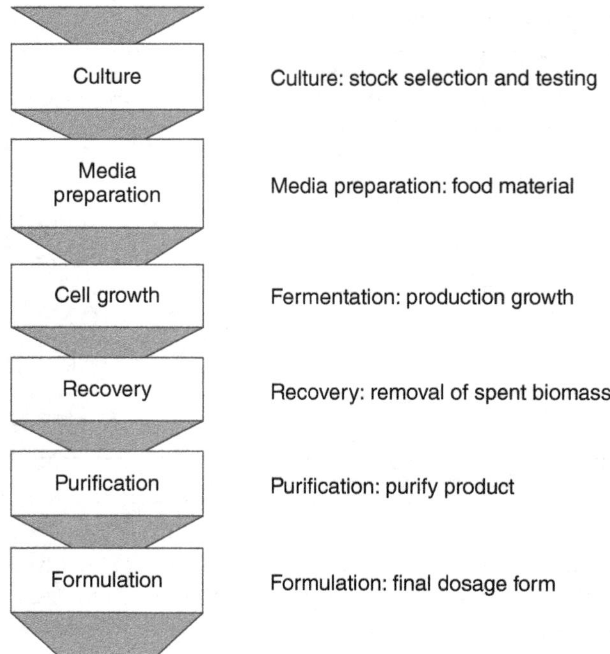

Figure 2.3 Basic biomanufacturing process steps.

normally require extensive purification in order to attain the protein/product target.

The basic cell growth process and the relationship to the manufacturing process are shown in Fig. 2.3.

In cellular and molecular biology, there are two types of cellular therapy products, extracellular and intracellular. Extracellular means "outside the cell," which usually means outside the plasma membrane of the cell. Proteins present in the extracellular space are stored outside the cells by attaching to various extracellular matrix components (Fig. 2.4a).

Intracellular means "inside the cell" where the protein of interest is found inside the cell membrane. This is important because it has a significant impact on how the manufacturing process is designed in order to obtain the protein of interest (Fig. 2.4b).

2.3 Understanding the Attributes

Regulatory policy is now focused on a quality by design approach where an understanding of product and process attributes is mandatory in providing a compliant facility.

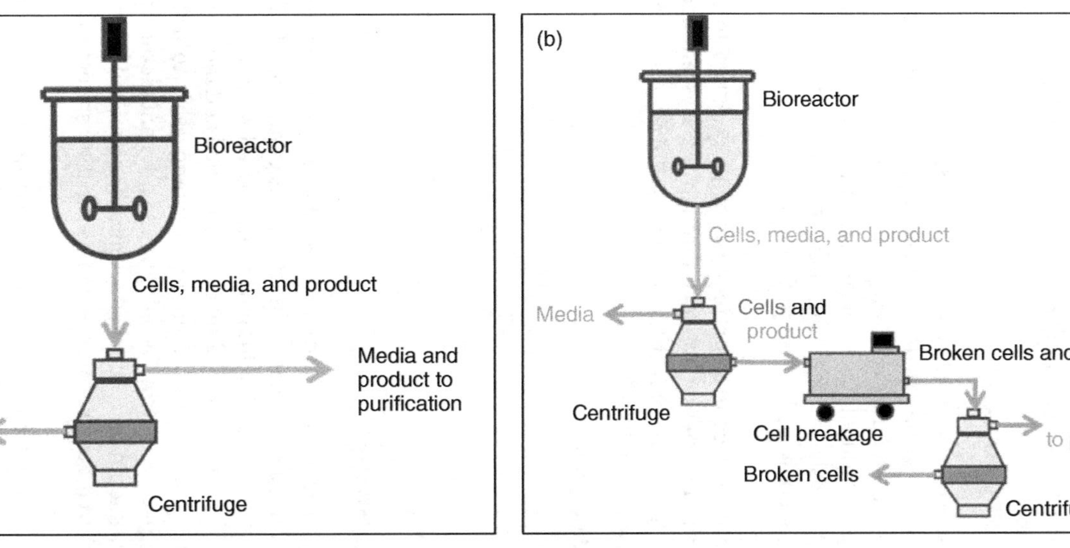

Figure 2.4 (a) Extracellular product diagram. (b) Intracellular product diagram.

- *Product Attributes.* Physical characteristics of products that influence design.
- *Process Attributes.* The materials, equipment, systems, methodologies, and techniques required to manufacture the product.
- *Facility Attributes.* The space allocations, environmental conditions, physical adjacencies, materials of construction, architectural finishes, and operating procedures required to accommodate product manufacturing.

2.3.1 Product Quality Attributes

ICH Q11 defines these critical quality attributes as "physical, chemical, biological, or microbiological properties or characteristics that should be controlled ... to ensure product quality" [2]. In the case of biologics, most of these attributes for the drug product are associated with the drug substance and are a direct result of the design of the drug substance or its manufacturing process. These attributes could include impurities (process or product related) and contaminants.

For these attributes, the control strategy has to be designed to ensure that product quality and consistency are met through a series of specifications. Specifications are chosen to confirm the quality of the drug substance. These specifications include the following [3]:

- Appearance and description
- Identity
- Purity
- Potency
- Quantity
- Microbial quality.

2.3.2 Process Parameters

The manufacturing process must be designed so that the product attributes previously referenced can be met. These critical process parameters (CPPs) should be determined by sound scientific judgment and based on research, scale-up, or manufacturing experience. They should be controlled and monitored to confirm that the impurity profile of the product is comparable to or better than the historical data that comes from development and manufacturing. All input process parameters should be controlled within a meaningful, narrow operating range to ensure that drug substance quality attributes meet their intended specifications [4].

Not all process parameters are critically based on the above discussion. There will be parameters that, while important to process performance, may not affect product quality. These may impact parameters such as process yield or the amount of time it takes to complete the specific unit operation in question. Being able to clearly define these two situations is critical to regulatory compliance.

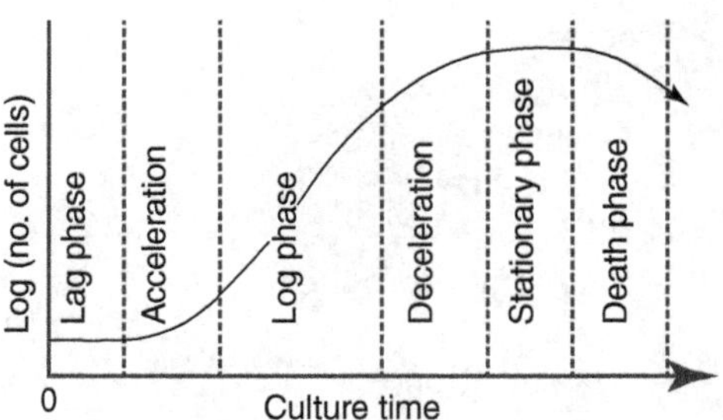

Figure 2.5 Cell culture growth curve.

An example of CPPs can be easily shown for cell culture operations. For a "typical" cell culture operation, the growth curve of the cell culture is shown in Fig. 2.5.

The CPPs for the cell culture growth phase would be as follows:

- Temperature
- Media conditions
- Composition
- pH
- Osmolality
- Oxygen content
- Mixing—agitation
- Energy transfer
- Mass transfer.

2.3.3 Facility Attributes

With the product and process attributes defined, the design of the facility must now be developed so that the current good manufacturing practice (GMP), as defined in USFDA 21CFR, part 211, is being met implementing the current Aseptic Manufacturing Guidelines. Some of the most common design considerations for facility attributes include the following:

- Process closure
- Biohazard levels
- Cleaning approach
- System definitions/operational approach
- Facility integration (segregation, flows, etc.).

Therefore, an example of the product–process–facility attribute synergy could be as follows.

- Product attributes
  - Cell line

- Process
  - Unit operations
  - Closure
  - Technology
- Facility
  - Single/multiproduct
  - Scale
  - Classification

2.4 Factors that Impact Facility Design

There are a number of integrated and independent factors that impact facility design.

1) Process considerations would include the following:
   - Cell line expansion method
   - Cell production platform
   - Recovery method
   - Purification approach
   - Characterization and stability
   - Scale of manufacturing
   - System closure
   - Biosafety level
   - Support utility requirements
   - Material handling and storage
2) Operational considerations would include the following:
   - Single, multiuse, or multiproduct operations
   - Process titer and yield
   - Equipment utilization
   - Segregation strategy
   - Technology platform(s)
3) Schedule/phasing
   - Coordination with clinical manufacturing/product launch
   - Future expansion in operating environments
   - Allowance for future shell space to accommodate new process development
   - Scale increases due to market demand, new technology platforms
4) Budgets and schedule
   - Capital expenditure limits
   - Cost-of-capital drivers
   - Fast-track project execution approach
   - Project delivery methodology

5) Industry standards
   - Non-GMP driven (ASTM, ANSI, ISO, etc.)
   - Global standard metrics
6) Site and code constraints
   - State and local codes
   - Building codes and zoning limitations (building height, wetlands, and esthetic requirements)
   - Safety
   - Environmental
7) Regulatory guidelines
   - Governing agency (country where product will be sold)
   - Classification of product
   - Single versus multiproducts

2.4.1 Facility Types

There are different types of facilities that are defined by specific activities that will take place within their manufacturing spaces. Each of these facility types has a unique set of characteristics that are driven by various manufacturing and regulatory considerations.

2.4.1.1 Product Development Facilities

The development phase is intended to establish parameters for nonclinical trial studies. These parameters are as follows:

- Absorption
- Distribution
- Metabolism
- Excretion/development.

ADME is an abbreviation in pharmacokinetics and pharmacology for "absorption, distribution, metabolism, and excretion" and describes the disposition of a pharmaceutical compound within an organism (Fig. 2.6). The four criteria influence the drug levels and kinetics of drug exposure to the tissues and hence influence the performance and pharmacological activity of the compound as a drug, which will influence the design of the process and the equipment used.

For a compound to reach a tissue, it usually must be taken into the bloodstream—often via mucous surfaces like the digestive tract (intestinal absorption)—before being taken up by the target cells. Factors such as poor compound solubility, gastric emptying time, intestinal transit time, chemical instability in the stomach, and inability to permeate the intestinal wall can all reduce the extent to which a drug is absorbed after oral administration. Absorption critically determines the compound's bioavailability. Drugs that

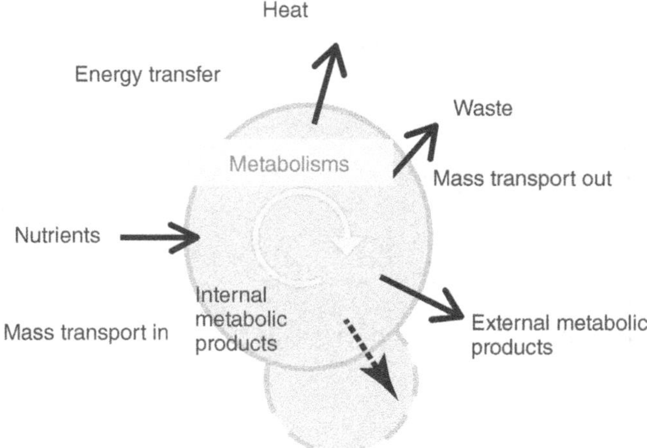

Figure 2.6 ADME components.

absorb poorly when taken orally must be administered in some less desirable way, like intravenously or by inhalation. The route of administration is also an important consideration in the design of the process.

Distribution is defined as the reversible transfer of a drug from one compartment to another. Some factors affecting drug distribution include regional blood flow rates, molecular size, polarity, and binding to serum proteins, forming a complex. Distribution can be a serious problem at some natural barriers such as the blood–brain barrier.

Compounds begin to break down as soon as they enter the body. The majority of small-molecule drug metabolism is carried out in the liver by redox enzymes, termed *cytochrome P450 enzymes*. As metabolism occurs, the initial (parent) compound is converted to new compounds called *metabolites*. When metabolites are pharmacologically inert, metabolism deactivates the administered dose of parent drug and this usually reduces the effects on the body.

Compounds and their metabolites need to be removed from the body via excretion, usually through the kidneys (urine) or in the feces. Unless excretion is complete, accumulation of foreign substances can adversely affect normal metabolism.

Product development facilities focus on how the proteins of interest will be manufactured based on these parameters. There are three primary biochemical functions that are the baseline of protein manufacturing:

- Replication—DNA ⇒ DNA
  - Provide for cell duplication
  - Retain genetic information

- Transcription—DNA ⇒ RNA
  - Transfer of information from genetics to protein-manufacturing mechanism
- Translation—RNA ⇒ Protein
  - Proteins from genetic information

What does this mean to the process engineers designing the facility?

- If you lose the structure, you lose the product.
- Loss of structure may be irreversible.
- You may not know it has happened.
- How can this happen? (How to design to prevent)
  - Temperature
  - pH
  - Shear
  - Ionic strength
  - Polarity
  - Enzyme activity
  - Incompatible materials
  - Aeration
- Process control, documentation, robustness, and repeatability are critical to assuring product integrity.

2.4.1.2 Pilot/Clinical

Pilot facilities often focus on scale-up and optimization of the manufacturing process. Pilot plants are sometimes referred to as the *playground of the process engineer*. In these facilities, the focus is often on an engineering approach to process development and scale-up that uses predictive models to develop the design criteria.

There are a number of challenges that impact pilot plant design:

- Biopharmaceutical scale-up is not linear; it is very dependent on the interactions of materials and processes.
- Trial and error methods are prone to failure.
- Reproducibility at larger scales may be difficult.
- Scale-up sometimes implies greater throughput than larger batch size.
- There may be difficulties in the application of "standard" engineering principles because of the sensitivity of the cells and the complexity of the reactions within the equipment at larger scales.

Clinical manufacturing facilities focus on the supply of materials that will be used in the different phases of the drug's clinical studies (phases I, II, and III). Each clinical phase has different requirements in terms of scale and needs, as discussed in Table 2.2.

Table 2.2 Clinical Trial Phases

	Phase I	Phase II	Phase III
Purpose	Side effects	Effectiveness	Larger population
Trial population	20–80 healthy subjects	100–300 patients	1000–5000 patients
Duration (years)	1	1–2	2–4
Cost	$100,000 to $1 million	$10–100 million	$10–500 million

Phase I clinical trials focus on determining any side effects associated with increasing dosage to the patient. During Phase I trials, early evidence of the drug's effectiveness will be determined. The patient population is small, thus the amount of material being manufactured is also small.

Phase II clinical trials determine the drug's safety and any side effects that may be seen in the patients. The phase II trial assesses the drug's effectiveness in treating the target disease or condition. It is important to know that less than one third of drugs entering phase II survive the clinical trial process to move into phase III. That is why companies are often reluctant to commit large sums of capital into these facilities; the risk is significant.

Phase III clinical trials are the final stage of the clinical process and confirm the drug's safety and effectiveness. The phase III trial will finalize the formulation and dosage of the drug product and will be where final information around the product's labeling and information will be confirmed. These trials occur at multiple testing sites and can have significant patient populations, and thus they are modeled around what will be the commercial manufacturing process. The FDA has to provide approval for a phase III trial to begin, and the expectation is that full compliance to the GMPs will be in effect.

One of the challenges with designing clinical manufacturing facilities is defining/designing what level of GMP compliance is required. The FDA has specific requirements for the manufacture of what are referred to as *investigational drugs* [5–7]. There are a number of relevant attributes:

- In early stages of preclinical and clinical development, the manufacturing process is often undergoing changes. Flexibility is necessary.
- There is no intended commercial use for clinical batches.
- The GMPs apply once the materials are introduced into human subjects.

Table 2.3 Clinical Production

Clinical production and product launch facility		
Product	Recombinant protein expressed in bacterial or mammalian cell line	
Function	Production of drug substance for phase III clinical trials and initial commercial launch with additional capacity for future expansion based on market growth	
Regulatory	GMP manufacturing	
Description	Estimates	
	Low	High
Operational scale		
Facility size (sq. ft.)	30,000	70,000
Cell culture trains	1	3
Bioreactor size (L)	2,000	10,000
Capture column size (m)	0.6	1.0
Wastewater generation (gallon/day)	35,000	120,000
Electrical power consumption (kWh/day)	12,000	35,000
Construction cost ($MM)	15	40
Project schedule (weeks)[a]		
Design	20	30
Construction	20	45
Commissioning	15	30
Total (years)	1.5	2.5
Personnel		
Lab	20	40
Manufacturing	20	45
Clerical, admin., management	10	20
Other (warehouse, validation, etc.)	15	25
Total personnel	65	130
Site factors		
Construction cost ($MM)	50	150
Land area (acres)	8	20
Regional / municipal factors		
	Relevance[b]	
Land cost	Moderate	
Labor cost	Moderate	
Highly trained labor force	High	
Proximity to university research centers	Moderate	
Proximity to government R&D and regional centers	Low	
Proximity to airports	Moderate	
Proximity to other biotech companies	High	
Proximity of support businesses[c]	High	
Public services infrastructure[d]	High	

a) Design, construction, and commissioning activities will overlap. Therefore, weeks are not additive.
b) Relevance, in this sense, is the level of importance attached to the listed factors. For instance, the cost of labor is more crucial to a manufacturing site than a R&D site where the labor cost is already expected be high.
c) Support businesses include suppliers of equipment and services. Examples are chemical reagent suppliers, hygienic piping welders, and lab instrument sales/service vendors.
d) Infrastructure items range from the quality of local water and sewer services, which have a direct bearing on the operation of the facility as well as indirect items such as the quality of roads and schools.

Table 2.4 Pilot Plant

Process Development / Pilot Plant		
Product	Recombinant protein expressed in bacterial or mammalian cell line	
Function	Technology transfer from lab to production environment. Limited production of drug substance for early stage clinical trials	
Regulatory	"GMP-like" manufacturing	
Description	Estimates	
	Low	High
Operational scale		
Facility size (sq. ft.)	15,000	35,000
Cell culture trains	1	3
Bioreactor size (L)	500	2,000
Capture column size (m)	0.3	0.5
Wastewater generation (gallons/day)	15,000	45,000
Electrical power consumption (kWh/day)	5,000	10,000
Construction cost ($MM)	10	25
Project schedule (weeks)[a]		
Design	15	25
Construction	25	35
Commissioning	12	20
Total (years)	*1*	*2*
Personnel		
Lab	15	40
Manufacturing	10	20
Clerical, admin., management	8	15
Other (warehouse, validation, etc.)	15	18
Total personnel	*48*	*93*
Factors		
Construction cost ($MM)	18	40
Land area (acres)	5	10
Regional / municipal factors		
	Relevance[b]	
Land cost	Low	
Labor cost	Low	
Highly trained labor force	High	
Proximity to university research centers	High	
Proximity to government R&D and regional centers	Moderate	
Proximity to airports	Moderate	
Proximity to other biotech companies	High	
Proximity of support businesses[c]	High	
Public services infrastructure[d]	Moderate	

a) Design, construction, and commissioning activities will overlap. Therefore, weeks are not additive.
b) Relevance, in this sense, is the level of importance attached to the listed factors. For instance, the cost of labor is more crucial to a manufacturing site than a R&D site where the labor cost is already expected be high.
c) Support businesses include suppliers of equipment and services. Examples are chemical reagent suppliers, hygienic piping welders, and lab instrument sales/service vendors.
d) Infrastructure items range from the quality of local water and sewer services, which have a direct bearing on the operation of the facility as well as indirect items such as the quality of roads and schools.

Table 2.5 GMP Commercial Manufacturing

GMP Production Facility		
Product	Recombinant protein expressed in bacterial or mammalian cell line	
Function	Large scale production of therapeutic protein for US/world markets. Campus setting with multiple buildings for manufacturing, laboratory and warehouse activities	
Regulatory	GMP manufacturing	
Description	Estimates	
	Low	High
Operational scale		
Facility size (sq. ft.)	100,000	300,000
Cell culture trains	3	8
Bioreactor size (L)	2,000	20,000
Capture column size (m)	0.5	2.0
Wastewater generation (gallons/day)	150,000	400,000
Electrical power consumption (kWh/day)	40,000	100,000
Construction cost ($MM)	100	220
Project schedule (weeks)[a]		
Design	36	60
Construction	50	120
Commissioning	40	60
Total (years)	2	3.5
Personnel		
Lab	20	40
Manufacturing	20	45
Clerical, admin., management	10	20
Other (warehouse, validation, etc.)	15	25
Total personnel	65	130
Site factors		
Construction cost ($MM)	120	400
Land area (acres)	20	50
Regional / municipal factors		
	Relevance[b]	
Land cost	Moderate/High	
Labor cost	High	
Highly trained labor force	Moderate	
Proximity to university research centers	Moderate	
Proximity to government R&D and regional centers	Low/Moderate	
Proximity to airports	Moderate	
Proximity to other biotech companies	Moderate/High	
Proximity of support businesses[c]	High	
Public services infrastructure[d]	High	

a) Design, construction, and commissioning activities will overlap. Therefore, weeks are not additive.
b) Relevance, in this sense, is the level of importance attached to the listed factors. For instance, the cost of labor is more crucial to a manufacturing site than a R&D site where the labor cost is already expected be high.
c) Support businesses include suppliers of equipment and services. Examples are chemical reagent suppliers, hygienic piping welders, and lab instrument sales/service vendors.
d) Infrastructure items range from the quality of local water and sewer services, which have a direct bearing on the operation of the facility as well as indirect items such as the quality of roads and schools.

2.4.1.3 Commercial Manufacturing

Commercial manufacturing facilities will supply the market with materials needed to address the needs of the patient population across the globe. The key points to consider for commercial facilities focus on the following:

- Production scale
- Markets where the product will be sold (drives compliance guidelines)
- Expansion/flexibility considerations
- Infrastructure requirements
- Schedule
- Cost.

2.4.2 Comparisons of the Facility Types

Each facility type has a different profile in terms of cost, schedule, operational resources, and other factors impacting the overall execution of the projects. Tables 2.3–2.5 provide some examples of case studies for the three types of facilities, defining estimated cost ranges, schedule durations, and operating resources.

References

1 FDA Glossary of Terms, US FDA, FDA.gov.
2 ICH Guideline Q11, EMA/ICH/425213, 2011, p 6.
3 Guidance for Industry, Q6B "Test Procedures and Acceptance Criteria for Biotechnological/Biological Products," US FDA, August 1999, p 12.
4 Guidance for Industry, Q8(R2) "Pharmaceutical Development," US FDA, November 2009, p 18.
5 Guidance for Industry, CGMP for Phase 1 Investigational Drugs, USFDA, July 2008.
6 Guidance for Industry, INDs for Phase 2 and Phase 3 Studies, Chemistry, Manufacturing, and Controls Information, USFDA, May 2003.
7 Guidance for Industry Q7A Good Manufacturing Practice Guidance for Active Pharmaceutical Ingredients, USFDA, August 2001.

Chapter 3

Regulatory Considerations of Biomanufacturing Facilities

Kip Priesmeyer

Kip Priesmeyer & Associates, LLC, St. Louis, Missouri, USA

3.1 Introduction

Before the architectural engineer puts pencil to paper and begins designing a multiproduct biopharmaceutical facility, he or she should have a clear understanding of the client's intended use of the facility. This includes knowing the types of products to be manufactured; their potential hazards and cross-contamination risks to other products, personnel, equipment, and to the facility; requirements for segregation and containment; the use of fixed equipment and/or single-use disposable systems; production demands in terms of product scale, output, changeover frequency; and the expected utilization and flexibility of the facility for the immediate and near future. These and many other issues must be identified, and risk-based decisions need to be considered to ensure that the completed design represents the most optimal use of the facility, while providing maximum protection for the products and related processes for which it will be used.

One of the more critical aspects that the architectural engineer must be cognizant of throughout the design process are the various global and regional regulatory statutes, related regulations, and expectations that are applicable to the facility's design, form, and intended function. Regulatory considerations must be clearly understood with respect to design and construction features, open and closed processing and related area classifications, plumbing and lighting, equipment suitability, and the flow of personnel, equipment, and materials. It is critical that the design architect understands the difference when regulators use the words "shall" versus "should," as well as the distinction between regulations and regulatory recommendations published in many guidance documents. The design engineer must have a full appreciation that the primary purpose for regulatory statutes and their enforcement is to

Process Architecture in Biomanufacturing Facility Design, edited by Jeffery Odum and Michael C. Flickinger.

promote and ensure consumer protection by preventing contamination and/or degradation of drug substances and drug products, or at least mitigating such hazards to safe and acceptable levels.

The intent of this chapter is to offer a proverbial regulatory "tool box," which the design team can consider from early conceptualization through commissioning of a multiproduct biopharmaceutical facility. In addition to an overview of US and European regulations and guidelines, this chapter focuses on the need to understand the client's quality systems, and control strategies will be discussed, along with bioburden control, to better comprehend how these elements can potentially impact facility design. A synopsis will be given on the Compliance Program guidelines used by FDA (Food and Drug Administration) investigators who inspect biopharmaceutical facilities, in addition to common FDA 483 observations, which include facility design issues. Lastly, design-related meetings with the FDA will be presented to emphasize the importance of interacting with the Agency early and often with respect to new technologies and/or novel design features and innovative use of multiproduct biopharmaceutical facilities. The information given in this chapter should not be considered as all inclusive, but rather as a basic foundation in the attempt to achieve regulatory compliance during facility design.

3.2 Regulatory "Uncertainty," A Two-Way Street

"Regulatory uncertainty" has been a major concern and criticism voiced by the pharmaceutical and biopharmaceutical industries for decades. "Uncertainty" is often expressed as attempting to understand the amount of scientific evidence needed to demonstrate that all patient-related safety concerns are addressed with respect to approval of innovative therapeutic products or delivery systems, novel concepts involving facility design and/or its flexibility of use, or perhaps existing technologies intended for a new application. The proverbial regulatory bar defining "safety" appears to be continuously raised, leaving the drug applicant to second guess how much data will satisfy the regulators' safety issues. Patient groups have also joined in voicing opposition that regulatory authorities may at times be taking an overly cautious approach with respect to safety, which has resulted in approval delays of effective medications [1]. In addition, uncertainty and the lack of uniformity have been often cited with respect to consistency of enforcement of existing regulations, particularly between the various FDA district offices and headquarter centers (CDER, CBER, and CDRH).

FDA has recognized the legitimacy of many of these concerns expressed by the industries it regulates and understands the global demands to promote new safe and effective medical products while operating under limited budgets and resources. For this reason, in part, the Agency launched the

Pharmaceutical cGMP Initiative for the twenty-first century in August 2002. The foundation of this initiative is a risk-based approach to identify and prioritize both FDA's and regulated industry's attention on critical areas and issues needing to be addressed to promote a more responsive approach to the consumer's health needs. This includes encouraging adoption and implementation of new technologies and concepts; using good science during regulatory review activities, as well as inspection and compliance assessments; and enhancing consistency and coordination of FDA's drug quality regulatory programs.

Uncertainty exists not only from an industry perspective but also from that of the regulators. For example, "uncertainty" is often revealed during review of biological license applications (BLAs) and related supplements submitted to the Agency for approval. Regulators understand the complexity of multiproduct biopharmaceutical facilities and recognize that no two facilities produce exactly the same products, under the exact same conditions, with the same variability and risks. Therefore, regulators expect the sponsor's BLAs not only include sufficient information describing processes, facility design features, and its strategy for operation but also identify and address sources of potential product contamination, degradation, and variability, including correlating methods of mitigation and control. This forms the basis for establishing a product protection strategy, which is of primary importance to regulators. Too frequently applications are deficient to provide such information, giving the regulator a level of uncertainty about the sponsor's product and process knowledge. In such cases, the regulator is obligated to request additional information to satisfy his or her interests and concerns.

Industry groups may often criticize FDA for being either too stringent or lax in the way the Agency administers its regulations, and most regulators are aware of the potential consequences to public health when either over or under regulating is exercised. By over regulating, the thought process may be to hold the sponsor/applicant to unreasonable and/or theoretical levels of absolute safety and effectiveness, which is impracticable and not realistically achievable. The effect can delay or deny regulatory approval of a needed safe and effective life sustaining drug, which could have dire results. At the other end of the regulating spectrum is failing to perform adequate inspections and document reviews to identify conditions that may have a high probability for patient harm. Regulators must be cognizant of the potential effects of both extremes and exercise the correct regulatory balance and expertise to ensure that good products are approved as quickly as possible and keep those from entering the market place where safety, purity, potency, and effectiveness cannot be fully demonstrated.

At an annual meeting of the Massachusetts Biotechnology Council, FDA Commissioner Margaret Hamburg, MD stated, "Regulatory science is the knowledge and tools needed to assess and evaluate a product's safety, efficacy, quality, purity, and performance. It involves the development of new methods, standards and models we can use to speed the development, review, approval and ongoing oversight of medical products." She went on to say that, "no matter how good your data and how important the unmet need, approval also hinges on whether you have a manufacturing facility ready to go and a plan in place for scaling up production so you can manufacture your new drug and do so in a reliable, high quality facility. Sadly, too many start-up companies fail to recognize this fundamental fact, and so approval is delayed." [2]. Dr Hamburg's message focuses on the need to reduce review and approval times for medical products, but not at the expense of jeopardizing safety and effectiveness. Along this same line, FDA Investigators and Product Reviewers are frequently reminded of the 1962 case involving Dr Francis Kelsey, an FDA medical officer, who denied approval for the US marketing of thalidomide tablets. Dr Kelsey was under tremendous industry pressure to approve this sleeping aid medication, but was adamant that the sponsor failed to provide sufficient safety data. Approximately a year after thalidomide was approved for use by German and Austrian authorities, clusters of adverse reports were received relating to a severe birth defect in thousands of infants of women who had taken this drug. Hands and feet projected directly from the body without limbs. Dr Kelsey's actions prevented such devastating birth defects in the United States, and this resulted in the 1962 Drug Amendments.

Regulatory uncertainty will always continue, to some degree; however, it is essential that regulatory authorities and the industry it serves have appreciation for each other's concerns. Absolute product safety may be unrealistic and should be approached on the basis of risk-benefit, and on the other hand, product and process understanding, knowledge, and experience are expected so that potential patient hazards can be identified and mitigated. Perhaps more common ground and better understanding can be gained between the two parties through communication improvement. One such avenue is presented later in this chapter with the discussion of FDA Type-C meetings.

3.3 Design with the Patient in Mind: Assess the Patient, Product, Process, and Plant

Whether designing a product, a process, or a facility, the first thought should be that with the patient in mind. It is explicitly understood that the medical industry exists primarily to provide safe and effective medications and medical devices to its customers, the patient, who places a valued trust in these products that are taken to manage, improve, and sustain quality of life. This trust is transferred to the companies that produce medical products, and more importantly to the individuals involved in developing, engineering, manufacturing, and controlling quality. As such, working in the health care industry brings personal rewards, challenges, and responsibilities with the understanding that small or large the contribution may be, anything and everything, developed, produced, approved, or distributed to the patient can significantly impact and alter their lives, for better or worse. Therefore, the pre-design phase of every project should consider taking time to acquire basic knowledge about the drug products to be produced in the facility, their intended use, and related production processes, as well as attain a general understanding about the facility's expected use and flexibility. It is also worth considering the personnel who will be required to operate the facility in a compliant state of control. The following is a nonexhaustive list of attributes to consider, during the pre-design stage of each project. It is recommended to assess these attributes as best as possible to better identify avenues of potential contamination risks associated with the product, process, facility, and its strategy for operation, while vigilantly keeping the patient's safety in mind.

Product (drug substances, intermediates, and final products)

1) Product classifications (protein therapeutics, monoclonal antibodies, viral and bacterial vaccines, and traditional pharmaceuticals)
2) Product scale/batch size
3) Drug substance: sterile or low bioburden
4) Dosage form and root of administration (e.g., sterile solution, suspension, IV, IM, and IP)
5) Container/closure system (e.g., vial, syringe, flex bag, and pressurized storage/shipping vessel)
6) Critical quality attributes, including drug substance and drug product release criteria
7) Toxicity and solubility considerations for product, media, buffer, and cleaning solutions
8) Inherent impurities attributed to cell line and/or viral seeds, including host cell proteins, viruses, microbial contaminants, and endotoxin
9) Stability considerations: in-process and final product requirements for hold time/expiration dating, temperature, photosensitivity, and humidity.

Process

1) Strategy for maximum process flexibility, including concurrent manufacturing of various class products
2) Sterile versus low bioburden
3) Bioburden reduction methods and profiles
4) Methods of sterilization and related process stage(s)
5) Pre/post viral inactivation/filtration and related segregation

6) Open versus closed processes and locations
7) Methods of segregation, closure, and containment, including localized protection
8) Common versus dedicated and/or single-use equipment, including equipment suitability
9) Campaign versus concurrent manufacturing and related frequency for product changeover
10) In-process testing by production personnel and required localized protection systems
11) In-process sampling systems and methods
12) Potential routes for microbial, viral, and endotoxin cross-contamination
13) Management/containment of hazardous waste.

Plant

1) Assess product protection design features in the existing facility and/or the lack thereof, including laboratories
2) Obtain overview of contamination history in existing facility, product, and environment-related
3) Compare number and types of products produced in the existing facility, including batch size and production frequency of each, with that intended to be manufactured in the proposed facility
4) Assess client's general operational strategies in existing facility against that for the proposed plant
5) Determine client's expectations for maximizing new facility's capacity and flexibility
6) Consider the potential hazards during transport of virally active samples and "dirty" equipment through the proposed plant.

It is recognized that a number of attributes listed above can be assessed only through interactions with the client's manufacturing and quality units. This is when the engineering company should enlist its regulatory compliance department to assist in interfacing and assessing design concepts with the client's existing GMP (good manufacturing practice) issues. Potentially critical regulatory concerns should be revealed, understood, and considered during pre-design discussions to better determine how the new facility and its operational strategies can best serve to correct, enhance, and not repeat the past.

3.4 Laws, Regulations, and Guidelines: Historical Background

It is often recognized that engineers may not be familiar with the laws and related regulations that are applicable to their fields of expertise, specifically the

Food, Drug, and Cosmetic Act (FD&C Act), Public Health Service Act (PHS Act), the relationship between these two Acts, and their related regulations published in the code of federal regulations (CFR). This often becomes evident during interactions such as GMP inspections, formal regulatory meetings (e.g., Type-C), and informal regulatory/industry exchanges when discussing specific aspects and interpretations of the regulations. Quality and regulatory personnel are typically responsible for interacting with regulatory authorities, with respect to knowledge, compliance, and enforcement of applicable regulatory requirements. The engineer often serves as a subject matter expert, during regulatory meetings, and may not be as versed in regulatory interpretations as his or her quality and regulatory counterparts. For this reason, this section serves to provide a historical background of some fundamental aspects of the FD&C Act, and its companion the Public Health Service Act and corresponding regulations with the purpose of either introducing or perhaps refreshing the engineers' understanding of the legal requirements. In addition, it is important to distinguish between regulatory requirements and recommendations, such as those presented in guidance documents generated by regulatory authorities, as well as organizations such as the International Conference on Harmonization (ICH) and Pharmaceutical Inspection Convention Co-operation Scheme (PIC/S).

The primary mission of regulatory health authorities is assuring consumer protection so the public have confidence in the products they purchase and use. In the United States, it is estimated that at least 20% of every consumer dollar spent is for products regulated by the FDA, which gives the Agency tremendous responsibilities for protecting and promoting public health and safety. FDA exercises this responsibility by enforcement of the FD&C Act and its related regulations. The Act was originally created in 1906 after more than 100 patients died taking a sulfanilamide elixir. It has been updated many times over the years, and the most salient amendments to the law effecting drug manufacturers were enacted as follows:

1938 The Federal FD&C Act was passed by Congress requiring drugs demonstrate safety before being marketed. In addition, this Act required establishing safe limits for unavoidable deleterious substance during manufacturing, as well as allowing FDA to conduct factory inspections.

1941 The Agency significantly changed manufacturing and quality control requirements when nearly 300 deaths and injuries resulted after taking sulfathiozole tablets, an antibiotic, combined with the sedative Phenobarbital. These changes eventually lead to the development of the GMP regulations.

1962 Congress passed the Kefauver–Harris Drug Amendments after Thalidomide was found to cause birth defects in thousands of babies born in Western Europe. This amendment required drugs be shown to be effective before marketing.

The FD&C Act [3] is comprised of various chapters, and Chapter 5, section 501(a)(2)(b), is of particular interest and importance to the pharmaceutical community as is defines adulteration of a drug as follows: *"if it is a drug and the methods used in, or the facilities or controls used for, its manufacture, processing, packing, or holding do not conform to or are not operated or administered in conformity with current good manufacturing practice to assure that such drug meets the requirements of this chapter as to safety and has the identity and strength, and meets the quality and purity characteristics, which it purports or is represented to possess."*

This definition does not distinguish between drug substance and drug product, it just states "a drug." The statute references "conformity with good manufacturing practice"; however, no regulations have been specifically promulgated for drug substances, only drug products (21CFR 211). The fact that the statute does not differentiate between drug substance and product gives FDA the authority to apply 21CFR 211 regulations to drug substances when appropriate. Therefore, it is important that the engineer be familiar with these CFR regulations and how they can apply to design projects to ensure that regulatory compliance is achieved. Otherwise, noncompliant conditions could result in the Agency having to enforce penalties that are described under Chapter 3 of the FD&C Act (Prohibited Acts and Penalties). Additional information about the FD&C Act can be found at: http://www.fda.gov/RegulatoryInformation/Legislation/FederalFoodDrugandCosmeticActFDCAct/default.htm.

In conjunction with the FD&C Act, biological products are also regulated under the PHS Act [4], which was passed by Congress in 1944. This Act requires companies that manufacture biologics to acquire an approved biologics license application (BLA). The license holder is required to comply with both Acts, as well as the respective regulations in 21CFR 21CFR 600 (Biological Products: General), 610(General Biological Standards), and 211 (cGMP in Manufacturing, Processing, Packaging, or Holding of Drugs and Finished Pharmaceuticals).

On October 1, 2003, FDA transferred the following therapeutic biological products from the Center for Biologics and Evaluation and Research (CBER) to the Center for Drug Evaluation and Research (CDER):

- Monoclonal antibodies for *in vivo* use
- Cytokines, growth factors, enzymes, immunomodulators, and thrombolytics
- Proteins intended for therapeutic use that are extracted from animals or microorganisms, including recombinant versions of these products (except clotting factors)
- Other nonvaccine therapeutic immunotherapies.

Companies wishing to obtain a BLA for these biological products are required to submit supporting data to CDER, while applications for human vaccines remain under CBER's regulatory jurisdiction.

Regulations of specific interest, including the design of multiproduct biopharmaceutical facilities, are listed in the following table, with the understanding that these should not be considered to be all inclusive and limiting.

211.28(a): personnel gowning
211.28(d): personnel health
211.42(b): flow of materials
211.42(c) : separate and defined areas
211.46(b): controlled air handling systems
211.48(b): air breaks installed in drain systems
211.50: disposal of waste in safe/sanitary manner
211.56(c): suitability of cleaning/sanitizing agents
211.65(a): equipment design and suitability
211.67(a): equipment cleaning and sanitization
211.80(b): component handling and storage
211.84(d): components liable to micro contamination
211.94(a–d): container/closure suitability
211.113: control of microbiological contamination
600.11(e): restrictions on building and equipment

It is worth noting that the word "shall" is used throughout the 21 CFR 211, as is with most regulations, to denote a mandatory requirement, whereas "should" is the regulators' method to suggest that an issue or topic be considered, but is not mandated. In other words, it is a nonbinding recommendation based on the Agency's current thinking, as found in regulatory/industry guidance documents, which will be discussed momentarily. Moreover, words such as "adequate," "appropriate," or "as necessary" are found within the regulations. As the regulations serve a highly diverse, changing and complex industry, FDA recognizes interpretation is sometimes required to ensure that regulatory flexibility is afforded across the pharmaceutical spectrum. Sections of the regulatory code containing such words provide the user with a degree of latitude for interpreting the regulation for their specific need and application. But be cautious in that regulatory interpretation (e.g., sample size, limits, and specifications) is frequently challenged by the authorities, who will not hesitate to criticize should there be events that can serve as examples of insufficient data supporting the company's interpretation and use. Such events are frequently cited as inspectional observations.

3.5 Global Guidance Documents

It has been estimated that 80% of the drugs marketed in the United States contain active drug substances that are manufactured and imported from abroad, and this is in addition to finished drug products that represent approximately 40% that are produced beyond US shores. For this reason, continued globalization of the pharmaceutical industry requires understanding the rules and regulations that are promulgated by regulatory authorities beyond the US boarders. For example, companies that market medical products in Europe must be familiar with Eurdralex vol. 4, which is the GMP guideline that governs medical products manufactured in the European Union, www.ec.europa.eu/health. This comprehensive document is not only closely aligned with many of the requirements published in FDA's CFR, 21 CFR 211, but also serves to centralize the regulatory requirements and guidance for topics such as that for active pharmaceutical substances (Part II), site master plans, ICH Q9 (Quality Risk Management, QRM) and Q10 (Quality Systems), and Mutual Recognition Agreements for batch certification (Part III). In addition, Eurdralex vol. 4 is a repository for numerous annexes (1–19) that include the production of sterile products (Annex 1), biological active substances for human use (Annex 2), computerized systems (Annex 11), manufacture of products derived from human blood or plasma (Annex 14), expectations for qualification and validation activities (Annex 15), and parametric release (Annex 17), and annexes that are specific to the production of veterinary medicinal products.

The pharmaceutical engineer will find Part II (Basic Requirements for Active Substances Used as Starting Materials) useful with respect to regulatory expectations related to facility design and construction, as well as that for utilities, water, containment, and management of sewage and refuse, which is often overlooked. European health authorities expect compliance with the regulations contained within these documents for medicinal products that are produced as well as imported in the European Union. Therefore, it would be wise for the pharmaceutical engineer to become familiar with Eurdralex vol. 4 and related annexes when approaching projects that impact products intended for the European market.

Continuing the discussion of international GMPs and standards, the Pharmaceutical Inspection Convention (PIC) and the PIC/S were established in 1970 and 1995, respectively. Although they were formed separately, today, they operate in parallel and are referred to as *PIC/S*. This international organization was founded on the principle for promoting harmonization between regulatory health authorities and the pharmaceutical industry with respect to technical standards, improve uniformity and quality assurance of regulatory inspections, exchange information while maintaining confidentiality, and provide a framework for continued training and learning that enhances consumer protection. Regulatory health authorities from approximately 19 countries, including the US/FDA, currently comprise the PIC/S organization. The pharmaceutical engineer should become familiar with the guidance documents listed on the PIC/S' website at www.picscheme.org.

Industry guidance documents are found on the web pages of most regulatory health authorities, including the FDA's. Those that are published by FDA are regulated under 21 CFR 10.115 (Good Guidance Practices) [5], and these documents represent the Agency's current thinking on a number of topics ranging from biopharmaceuticals and biosimilars to generic drugs, changes to approved applications, and many other subjects relating to the manufacture, testing, packaging, and distribution of regulated products, as well as inspection and enforcement policies. It should be understood that these guidance documents do not represent regulation, but rather function to recommend and promote regulatory compliance based on the Agency's interpretation of new technologies and best practices. Industry is not legally bound to follow or comply with guidance documents when alternative approaches are taken that ensure compliance with statutory requirements. However, it should be understood that the Agency may enforce the concepts of any guidance document that is incorporated into a company's standard operating procedures, when those procedures may be subsequently violated. It should be also noted that 21 CFR 115 (f) permits regulated industry to participate in the creation of guidance documents, as well as recommend for revisions as needed and/or withdrawal of documents when justified. Industry Guides for CDER and CBER regulated products may be found at www.fda.gov/Drugs/GuidanceComplianceRegulatoryInformation and www.fda.gov/BiologicsBloodVaccines/GuidanceComplianceRegulatoryInformation/Guidances.

Health authorities and the industries they regulate recognize that "regulatory uncertainty" is best mitigated when there is a common process for harmonizing the regulatory requirements and expectations that frequently exist between the various inspectorates. This recognition began taking shape in 1990 during a meeting between regulatory authorities representing Europe, Japan, and the United States and various pharmaceutical industry associations. It became obvious that harmonization of technical requirements was necessary to meet regulated industry's globalized challenges. It was also recognized that alignment could be best achieved through a joint regulatory-industry initiative, and thus the ICH was born. Soon afterwards, the topics of safety, quality, and efficacy were selected to focus harmonization efforts for approving medical products, and committees were formed to generate technical guidance documents in each of the three respective topics.

Though the ICH guidelines may have not been specifically written with the design engineer in mind, it is highly recommended that the engineer becomes familiar with at least several of the ICH Quality-related documents, such as

Q7A (GMPs for Active Pharmaceutical Ingredients), Q8-R2 (Pharmaceutical Development), Q9 (QRM), Q10 (Pharmaceutical Quality Systems), and Q11 (Development and Manufacture of Drug Substances). These five documents address topics that pharmaceutical engineers will likely need to consider for each design project, such as the GMPs that are applicable to design, construction, and containment discussed in chapter four of Q7A [6], as well as guidance specifically related to API's produced by cell culture/fermentation found in Chapter 18. Although the intent of Q8-R2 [7] is focused on product development, one of the primary messages of this guide is to build quality into product and process design by defining design space and understanding its relationship to scale, equipment, and operational flexibility. QRM is presented in the Q9 guidance document [8], and it not only promotes the concept of consistent, science-based decision making but also offers systematic principles as to how the management of risks can be integrated into decision processes that may impact product and process quality. There are a number of documents that define Quality Systems (e.g., ISO 9000, 9001, 9004, FDA's draft guidance on Quality Systems, ICH Q7A), and Q10 (Quality Systems) [9] was created to bridge and harmonize regional regulatory differences that may exist. This guideline integrates the concepts of Quality Systems with QRM, which if applied correctly, may reduce process variability and result in improvements in-process robustness and product quality. Lastly, Q11 (Development and Manufacture of Drug Substances) [10] takes many of the concepts and principles addressed in Q8, 9, and 10 and shows how they are applied with respect to development and manufacturing. These and other ICH guidelines can be found on the ICH website, www.ich.org. It should be noted that FDA has adopted these guidance documents, which can also be found at www.fda.gov/regulatoryinformation/guidances.

The guidance documents discussed in this section focus to build quality into pharmaceutical product and related processes by emphasizing critical quality product and process attributes be accurately defined, unacceptable variability identified and reduced, use of risk management to make sound and effective science-based decisions, while reducing regulatory uncertainty as best as possible. It is incumbent that the pharmaceutical engineer be familiar, and considers the information contained in these guidelines, during each and every design project. Facility form and function serve to ensure that not only manufacturing quality and consistency are achieved but also product protection is guaranteed throughout the product lifecycle.

3.6 Quality Systems and Risk Management

The preamble to the 1978 amendment of the GMPs emphasizes that "quality should be built into the product, and testing alone cannot be relied on to ensure product quality" [11]. The current concept of quality systems has its roots in FDA's October 7, 1996 Federal Register announcement, which was subsequently promulgated under 21 CFR 820, better known as the *Medical Device Quality System Regulation*. The need for this regulation was in part a response to design defects that were responsible for many medical device recalls experienced between 1983 and 1989, and its corresponding European counterparts are found in the Quality Management Systems guidelines published as ISO 9000, 9001, and 9004. The regulation recognizes that not only procedures, processes, and resources but also control and monitoring of product design are necessary to ensure a product's fitness-for-use, as well as a defined and committed organization structure is required to implement a functional quality system. Although aspects of the quality system already existed in the pharmaceutical GMPs, the Pharmaceutical cGMP Initiative for the twenty-first century, launched in 2002, recognized that industry technologies are rapidly changing and a modern regulatory quality system is necessary to promote and keep pace with these changes. The 1978 amended GMPs focused on quality control, and a more modern system was needed that included quality management, quality assurance, and the use of risk management tools. In 2006, FDA published the *Guidance for Industry Quality Systems Approach to Pharmaceutical cGMP Regulations* [12] (www.fda.gov/downloads/drugs). This document describes the philosophy and elements of a modern quality system, and it considers existing GMP regulations to achieve regulatory compliance. The content of this guidance document is similar to that found in ICH Q10 (*Pharmaceutical Quality System*), published in 2007 (www.ich.org).

An effective Quality System is designed to monitor, detect, and appropriately respond to an operation's performance, and if properly managed, it will be an invaluable resource for continued product and process knowledge and improvements, and assurance of the operation's state of control. One of the first considerations is obtaining an objective realization of the current state of control, which is necessary for implementing a control strategy. You have to understand where you are before knowing where you need and want to go. This journey begins with interpreting and complying with cGMP requirements such as investigating and determining root cause for unexplained discrepancies and deviations related to "the failure of a batch or any of its components to any of its specifications" (21 CFR 211.192) [13]. Thorough and complete investigation of product and process-related deviations generally provides information and knowledge that was previously unknown or not understood. Once root cause is identified, and appropriate corrective and preventive actions (CAPA) are applied and verified, repetitive events can be avoided and continued product and process knowledge is gained.

Other cGMP requirements that should be included, but not limited to, an effective quality system, include investigating complaints (21 CFR 211. 198) [13]; returned drug products (21 CFR 21.204) [13]; annual review of records that document "quality standards of each drug product to determine the need for changes in drug product specifications or manufacturing or control procedures" (21 CFR 211.180e) [13]; and any event requiring product recall, reject, rework, reprocessing, revalidation, and related changes to correct and prevent recurrences. If deviation events are properly and adequately investigated, the outcome will typically result in increased knowledge and understanding of process performance, limitations, and variability, which can lead to product and process improvements if properly addressed and corrected. This ultimately enhances the effectiveness of the quality system, including documented evidence of an acceptable state of control. Only then will the manufacturer be able to ensure that the products produced are fit for intended use and the patients' needs are met.

It should also be mentioned that quality systems are one of the central areas reviewed during FDA inspections, and approximately 80% of the Warning Letters issued by the Centers for Drugs (CDER) and Biologics (CBER) include citations related to deficiencies found in the management of this system. More will be said about FDA inspections later in this chapter.

QRM is an integral component to an effective quality system, and it must be considered to ensure that the current state of control is factual and realistic and that a control strategy has been adequately defined and implemented to increase the understanding of product and process. Both FDA and ICH (Q9) have published guidelines on this topic, which are respectively found at the websites listed earlier.

Risks are common and expected in all areas of our culture, such as business, government, and our personal lives, and they are the result of having incomplete information or knowledge. Risks cannot be avoided, only managed as best possible by first identifying potential risks and consider what could go wrong, then understanding the probability of it happening, and finally assessing the possible impact or harm that could result should it occur. Effective risk management includes the use of recognized methods to assess levels of risk to determine if acceptable or not. These methods include, but are not limited to, failure mode effects analysis (FMEA), hazard analysis and critical control points (HACCP), fault tree analysis (FTA), or any number or combination of the valid approaches listed in the QRM guidelines published by FDA and ICH. First, it must be understood that persons who use these risk-related tools should be qualified in their use before applying them. Otherwise, the results could be incomplete and/or inaccurate, which would jeopardize the credibility of the risk assessment outcome. When properly used, the information gained using a valid risk assessment method can assist in making better and more accurate decisions, which can ultimately lead to risk mitigation.

Pharmaceutical engineers are frequently challenged by clients to develop proposals relating to new and novel designs and concepts to increase facility flexibility for use, whereas assuring product protection is guaranteed and will be accepted by regulatory authorities. Examples of such challenges may include assessing the potential risks of producing viral and bacterial vaccines in the same facility; starting upstream manufacturing of a monoclonal antibody concurrently with downstream production of a therapeutic DNA recombinant product; or supporting the concept of open "ballroom" manufacturing for pre/post viral inactivation, while reducing area classification and related environmental monitoring. These engineering challenges are real and have obvious product and process risks associated with them that must be objectively and scientifically assessed to determine feasibility of the project. As part of the assessment, it is incumbent that the engineer has an understanding of the client's current state of control, including regulatory issues that may need consideration as part of assessing proposed risks. From a regulator's perspective, there may be no justification for adding additional challenges and risks to a facility and/or a process until existing regulatory issues be satisfactorily remediated and/or unless the intent of the project is to correct problems that have been identified. For this reason, it is incumbent that the engineer interacts with the client's Quality/Compliance Unit, which will hopefully provide a candid overview of current regulatory challenges that need to be considered as part of the project's risk assessment. In doing so, the engineer needs to be familiar with the principles of root cause analysis, CAPA effectiveness, and change management to effectively communicate those interests that may potentially impact the proposed project at hand.

It is expected that risk assessments be performed and documented by a cross-functional team representing, but not limited to, the engineering company as well as the client's engineers, Quality, Manufacturing, Maintenance, and, if necessary, Development and Regulatory Affairs. Real and potential risks associated with the project need to be identified and scored according to severity/harm, probability of occurrence and ability for detection, then ranked according to criticality (high, medium, and low). Criteria must be established for ranking risk criticality levels, as well as documentation maintained that supports scoring for each identified risk. It has been said that a risk score is not the risk, but rather what actions are taken to reduce risks that are identified, especially those ranked as "high" or "medium."

FDA currently does not require risk assessments be performed, but when they are, investigators may likely review and challenge the related documents to determine if such assessments are properly used. If a risk assessment is believed to be a misuse to justify a poor practice and/or to circumvent cGMP requirements, the regulator may cite this as an observation on an FDA-483, List of Inspectional Observations, and reviews of other risk assessments may be done so under suspicion. It should be understood from the regulator's perspective that to validate a poor practice is a poor practice, and should be avoided.

The Quality System and QRM go hand-in-hand and are indispensable to the design engineer. Understanding the elements of the quality system allows the engineer to better assess the client's current regulatory compliance and control strategies to determine if they are aligned with the proposed engineering project. This can be then confirmed, if necessary, using risk management tools that can proactively identify various levels of risks so better informed decisions can be made to determine how such risks will be managed and remediated to ensure success of the project. It is a regulator's nature to be skeptical of novel engineering designs and concepts that he or she has little or no prior experience. This chasm can often be successfully bridged with a scientifically sound risk assessment that considers current quality aspects with sound engineering principles and recommendations that enhance the client's control strategy, and in doing so, enhances protection of products and processes as well.

3.7 Product Changeover and Regulatory Assessment of Cleaning Validation

Optimal flexibility is one of the primary objectives to manage and operate a modern multiproduct facility, and the multipurpose use of shared equipment, process systems, space, and personnel are common practices necessary to increase production efficiency and reduce costs. Regulatory authorities understand the expense of doing business and the need to maximize the use of space, equipment, materials, personnel, and time. However, they are keenly aware of the increased risks and potential consequences of cross-contamination, especially in facilities that manufacture complex biological products that use living organisms that produce protein-based drug substances that can be viscous and tacky, difficult to clean and remove from equipment, and highly susceptible to adventitious agents (bacterial, viral, and mycoplasma). Cross-contamination has been responsible for facility shutdowns, and lengthy, expensive remediation activities that can lead to product shortage. For this reason, in part, FDA specifically requests methods of containment and contamination control be included in BLAs, and guidance can be found in the Agency's Chemistry, Manufacturing, and Controls guidance for Therapeutic Recombinant DNA-Derived Product or a Monoclonal Antibody for *In Vivo* Use [14].

FDA and EU regulations contain specific requirements that address equipment cleaning, maintenance, and related documentation. FDA requirements are found under sections 21 CFR 211.67 (a–c) [13], which state, in part, "*Equipment and utensils shall be cleaned, maintained, and sanitized at appropriate intervals to prevent malfunction or contamination that would alter the safety, identity, strength, quality, or purity of the drug product beyond the official or other established requirements.*" This is echoed by EU guidance 3.37, which states, "Washing and cleaning equipment should be chosen and used in order not to be a source of contamination." [15]. One of the common themes between these two regulations is contamination prevention, as regulators clearly understand that the level of risks for potential cross-contamination increases in a multiproduct facility, especially when shared equipment is employed.

Cross-contamination between different products produced in the same facility is often attributed not only to inadequate changeover procedures, but more so to cleaning validation of equipment and the facility, especially equipment intended for multiproduct use. Cleaning validation is the foundation for product/process changeover activities. The strategies, procedures, and policies employed in cleaning and sanitization must demonstrate compliance with defined critical process parameters and quality attributes before a regulatory agency will approve a facility for multiproduct use. For this reason, regulatory assessments of product and process changeover procedures often begin with review of the cleaning validation program and its master plan, because serious deficiencies found in this area can provide regulatory authorities with sufficient evidence and justification for objecting to a company's intended multiproduct use of a facility and/or the equipment intended for multiproduct use. Should this occur, it could jeopardize the approval of the BLA until the company is sufficiently able to rectify the regulator's concerns and objections.

Regulatory assessment of a company's cleaning validation program typically begins with first understanding the following basic information before reviewing and challenging related documentation:

1) Production strategy with respect to campaign versus concurrent manufacturing
2) Identification of all products intended to be manufactured within a given facility, including, but not limited to, drug substances, intermediates, drug products, investigational new drugs, buffer and media solutions, reagents, and that which are microbiologically and/or virally active, and location within the process (e.g., cell culture lab)
3) Identification of products intended to be sterile, nonsterile, and/or low bioburden, including that designated at each unit operation for each respective product
4) List of product dedicated versus shared, multiproduct use equipment, including the identification of related contact materials, and equipment surface types (e.g., stainless steel, glass, plastic/polymers, and Teflon)
5) Determination of soil types, sources, and respective solubility, toxicity, and potency, as well as identification of "worst-case" soils that are most toxic and difficult to remove at the maximum "dirty" hold time established
6) Risk assessments conducted to identify potential avenues and levels of risks (high, medium, and low) associated with cleaning (e.g., equipment washers

that concurrently clean glassware and utensils used in the production of different products)

7) List of cleaning, sanitizing, and disinfecting agents, including their intended use (e.g., bactericide, sporicide, and viralcide), concentration and contact time, and equipment surface types applied
8) Equipment cleaning methods (CIP, SIP, COP, and manual) and justification for critical cycle parameters (time, temperature, rinse volumes, cleaning/sanitizing agents, etc.)
9) Sampling locations and methods, tests performed, and related acceptance criteria
10) Justification of acceptance criteria for maximum allowable product carry-over (batch to batch of same product and between different products)
11) Development and justification of cleaning validation matrices/grouping based on either product and/or equipment design and surface types
12) Personnel training and qualification, specifically that applicable to manual cleaning operations.

Biopharmaceutical manufacturing requires the use of numerous and varied types of equipment having varying sizes, configurations, geometries, product contact surface types, and intended uses that must be cleaned according to validated processes. Developing and validating cleaning cycles for individual pieces of equipment is time consuming, costly, and impracticable in many cases, which is why regulatory authorities acknowledge adopting a matrix or bracketed approach to cleaning can be justified when sufficient evidence exist. Such evidence is based in part from the above information that can support grouping products and/or pieces of equipment into individual families that employ a common cleaning method. The cleaning cycle defined for a particular grouping is based on worst-case conditions, which include, but not limited to, least soluble, most toxic soils to remove from the most difficult equipment to clean, based on size, configuration, and surface type, and includes consideration for "dirty" hold time. Developing individual matrices must be scientifically justified and documented using a risk-based approach that evaluates the above variables against individual pieces of equipment within each family grouping. Predefined acceptance criteria (e.g., conductivity/resistivity, TOC, micro/endotoxin, protein, and visual inspection), for each defined cleaning cycle, must be achieved for that equipment judged to be worst-case in each matrix group. Only then will a regulator develop a level of confidence that a defined cleaning cycle is valid, as well as valid to a related equipment matrix.

Biopharmaceutical manufacturing processes employ many complex and varied pieces of equipment, such as bioreactors; product transfer lines; harvesting, inactivating, and pooling vessels/tanks; process chromatography resins and columns; ultrafilters and related housings; and numerous in-line filters, sensors, probes, gaskets, and connectors. Some equipment is assigned to be product dedicated (e.g., chromatography resins and UF/DF cassettes), some is intended for single use (e.g., filters, bags, silicone tubing, and connectors), whereas other equipment may be meant for multi-use purposes (e.g., bioreactors, vessels, and transfer lines). Campaign manufacturing is a common practice for producing biopharmaceuticals; whereby, a product is produced for either a defined period of time and/or a defined number of batches before the campaign ends and production suites, and equipment are prepared/changed-over for campaign manufacturing for a different product. This approach requires a defined strategy, supported by detailed policies and written procedures that ensure that avenues of potential cross-contamination between the two products and processes are identified and precluded.

There are a few basic steps that should be considered when conducting product changeover activities.

1) At the end of a production campaign, procedural controls should be implemented to ensure that manufacturing areas and related equipment are placed in a decommission, hold status, and labeled/identified accordingly, making further manufacturing operations inaccessible until changeover activities are satisfactorily completed and the areas are released back to the Production Department by the Quality Unit.
2) Changeover activities, operations, and instructions should be detailed in an SOP and documented by a changeover batch record or checklist. Personnel performing changeover activities must be trained and/or qualified.
3) All equipment, materials, solutions, and documentation not required for the next production campaign are labeled with the appropriate status and removed from the area.
4) Manufacturing equipment and production areas should be cleaned and sanitized, before disassembly activities, using validated cleaning procedures. Disassembly of multi-use, shared equipment begins with removal of equipment-related filters, gaskets, o-rings, sensors, and so on. Should a company elect not to remove and replace the equipment's soft rubber parts, regulators will likely confront this decision by challenging cleaning processes for effectiveness to remove product residues and potential microbial contaminants immediately around and behind these parts. For the production of vaccines, particularly viral vaccines, FDA/CBER typically requires all soft rubber parts, related to equipment used in virally active processes, be replaced and not reused.
5) Once equipment and related parts are cleaned and reassembled, a second cleaning is performed, using the validated cleaning process, which is followed by sampling (e.g., rinse water and swabs) and testing to determine if product-carryover acceptance limits are met. Failing results must be investigated.

6) The Quality Unit is responsible to ensure that all changeover activities are satisfactorily performed, including compliance with cleaning acceptance criteria, before process equipment and the production area are returned to the Manufacturing Department for the start of a campaign for a different product.

3.8 Control Strategy

A control strategy begins with building quality into the product and its related manufacturing processes. This starts with sufficient knowledge and understanding that are required to define and assign appropriate quality attributes that define and characterize the product and process. It continues with improvements throughout the product's life cycle by making valid scientific decisions using the principles presented in ICH Q8 (*Pharmaceutical Development*), ICH Q9 (*QRM*), and ICH Q10 (*Pharmaceutical Quality Systems*). In addition, a valid control strategy must be based on clear understanding, interpretation, and execution of cGMP regulatory requirements, as well as the current expectations of regulatory authorities (e.g., FDA, EMEA, and MHRA).

As will be discussed later in this chapter, the regulatory inspection is one of the primary tools used by regulatory authorities to assess a company's compliance status and state of control. ICH 10 defines "state of control" as, "*A condition in which the set of controls consistently provides assurance of continued process performance and product quality.*" FDA's definition of "state of control" is found in Compliance Program Guide 7356.002 [16], and it is similar to that defined by ICH, but also includes a reference to compliance with cGMP regulations, as well as criteria for when a company and a system is found to be "out of control," as follows:

> *A drug firm is considered to be operating in a state of control when it employs conditions and practices that assure compliance with the intent of Sections 501(a)(2)(B) of the Act and portions of the CGMP regulations that pertain to their systems. A firm in a state of control produces drug products for which there is an adequate level of assurance of quality, strength, identity and purity.* ***A firm is out of control if any one system is out of control. A system is out of control if the quality, identity, strength and purity of the products resulting from that/those system(s) cannot be adequately assured.*** *Documented CGMP deficiencies provide the evidence for concluding that a system is not operating in a state of control.*
>
> (*CP 7356.002*)

To be in a "state of control" essentially means that the elements of a product's life cycle, defined in ICH Q10 (e.g., development, tech transfer, manufacture, and discontinuation) are stable, predictable, and consistent with predefined acceptance criteria. However, before a state of control can be assured, a strategy must be developed that is able to detect and monitor the level and progress of control that is required at each stage of a product's life cycle. ICH Q10 defines "control strategy" as, *"A planned set of controls, derived from current product and process understanding that assures process performance and product quality. The controls can include parameters and attributes related to drug substance and drug product materials and components, facility and equipment operating conditions, in-process controls, finished product specifications, and the associated methods and frequency of monitoring and control."*

The key words in the above ICH definition are *"set of controls, derived from current product and process understanding that assures process performance and product quality."* The first requirement for ensuring that regulatory "control" is having a robust "set of controls" that can objectively detect, monitor, assess, and respond to deficiencies and discrepancies that are experienced in the course of manufacturing biopharmaceuticals. The elements of such controls were designed into the 1978 GMP amendment with the inclusion of the following 21CFR regulations:

1) 211.22 (Responsibilities of the quality control unit)

> *(a) There shall be a quality control unit that shall have the responsibility and authority to approve or reject all components, drug product containers, closures, in-process materials, packaging material, labeling, and drug products, and the authority to review production records to assure that no errors have occurred or, if errors have occurred, that they have been fully investigated. The quality control unit shall be responsible for approving or rejecting drug products manufactured, processed, packed, or held under contract by another company.*

2) 211. 194 (Production record review)

> *...the failure of a batch or any of its components to meet any of its specifications shall be thoroughly investigated, whether or not the batch has already been distributed. The investigation shall extend to other batches of the same drug product and other drug products that may have been associated with the specific failure or discrepancy. A written record of the investigation shall be made and shall include the conclusions and follow-up.*

3) 211.180e (Records and reports, general requirements)

Written records required by this part shall be maintained so that data therein can be used for evaluating, at least annually, the quality standards of each drug product to determine the need for changes in drug product specifications or manufacturing or control procedures. Written procedures shall be established and followed for such evaluations and shall include provisions for:
(1) *A review of a representative number of batches, whether approved or rejected, and, where applicable, records associated with the batch.*
(2) *A review of complaints, recalls, returned or salvaged drug products, and investigations conducted under 211.192 for each drug product.*
(f) Procedures shall be established to assure that the responsible officials of the firm, if they are not personally involved in or immediately aware of such actions, are notified in writing of any investigations conducted under 211.198, 211.204, or 211.208 of these regulations, any recalls, reports of inspectional observations issued by the Food and Drug Administration, or any regulatory actions relating to good manufacturing practices brought by the Food and Drug Administration.

4) 211.198 (Complaint files)

……… *any complaint involving the possible failure of a drug product to meet any of its specifications and, for such drug products, a determination as to the need for an investigation in accordance with 211.192. Such procedures shall include provisions for review to determine whether the complaint represents a serious and unexpected adverse drug experience which is required to be reported to the Food and Drug Administration in accordance with 310.305 and 514.80 of this chapter.*

5) 211.100 (Written procedures; change control)

There shall be written procedures for production and process control designed to ensure that the drug products have the identity, strength, quality, and purity they purport or are represented to possess. Such procedures shall include all requirements in this subpart. ***These written procedures, including any changes, shall be drafted, reviewed and approved by the appropriate organizational units and reviewed and approved by the quality control unit.***

Regulators clearly understand and appreciate the many challenges involved in the daily operations of a multiproduct biopharmaceutical facility that produces and tests complex biological products. They understand that these products are often sensitive to their immediate environments and related process components (e.g., temperature, pressure, pH, nutrients, contact surfaces, and contaminants), and recognize that all aspects associated with a batch may not meet defined acceptance criteria if appropriate procedures and operating controls are not established and properly followed. Should this occur and result in product/process-related deviations and discrepancies, a thorough and complete investigation is required according to 21CFR 194, as stated earlier.

A thorough and complete investigation will typically result in gaining knowledge and understanding about a product, process, facility, and/or procedures that were previously unknown or not recognized. Appropriate corrective actions are expected to be implemented to ensure that repetitive events do not recur. Identifying and correcting product- and process-related deficiencies represents a reactive approach to regain a desired level of GMP control. However, once this is achieved, the information and knowledge gained should contribute to develop a more robust, proactive control strategy that will effectively reduce deviations, complaints, rejected materials, and product recalls by refining and improving product quality. This information is a source of experience and understanding, which can contribute to future risk assessments, so better informed decisions can be made proactively rather than reactively. This is the goal of the quality system and path to increase protection for patients and the products they rely on.

3.9 Contract Manufacturing Organizations

Contract manufacturing has become commonplace in the biopharmaceutical industry over the past decade, and trends indicate that this will continue for the foreseeable future, as growth in the industry continues to expand and new milestones are reached each year with new products entering the market. Outsourcing services and activities have become more specialized, and logistically more complex for the BLA owner, than 10–15 years ago when they were somewhat limited (e.g., fill/finish, packaging, and labeling). Today, outsourcing can involve upstream manufacturing of a monoclonal antibody at one facility, which then ships the harvested material half a globe away to a contractor that performs downstream purification, before transporting the finished drug substance to a contractor that aseptically fills, lyophilizes, and visual inspects the dosage form, then sends it to a company that packages, labels, and cartons the finished drug product. Each of these contractors will be required to collect and submit in-process and finished drug product samples to one or more contract laboratories offering analytical services specializing in complex biopharmaceutical testing. These laboratories may offer other services such as storing and testing stability samples and/or maintaining drug product retains. Documents generated by each contract service, including deviation investigations, are submitted to the BLA license owner for review and approval before product release.

The above scenario is real for many companies that are licensed to market biopharmaceuticals, as outsourcing provides options to reduce capital costs and expenses required to operate and manage complex, manufacturing facilities. Outsourcing allows a company to focus its internal resources on core competencies that can be utilized to increase efficiency and effectiveness of its operations, while reducing overhead expenses. However, such flexibility does not come without risks when placing a product's safety, quality, purity, and efficacy in the hands of an outside company, whose quality culture may not be fully understood or aligned with that of the license holder. Biopharmaceutical manufacturing is a complex process requiring strict controls, detailed procedures, thorough documentation, and well-trained/qualified personnel to perform the tasks at each unit operation.

Regulatory authorities are well aware of the complexities and risks involved in these activities and are ever more cognizant when multiple companies are engaged in the single effort to produce a marketed biopharmaceutical. Outsourcing requires a highly coordinated effort on the part of the license holder, who has full responsibility for the marketed product and must ensure that all contract parties are operating in a state of control and cGMP compliance before qualifying and approving their services. The contractor and license holder then enter into a contractual quality agreement, which includes identifying responsibilities of each party. These agreements must be thorough and complete; otherwise, gaps can result in the technical transfer of processes, methods, and related documentation.

The Quality Agreement is one of the first documents requested by FDA Investigators when conducting inspections of contract manufacturers, so responsibilities shared between the two parties are identified and connected with the appropriate company being inspected. Though there is no regulatory requirement for quality agreements per se, ICH Q7 (*Good Manufacturing Practice Guidance for Active Pharmaceutical Ingredients*) recommends formal agreements be established that identify responsibilities of each party. In addition, 21 CFR 211. 22(d) states, "*The responsibilities and procedures applicable to the quality control unit shall be in writing; such written procedures shall be followed.*" When the responsibilities for any part of manufacturing, testing, packaging, and/or holding of a drug are contracted to an external service, these respective responsibilities must be identified to ensure that the activity is performed according to the procedures reviewed and approved by the license holder's Quality Unit.

There are a number of elements that need to be considered when developing quality agreements, which delineate responsibilities, and this information can be found in FDA's *Guidance for Industry Contract Manufacturing Arrangements for Drugs: Quality Agreements* [17]. The guidance document recommends that quality agreements established between two companies identify responsibilities and procedures that address, but not be limited to, the following:

1) Product release is the responsibility of the license holder and cannot be delegated to the contract service provider.
2) A written plan for communicating deviations, which may be experienced during manufacturing, including the investigation, documentation, and resolution of such events.
3) The license holder should have the authority to periodically audit the contract facility, as well as perform "for-cause" audits when necessary to evaluate the contractor's cGMP compliance status.
4) The Agreement should include a section on change control to ensure that each party advises the other of proposed changes (e.g., raw material suppliers, specifications, equipment, procedures, and validation) before their implementation.
5) The Agreement should also have a section that addresses preventing cross-contamination, when a contract facility manufactures and/or performs testing for multiple clients.

The FDA guidance document recommends that a quality agreement includes information that lists the following categories and identifies responsibilities for those activities under each as follows:

6) Facilities and equipment: validation, qualification, and maintenance activities, including that for utilities and automated systems, as well as for performing environmental monitoring and assigning room classifications.
7) Materials management: responsibilities for establishing specifications for raw materials and components, including auditing and approving suppliers, and an appropriate inventory system for management and accountability of materials.
8) Product-specific considerations: product and process-related specifications, batch numbering system, expiration and retest dating, storage and shipment of products, process validation, and technical transfer of product and process information.
9) Laboratory controls: procedures for sampling and testing, method validation and the transfer of methods to the contract facility, qualification/calibration/maintenance of laboratory equipment, as well as timely communication of results, and responsible party for investigating test result failures and deviations.

10) Documentation: the license holder should review and approve documents generated by the contract facility that are related to the product produced and/or service provided, such as procedures, production and testing records, and deviation investigations, if required.

Having a clear and documented description of each party's responsibilities and a defined process for communicating changes, product/process-related problems, activities, and expectations contributes to a better understanding and coordination between the license holder and contractor that will enhance safety and quality of the marketed drug product.

3.10 FDA Inspections of Biopharm Facilities and Regulators' Priorities

As discussed earlier in this chapter, the FDA is a consumer protection agency that enforces the FD&C Act, specifically section 501(a)(2)(b), as it relates to the GMPs for drugs. There are several methods of enforcement, but the factory or plant inspection is the primary tool used by the Agency to assess a company's compliance status with regulations and established procedures. Biopharmaceuticals are regulated by FDA's CDER, as well as the CBER. Those biopharmaceuticals regulated by CDER include therapeutic synthetic peptides of 40 or fewer amino acids, therapeutic DNA plasmid products, monoclonal antibodies for *in vivo* use, and therapeutic recombinant DNA. Those regulated by CBER include fractionated blood products, antitoxins, allergenic extracts, human vaccines, products from manipulated human cells, and that based on gene therapy intended to replace defective or missing genetic material.

Inspections include both domestic and foreign manufacturers that file BLAs with one of the two FDA Centers. Inspections are coordinated and conducted by the division manufacture product quality (DMPQ), which reside within each of the two respective Centers, as well as by field Investigators who are located in FDA district offices throughout the country. In addition, inspections of CBER-regulated products are performed by Team Biologics, which is a specialized group of investigators that report to FDA's Office of Regional Operations, which is within the Office of Regulatory Affairs, as depicted below. This team was formed in 1997 to conduct biennial inspections of CBER-regulated products, whereas CBER/DMPQ continues managing the preapproval inspection program.

Since 2002, FDA has used a risk-based approach for conducting inspections, which was done in response to the provisions in the FDA Modernization Act of 1997. Before that time, inspections were conducted by reviewing product profile classes (e.g., injectables, solid dosage, bulk drug substances, and liquids), and it was decided to revise the inspection method by organizing typical pharmaceutical production and related activities into six quality systems for better efficiency and effectiveness. These six systems are summarized as follows:

1) *Quality System.* This is the only one of the six systems that is required to be covered during each and every inspection, which is an indication of the priority FDA places on its review. This system includes the quality unit's responsibilities for establishing and managing subsystems that address customer complaints, change control, discrepancy and failure investigations, CAPA, Biological Product Deviation Reports (BPDR), batch release, annual record review, product rejects, rework/reprocess, recalls, returns, quarantine materials, stability failures, validation/revalidation (e.g., computer, manufacturing process, and laboratory methods), employee training/qualification, and product improvement projects. Should significant deviations of this system be discovered during the course of an inspection, regulatory actions will likely be recommended, as this system is a direct reflection of a company's overall status of compliance with cGMPs. For example, repetitive deviations may indicate that either root cause investigations are inadequate and/or the corrective actions assigned are inappropriate and/or ineffective. Environmental monitoring trends that reflect an increase in microbiological contaminations could suggests poor employee practices, which may be indicative of an ineffective training/qualification program or perhaps an signal that increasing production activities are beginning to adversely challenge the production environment and its capabilities. The Quality System could be judged as deficient and ineffective should trends in customer complaints begin to rise for no apparent reasons, giving thought that the design and/or management of this system is incapable of early warning detection of drift to a noncompliance status.
2) *Facilities and Equipment System.* This system includes buildings and facilities along with maintenance, equipment qualifications (installation and operation), equipment calibration and preventative maintenance, cleaning and validation of cleaning processes, and utilities that are not intended to be incorporated into the product such as HVAC, compressed gases, steam, and water systems.
3) *Materials System.* This system includes activities to control finished products, components, container/closure systems, validation of computerized inventory control processes, drug storage, and distribution controls, as well as water or gases that are intended to be incorporated into the product.
4) *Production System.* This system includes activities to control the manufacture of drugs and drug products including batch compounding, dosage form production, in-process sampling and testing, and process validation.

It also includes establishing, following, and documenting performance of approved manufacturing procedures.
5) *Packaging and Labeling System.* This system includes activities that control the packaging and labeling of drugs and drug products. It includes written procedures, label examination and usage, label storage and issuance, packaging and labeling operations controls, and validation of these operations.
6) *Laboratory Control System.* This system includes activities related to laboratory procedures, testing, development and validation of analytical methods, and the stability program.

The frequency and depth of inspections are determined by several factors such as product/patient criticality (e.g., sterile injectable products and low therapeutic pediatric dose), as well as the company's GMP compliance history and technologies that are employed (e.g., barrier isolation), which dictate the level of inspection (I or II) that will be considered.

Level I, also referred to as a *Full* inspection, is performed when there is either little or no prior information about the company's compliance status, such as an initial inspection, or when an unfavorable compliance history is known to exist from previous inspections (e.g., Warning Letter and Consent Decree), or significant changes have occurred since the last inspection. In such case, FDA Investigators are instructed to select four systems to cover during the inspection, one of which must include the Quality System. For biological products, Investigators are instructed to include the production system as well.

Level II inspections, also known as *Abbreviated*, are conducted when a satisfactory inspection history exist where at least two consecutive inspections resulted in no regulatory actions or one of the two previous inspections was Level I. This inspection strategy requires coverage of the Quality System and one of the other six systems, preferably one not assessed during the previous, most recent inspection.

Both inspection levels I and II require evaluation of the Quality System, which is the Investigators' proverbial "window" for making a quick assessment of a company's compliance status and state of control. FDA observations frequently found with the system include incomplete investigations of product/process failures, as well as customer complaints (e.g., product quality impact assessment and root cause analysis), repetitive events with inadequate corrective actions, changes made but not documented, including changes to BLA commitments without notifying FDA. Should this system be found out of compliance, regulatory action (e.g., Warning Letter) would likely be recommended. As mentioned earlier, approximately 80% of Warning Letters issued include one or more citations referencing the Quality System. Section V of CP 7356.002 M provides examples of significant cGMP deviations, which could also support regulatory action.

The "for-cause" inspection is another type of inspection at the FDA's disposal, and it is used when the Agency has documented and/or credible circumstantial information that serious GMP violations (e.g., data integrity/falsification) have occurred. When this happens, an inspection team of experienced, senior Investigators is assembled to develop and execute an inspection approach that may appear to company management as routine (e.g., Level I inspection), but it is designed to focus on the credibility of the information reported to the Agency, such as that from an informant. Should the allegations be confirmed and documented, they will be assessed to determine if violations are limited to cGMP and/or if potential violations against Title 18 occurred, which is the criminal and penal code of the federal government. Should the inspection reveal potential violations of Title 18 (e.g., submitting false data to FDA and/or making false statements to a federal official), FDA may decide to refer the case to its Office of Criminal Investigations (OCI) for review and follow-up. These investigations can result in the most serious consequences involving the company and persons responsible for Title 18 violations.

Regulatory authorities expect manufactures to be "inspection ready" at all times, meaning that companies be prepared to present verbal and documented evidence for those activities and decisions made since the previous inspection, including problems that have been experienced over that time. One of the first priorities regulators address is to determine if all product batches released for commercial distribution met their respective quality standards and criteria at the time of release, and continue doing so over the defined shelf life (stability). Should this not be the case, the company must be fully prepared to explain and justify the actions and decisions taken for leaving such product in the market place. Another priority is determining how product/process-related problems and deviations are managed by assessing if appropriate responses and corrective actions were taken. Investigators clearly understand that product and process problems are a fact of life in a normal manufacturing environment, and they want to evaluate how such problems are captured, escalated to management, and resolved to prevent repetitive events, as well as no finished product is released before all related deviation investigations are fully investigated and closed. Companies must be prepared to adequately discuss the rationale and outcome of investigation activities, which the regulator will assess. New facility designs, modifications, and changes, including clean utilities, area classifications, and new equipment are typically requested for review during inspections, and the company must be prepared to present this information along with any related FDA notifications for such changes, particularly changes made that the company felt did not require prior notification to FDA.

These are only a few examples of "inspection readiness," as well as a regulator's inspectional priorities. FDA Compliance Program Guide CP7356.002 M (Inspection of Biological Licensed Therapeutic Drug Products) includes examples of significant cGMP deviations that every company should assess to against their current state of inspection readiness [18].

3.11 Regulatory Meetings

As discussed earlier in this chapter, optimal facility flexibility is the desired outcome with respect to designing and operating a multiproduct biopharmaceutical facility. However, it should be recognized that increased flexibility often equates to increase levels of risks. Although there are tools to assess known and perceived risks (see ICH Q9, Annex 1), there may be times when the boundaries of acceptable risks may be on the borderline of being unacceptable. Should any questionable risks or regulatory uncertainty be related to issues involving the approval of a BLA, such as the design of a new production facility, technical transfer of a production process, or major changes that may require a preapproval supplement of an existing BLA (e.g., facility remodeling to improve GMP compliance), it is highly recommended to reach out to FDA as early as possible and engage the Agency in discussion of such issues. Doing so provides industry officials with the opportunity to receive comments, questions, and sometimes advice, when necessary, from FDA officials who may have direct involvement in the subsequent review of the BLA and/or preapproval facility inspection. Receiving FDA feedback on issues presented in these meetings, before the BLA submission and facility inspection, provides useful information and guidance, which typically results in successful outcomes. Regulatory authorities do not appreciate last-minute surprises that may potentially impact the approval process, and could have been presented, addressed, and resolved at an earlier time. "Surprises" often lead to costly redesign, modifications, and corrective actions, which can result in delay of regulatory approval.

FDA offers a formal process to meet with industry officials to discuss matters of actual or perceived regulatory uncertainty that require more clarity. The types of meetings that are offered, and related criteria, are found in the *Guidance for Industry: Formal Meetings Between the FDA and Sponsors or Applicants* [19]. This document discusses the following three types of meetings:

1) *Type A*. These meetings are intended for product development issues, which include clinical protocols and testing, as well as related dispute resolution between FDA and drug sponsors.
2) *Type B*. These meetings are designed to address pre-investigation new drug/BLAs, as well as end of phases 1, 2, and pre-phase 3 issues.
3) *Type C*. These meetings are for all other matters not addressed in type A and B meetings, including, but not limited to, design of new and/or remodeled facilities, flexible and optimal facility use, new/novel technologies, and their application to related processes and products. These are the most common meetings requested by industry officials who wish to have the Agency's

review and response on issues being considered before either starting or completing projects that may require FDA approval.

Meeting requests are submitted to the Agency in writing and must clearly state the type of meeting being requested, its purpose and objectives, a proposed agenda, and questions that the applicant or sponsor wishes FDA to address. Other criteria for these meetings can be found in the guidance document referenced earlier. Meetings can be face-to-face, conference calls, or by teleconferencing, and the Agency will respond within a defined time period whether or not meeting requests are accepted, and reasons if a request is denied. FDA documents meeting minutes, which become the official record that is issued to attendees within 30 days. The sponsor/applicant has an opportunity to dispute the accuracy of the minutes if necessary, but not the Agency's position taken during the meeting.

There are many benefits to meeting with regulatory authorities, as are seen by the number of industry conferences FDA representatives are invited to attend each year. Formal meetings with Agency officials are industry's opportunity to acquire regulatory clarity on issues that are not necessarily addressed by the regulations, guidance documents, or published literature. Such meetings that involve new technologies and novel approaches to GMP compliance can often be learning experiences to FDA Investigators who may subsequently review this information during inspections. Regulatory authorities must educate before they regulate, and formal FDA/industry meetings can offer this opportunity so that industry can provide new information to the Investigator to consider before he or she must make a regulatory/compliance decision. It is much easier having a discussion, which may include disagreements, between two parties having similar technical understanding than one where party A must take the time to educate party B before they are both on common ground to appreciate each other's viewpoint.

3.12 Conclusion

This chapter has attempted to provide a regulatory perspective for the design engineer, who may not be familiar with the regulatory requirements and expectations that directly and indirectly impact the design, operation, management, and optimal use of a multiproduct biopharmaceutical facility. A brief historical overview of the FD&C Act was intended to introduce the engineer to the foundation on which the cGMP regulations are based and were developed over time, often owing to known and unknown risks that resulted in tragic consequences impacting the safety, health, and welfare of the American consumer. Regulatory expectations need not always be codified by specific regulations, and for this reason, numerous guidance documents, which of some were discussed in this chapter, have been published over the past two decades by FDA, the European Health authorities, and ICH. These documents are intended to share and recommend regulating authorities' current thinking on a wide range of topics designed to promote and encourage regulatory compliance. Although the contents of these guidelines are not legally enforceable, regulators have used them as a compass to lead them to noncompliance of regulatory requirements (cGMPs).

Discussion of the quality systems, which was presented in this chapter, recognizes that the engineer may have limited experience with these systems and/or exposure to the client's Quality Unit during a design project. Nevertheless, it is critical that the engineer must be familiar with elements that comprise the quality system (e.g., deviation investigations, complaints, and rejected products and materials) in an effort to identify potential areas of regulatory concern that may need to be factored into and addressed as part of a project that is intended to ensure protection of both the product and the related processes. The current compliance status of existing operations can be a valuable source of information, which should be considered in every design project, and a functional, properly managed quality system can be the Rosetta stone that leads the designer to a successful outcome with the client.

Before a regulatory authority approves a facility for multipurpose use, one of the more critical elements that must be demonstrated is prevention of cross-contamination during changeover of products and processes. This chapter emphasized the importance of cleaning validation and discussed a number of background topics that regulators inquire about while assessing cleaning validation programs. It is incumbent that the design engineer considers this information as it can be a resource applied to the design and functionality of a facility to ensure that the proposed engineering project sufficiently contributes to the concept of product protection. Along this same line is having a clear understanding of not only understanding the client's control strategy for operating a multiproduct, biopharmaceutical facility, but more importantly knowing the client's current state of control, which is one of the essential raw materials necessary to gauge if a proposed facility design project adequately contributes to regulatory compliance.

An overview of FDA's approach to inspecting biopharmaceutical facilities was discussed for the purpose of providing insight to the design engineer, who rarely has the opportunity to participate in the regulatory inspection process. This insight was intended to give the engineer an appreciation of the regulator's focus and priorities when conducting inspections, especially as they relate to coverage of the quality systems. The cGMPs require product-, process-, and facility-related problems be thoroughly investigated and documented, and appropriate corrective actions be taken to prevent recurrences. Repetitive events are indicators of ineffective and/or inappropriate corrective actions, and the engineer needs to be aware of those related to the facility and utilities to properly assess potential design deficiencies that may have lead to noncompliance. This requires good communications with the Quality Unit organization, which is responsible for managing and monitoring the company's quality system program.

As discussed earlier, contract manufacturing organizations (CMO) continue to play an ever important role in the biopharmaceutical industry and are critical partners with their respective BLA license owners. Regulatory authorities appreciate the many complexities involved and the skill sets required to manufacture, test, package, and ship protein derivative products, and they recognize the need to employ the right services that specialize in these activities. In doing so, regulators also expect quality agreements be established between the license owner and its contract service organizations, and that these agreements clearly define responsibilities for each party. Regulators consider the CMO as an extension of the license owner's operations, and they hold the owner ultimately responsible for monitoring and ensuring that the service provider not only remains in compliance with the quality agreement but more importantly with all applicable cGMP regulations as well. It is therefore recommended that contract engineering companies also consider performing regulatory compliance assessments for projects being proposed before presenting to the client.

As the biopharmaceutical industry continues to grow, and new technologies are discovered and introduced, and manufacturing facilities become more complex and flexible in operations and activities, regulatory uncertainty will remain a certainty, in some respects. The principles of risk management and its related tools may be able to identify some risks, but not all risks, particularly where there may be no existing regulatory requirements or guidance available. Introducing a novel facility design or an innovative use of technology that may challenge traditional methods of cGMP compliance may be considered "leading edge," but the leading edge can often become the "bleeding edge" should regulatory authorities consider potential product risks to be unacceptable. This could result in activities involving the redesign of facilities, replacement of equipment, recommission utility systems, and/or revalidation of processes and related controls to comply with regulatory expectations, which, again, may not be published in existing regulations or guidance documents. Should any of these events occur, it can likely compromise and delay approval of the BLA or a preapproval supplement, and companies must clearly understand when such risks exists.

For this reason, FDA encourages industry representatives to request formal regulatory meetings, which give companies the opportunity to meet with Agency officials to present and explain ongoing projects and proposed plans that may represent regulatory risks or uncertainty. Should the regulators disagree with a company's strategies, designs, or concepts, it is better receiving this feedback as early as possible so that regulatory recommendations can be considered before too much investment in capital and resources are committed. Regulators do not appreciate a "last-minute surprise," which was not brought to their attention before a preapproval inspection, as it may be perceived as an intentional lack of transparency that could compromise the inspection outcome.

As stated earlier in this chapter, the pre-design stage of every engineering project should include acquiring as much information and knowledge about the client's products, processes, and operating strategies, as well as the client's current regulatory compliance status, to conceptualize how best to design a facility that contributes and provides optimal product protection. This is the goal of the pharmaceutical engineer to design with the patient in mind, as we are or will all become "the patient" one day.

References

1 Wechsler J. Added responsibilities and outside concerns prompt overhaul of agency's structure. *BioPharm Int* 2011;**24**(12):12–15.

2 Hamburg MA. Annual Meeting of the Massachusetts Biotechnology Council. M.D. Cambridge, Massachusetts, March 15, 2013.

3 The Food, Drug, and Cosmetic Act. http://www.fda.gov/regulatoryinformation/legislation/federalfooddrugandcosmeticactFDCAct/default.htm. Accessed 2014 June 2.

4 Public Health Service Act (PHS Act). http://www.fda.gov/regulatoryinformation/legislation/ucm148717.htm. Accessed 2014 June 2.

5 21 CFR 10.115 (Good Guidance Practices). http://www.gpo.gov/fdsys/granule/CFR-2012-title21-vol1/CFR-2012-title21-vol1-sec10-115. Accessed 2014 June 2.

6 ICH Q7A (Good Manufacturing Practices for Active Pharmaceutical Ingredients). fda.gov/BiologicsBloodVaccines/GuidanceComplianceRegulatoryInformation. Accessed 2014 June 27.

7 ICH Q8-R2. http://www.fda.gov/downloads/Drugs/Guidances/ucm073507.pdf. Accessed 2014 June 27.

8 ICH Q9 (Quality Risk Management). http://www.fda.gov/downloads/Drugs/…/Guidances/ucm073511.pdf. Accessed 2014 June 27.

9 ICH Q10 (Quality Systems). http://www.fda.gov/downloads/Drugs/Guidances/ucm073517.pdf. Accessed 2014 June 27.

10 ICH Q11 (Development and Manufacture of Drug Substances). http://www.fda.gov/downloads/Drugs/GuidanceComplianceRegulatoryInformation/Guidances/UCM261078.pdf. Accessed 2014 June 27.

11 Preamble to GMPs Federal Register docket #75N-0339. http://www.fda.gov/downloads/Drugs/DevelopmentApprovalProcess/Manufacturing/UCM206779.pdf. Accessed 2014 June 27.

12 Guidance for Industry Quality Systems Approach to Pharmaceutical cGMP Regulations. http://www.fda.gov/downloads/Drugs/…/Guidances/UCM070337.pdf. Accessed 2014 June 27.
13 Code of Federal Regulations, 21 CFR 211. http://www.gpo.gov/fdsys/granule/CFR-2011-title21-vol4/CFR-2011-title21-vol4-part211/content-detail.html. Accessed 2014 June 2.
14 FDA/Chemistry. Manufacturing, and Controls guidance for Therapeutic Recombinant DNA-Derived Product or a Monoclonal Antibody for in-vivo Use. http://www.fda.gov/downloads/biologicsbloodvaccines/guidancecomplianceregulatoryinformation/guidances/general/ucm173477.pdf. Accessed 2014 June 27.
15 European Union Guidance on Good Manufacturing Practice. The Rules Governing Medicinal Products in the European Union, Volume 4, EU Guidelines for Good Manufacturing Practice for Medicinal Products for Human and Veterinary Use (year?).
16 FDA Compliance Program Guide CP7356.002 (Inspection of Drug Manufacturers). http://www.fda.gov/downloads/ICECI/ComplianceManuals/ComplianceProgramManual/UCM125404.pdf. Accessed 2014 June 27.
17 FDA's Guidance for Industry Contract Manufacturing Arrangements for Drugs: Quality Agreements (Draft). http://www.fda.gov/downloads/Drugs/GuidanceComplianceRegulatoryInformation/Guidances/UCM353925.pdf. Accessed 2014 June 27.
18 FDA Compliance Program Guide CP7356.002M (Inspection of Biological Licensed Therapeutic Drug Products). http://www.fda.gov/downloads/ICECI/ComplianceManuals/ComplianceProgramManual/UCM125422.pdf. Accessed 2014 June 27.
19 FDA Guidance for Industry: Formal Meetings Between the FDA and Sponsors or Applicants. http://www.fda.gov/downloads/Drugs/Guidances/ucm153222.pdf. Accessed 2014 June 27.

Chapter 4

Biopharmaceutical Facility Design and Validation

Jeffery Odum

NNE, Durham, North Carolina, USA

4.1 Introduction

The manufacture of biotherapeutic drug substances is based on a synergy between science, product, process, and facility. This relationship is often difficult for many outside the industry to understand; manufacturing drug products from living organisms is a complex effort that must focus priority on patient safety. The guidelines that provide the foundation of compliance and regulatory oversight are supported by a number of guidance documents that provide the basis for validation, the precursor to product licensure. Facility validation is recognized as the documented process that supports the verification that the facility and its systems, equipments, and features meet the established specifications and quality attributes defined by the manufacturing organization.

Biomanufacturing facilities are based on a simple, yet complex principle: protect the product. As the manufacturing process requires the introduction of living organisms into a mechanical-based process manufacturing system, the focus on product protection and environmental control is one of the key attributes of facility design.

There are a number of basic considerations of biomanufacturing facility design. There is no "universal" standard for good manufacturing practice (GMP). Each product, process, and facility has unique attributes that influence regulatory compliance as defined by the GMPs. As we will see, defining these attributes is a critical aspect of the validation process and one that regulators have placed a greater focus on today than in the past.

The product and manufacturing process have a great influence on GMP interpretation. The guidelines that are defined in 21 Code of Federal Regulations (CFR), parts 210 and 211 [1], are open to interpretation and influenced by a number of global guidance documents. For products

Process Architecture in Biomanufacturing Facility Design, edited by Jeffery Odum and Michael C. Flickinger.

marketed in countries outside the United States, compliance to global guidelines from Europe, Asia, and other regions, having a familiarity with the validation requirements, will be critical in the validation process.

While the majority of biologic drug substances are not sterile but controlled to a low, measurable level of bioburden, the principles of aseptic processing as defined by the US Food and Drug Administration (FDA)'s 2004 Aseptic Guidance [2] are generally followed. This higher level of manufacturing control is also one of the reasons that biologic products are more complex and costly to manufacture. The convergence of unique product attributes, complex process unit operations, fragile cell lines, and interpretive regulatory requirements becomes the backdrop of the validation challenge.

4.2 Designing for Compliance

From the beginning of the facility design process, the focus on regulatory compliance leading to successful validation and product/facility licensure must be maintained. Every design activity will eventually have some form of impact on the validation process. Commissioning, qualification/verification, and process validation (PV) can all be influenced by what may seem the most insignificant decision at the time.

The FDA utilizes a systems-based approach for regulatory review [3]. The six systems are quality, production, facilities, materials, packaging and labeling, and laboratory controls.

The routine inspection process will either be a full inspection of the quality system and three of the other defined systems or an abbreviated inspection of the quality system plus one other system. The fact that the facility is considered a critical system defines the importance of the relationship between facility design and validation.

Regulatory guidance for how drug manufacturing facilities will be reviewed and inspected comes from a number of different documents. In addition to the previously referenced current good manufacturing practices (cGMPs), the US FDA has issued Q7A that defines many of the requirements for facility and equipment compliance to the cGMPs [4].

The European Union facility requirements can be found in Annex 1 [5].

The current global guidance that is a reference foundation document has been issued by the International Committee on Harmonization (ICH) under its guidance Q7 [6]. This document provides the base compliance requirements that have been accepted by all members of the ICH working groups, including the United States, Europe, and Japan.

4.2.1 Facility Considerations

Key to a successful validation effort is a well-defined manufacturing process that is seamlessly integrated with the facility. In defining the manufacturing process, there are a number of drivers that will have a significant impact on the facility design. The synergy between product–process–facility will become apparent.

The large majority of biomanufacturing operations are not designed to produce sterile drug substance. Even if aseptic manufacturing standards are implemented, the focus is on bioburden control. In bioburden-controlled processing, the distinction between upstream and downstream operations is important. The upstream operations (seed inoculation and cell culture fermentation) are commonly designed as axenic operations that imply the need for a greater level of product protection. Downstream operations (recovery, purification, and bulk filling) are operated as bioburden-control processes, again necessitating the need for contamination control, but to predetermined levels based on the product/process attributes.

Biomanufacturing process systems will be either defined as open or closed [7]. Process definition in this case relates to how the various manufacturing unit operations are executed based on their ability to protect the product. Open systems require special design attributes that will allow for the reduced risk of product contamination due to exposure to the manufacturing environment. Open systems are allowed under the GMP regulations but must be integrated into the facility in a manner that minimizes risk in accordance with current regulatory expectations.

Closed systems are designed in a manner to ensure product protection by removing the environmental risk of exposed product. While the concept of a closed manufacturing system may seem like a simple undertaking, it is often very difficult to deliver. A simple example is the process of taking in-process samples. Many current sampling systems require the system to be placed in an "open" state to remove the sample from the system. If overlooked, GMP compliance is not only unmet but product safety may be compromised.

The GMPs make reference for the need to have an orderly flow of manufacturing operations to minimize product contamination risks and allow for an orderly flow of material, personnel, and equipment through the facility during manufacturing operations [8]. There is also a need for "logic" in terms of room and area adjacencies to ensure manufacturing efficiency and provide for room/area segregations that also meet GMP requirements. The importance of these facility attributes is seen in the documentation required for licensure. FDA and other regulatory agencies require companies to submit specific documents as part of the application, including layouts and flow diagrams [9].

As facility design concepts are developed, the focus on functional adjacency becomes a critical aspect of GMP compliance. This concept includes the definition of layouts so that flows are unidirectional, segregation of clean and dirty operations is maintained, area environmental classifications support open/closed manufacturing operations, and air flows and room pressures support both product and environmental protection. Examples of these documents are given in Figs. 4.1–4.5.

There may also be an impact of facility validation that is related to the biosafety classification of the product/process. While the current GMPs do not specifically mention biosafety-related issues around containment, the attributes of the facility design and thus, the validation of support systems will be affected. For example, common biosafety level 3 design requires the building ventilation systems utilize segregated, single-pass air flow with negative room pressures to "contain" any particles that may escape to the environment that can prove detrimental to the workers in the facility. For this type of systems, the qualification protocols must be written to specifically address the issues around air flow volume, direction, and quality.

Another facility consideration that affects validation is the gowning philosophy/approach that is implemented. The definition of personnel flows will have a major impact on how gowning philosophy is translated into physical room layouts and air lock configurations. Figure 4.6 provides some different examples of how the questions around air lock configuration to meet defined gowning requirements can be addressed.

Facility considerations also include design aspects such as materials-of-construction selection, finishes that support cleaning in accordance with the GMPs [10], and safety aspects based on personnel and/or environmental aspects. These include containment requirements for biosafety level compliance [11], hazardous material handling and storage, and unique cleaning requirements for equipment. Gowning approach and requirements will also affect facility design and be a key element in the overall compliance strategy.

4.2.2 Product–Process–Facility Integration

Designing for compliance in the manufacture of biological drug products is based on an industry continuum that recognizes the synergy between product and process. This synergy is aptly discussed in Bioprocess design, computer-aided of the Biopharmaceutical Facilities Baseline Guide [12]. For a facility to be in compliance with the cGMP guidelines, there must be a well-defined link between the product attributes, the process unit operations, and the facility attributes. This link is discussed in more detail later.

4.2.3 The Role of Quality by Design

Quality by design (QbD) is defined in ICH Q8 as a systematic approach to product development that begins with predefined objectives and emphasizes product and process understanding and process control based on sound science and

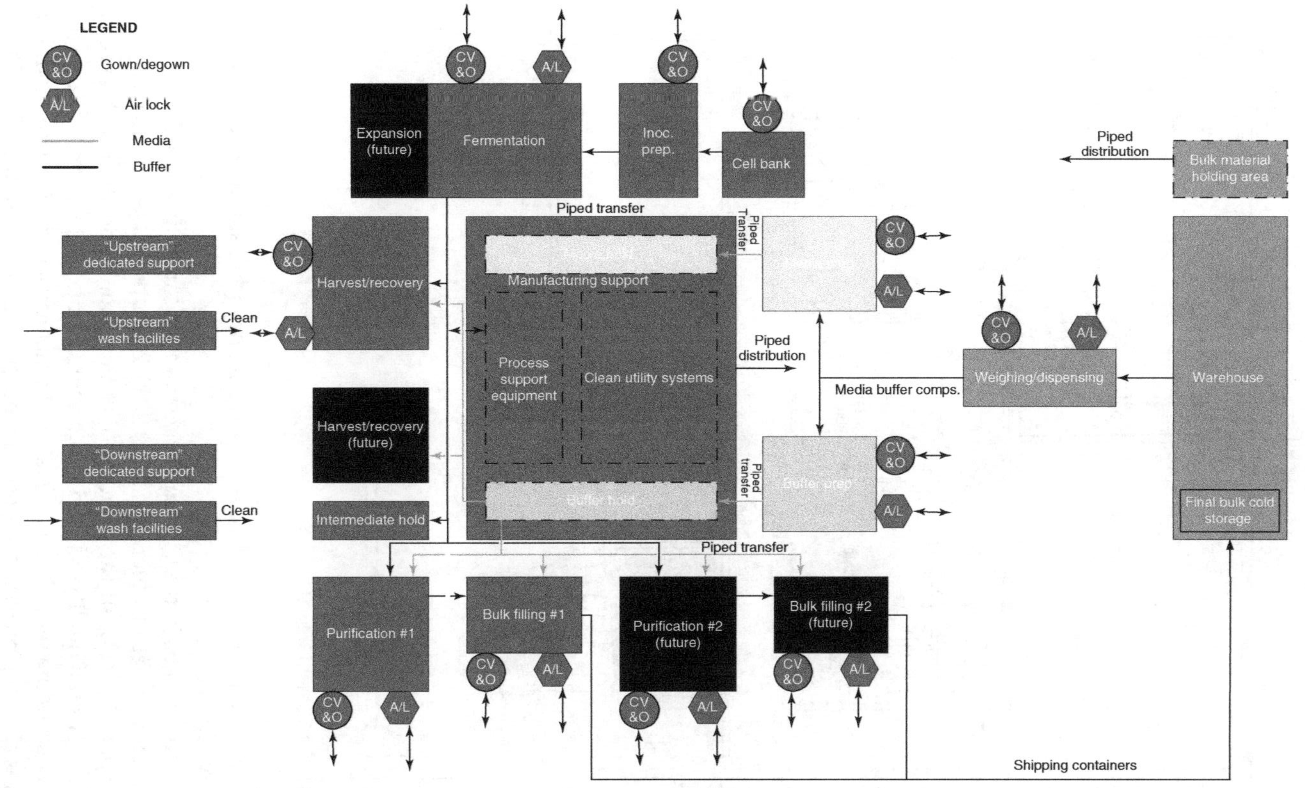

Figure 4.1 Block flow adjacency diagram.

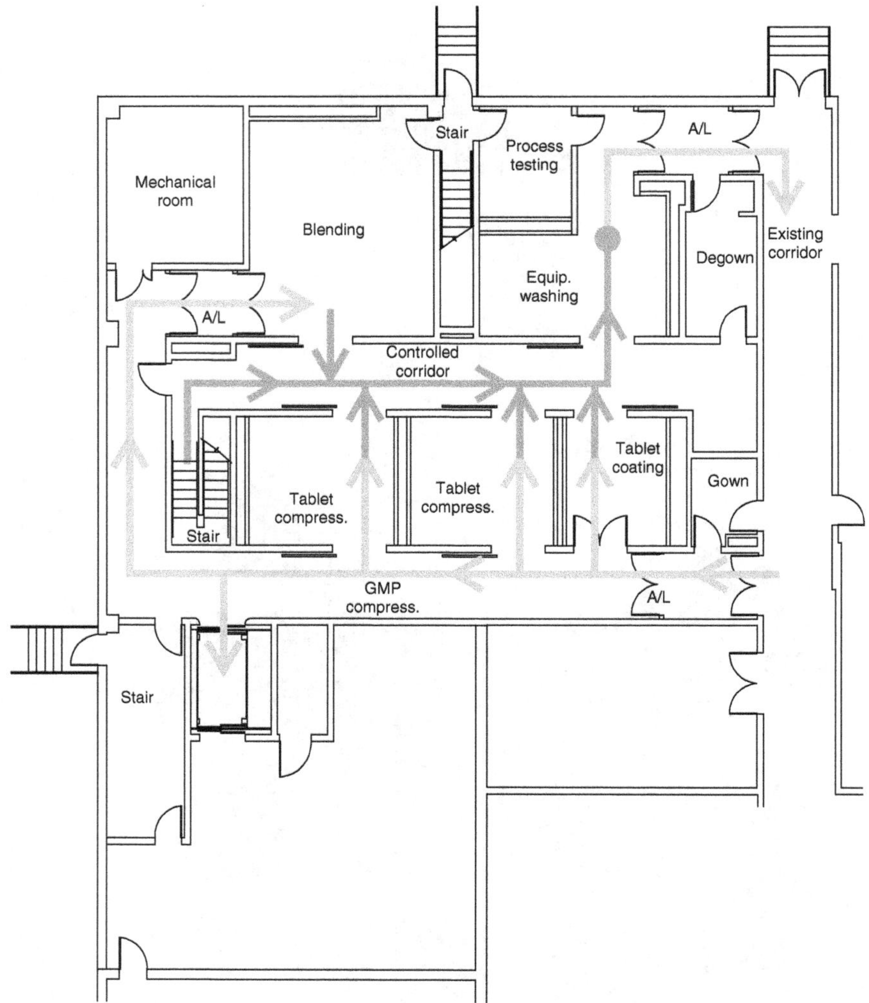

Figure 4.2 Equipment flow diagram.

quality risk management [13]. Current regulatory philosophy focuses on the implementation of QbD to support PV.

There are six steps in the QbD approach and are shown in Fig. 4.7.

The quality target product profile (QTPP) is a prospective summary of the quality characteristics of the drug product that should ideally be achieved to ensure the desired quality, taking into account product safety and efficacy. The QTPP includes the intended use, quality characteristics, delivery method, and attributes affecting pharmacokinetic attributes. A simple example of a QTPP is shown in Table 4.1.

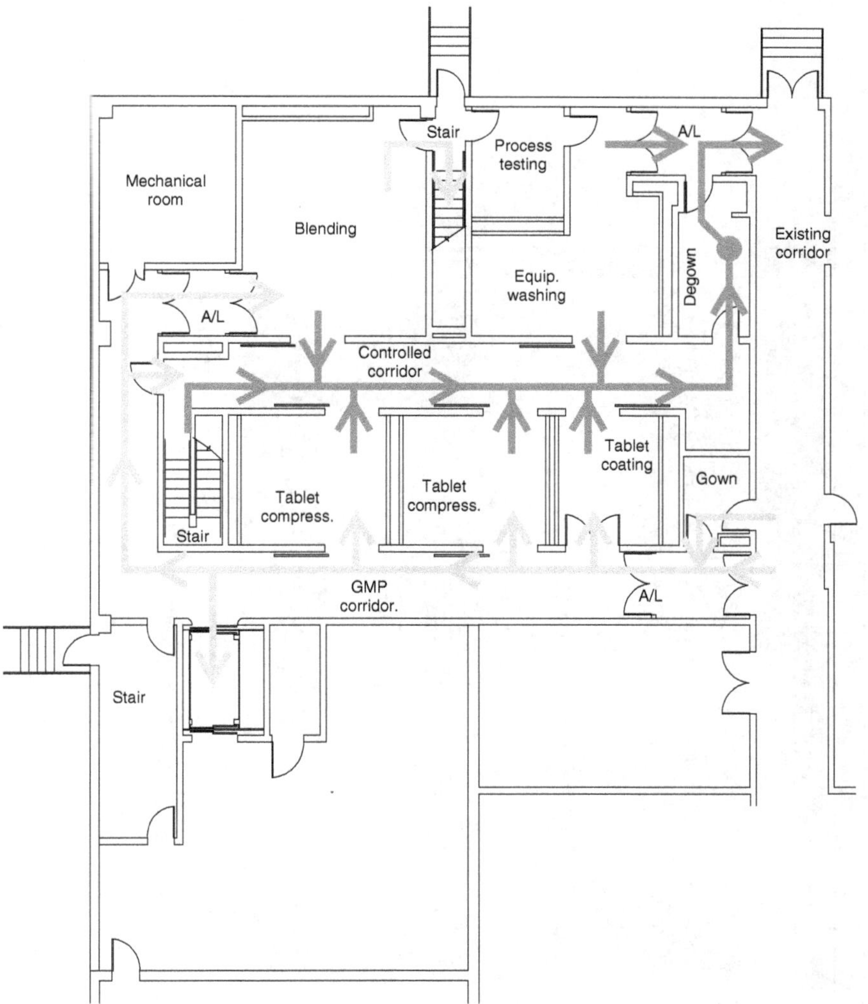

Figure 4.3 Personnel flow diagram.

Critical quality attributes (CQAs) are any physical, chemical, biological, or microbiological property or characteristic of a material that should be within an appropriate limit, range, or distribution to ensure the desired product quality. With respect to QbD, the need is to focus only on those attributes and the impact of those attributes across their ranges on safety and efficacy. For biological products, some common CQAs would include chemical purity, qualitative and quantitative impurities, and microbial quality.

Critical process parameters (CPPs) are the independent parameters of the manufacturing process that are most likely to affect the CQAs of the product.

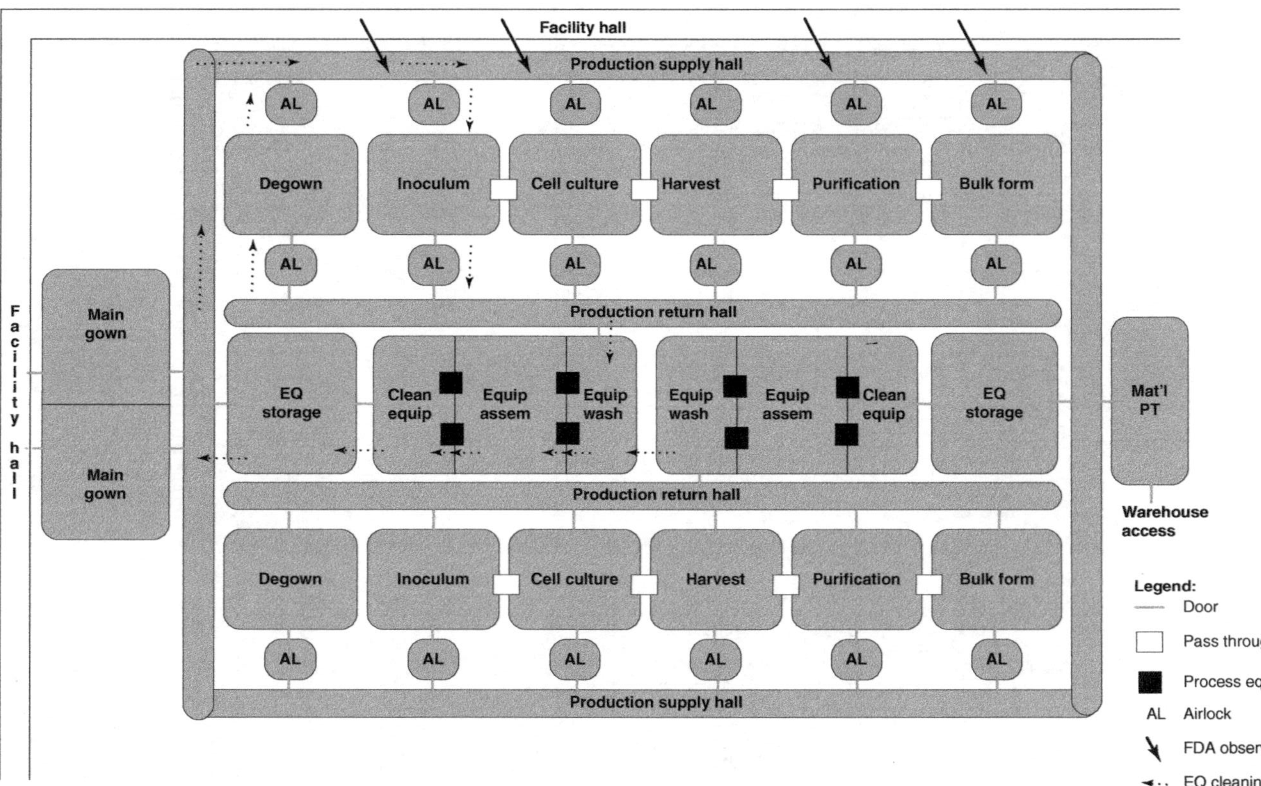

Figure 4.4 Segregation diagram. [courtesy of International Society of Pharmaceutical Engineering.]

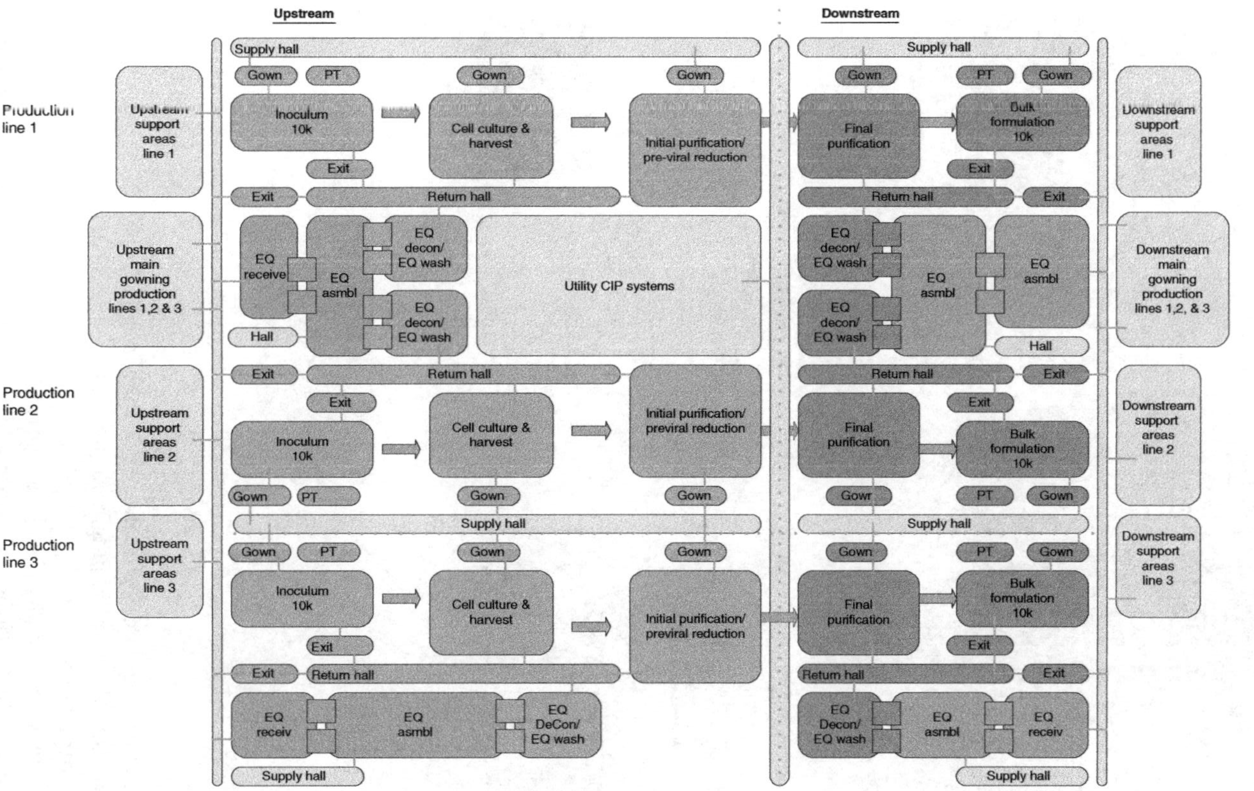

Figure 4.5 Facility layout adjacency diagram with flows.

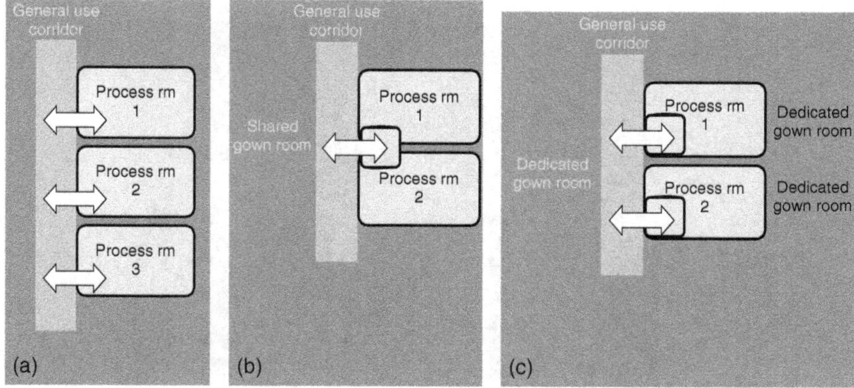

Figure 4.6 Air lock configurations.

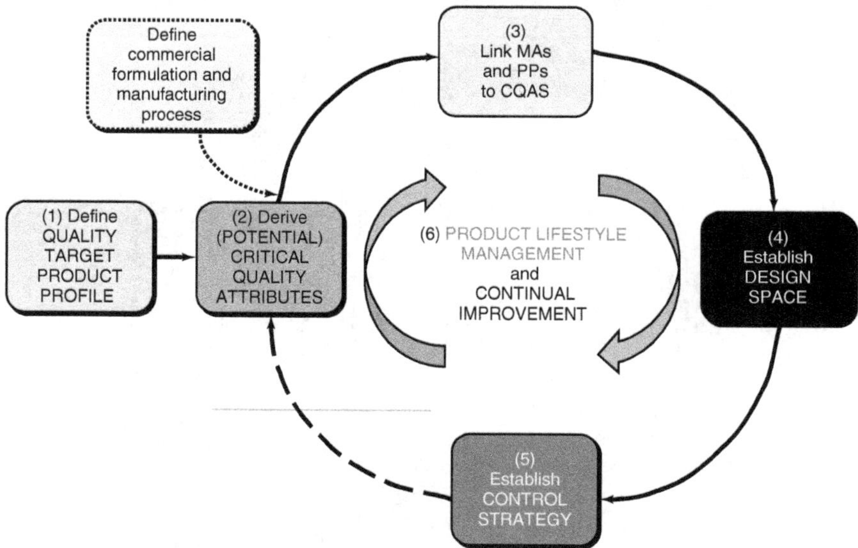

Figure 4.7 QbD approach steps.

Under current regulatory guidance as defined in ICH Q8, CPPs must be determined by sound scientific judgment and based on research, scale-up, and/or manufacturing experience.

CPPs must be controlled and monitored to confirm that the product profile is comparable to or better than the historical data from both development and manufacturing operations. In terms of the manufacturing process, these parameters should be controlled within a meaningful, narrow operating range

Table 4.1 QTPP Example

Variable	QTPP Design Specs (Must-Haves)
Indications	Prostate cancer
Presentation(s)	100-mg/mL vial
Form(s)	Lyophilized product
Dose(s)	5 mg/kg/dose
Dosing regimen(s)	Q3
Storage conditions	24 months @ 2–8°C
Delivery system(s) and devices	IV infusion
Placebo required?	Yes

to ensure that specifications are met. Common parameters would include temperature, pH level, oxygen content, or agitation speed.

The CQAs and CPPs are the key elements that are used to determine a product's design space, as defined in ICH Q8. Design space is the multidimensional combination of interaction of input variables and process parameters that have been demonstrated to provide assurance of product quality. The design space is proposed by the applicant (product manufacturer) and is subject to regulatory assessment and review.

Design space is a region where quality product can be produced. It includes material attributes and process parameters and includes an evaluation of scale and the equipment that is used in the manufacture of the drug product. It is arrived at by an iterative application of risk assessment and experimental design to the knowledge space. The steps in the development of a design space include (i) defining the CQAs, (ii) using prior knowledge and risk assessment to determine CPPs, (iii) understanding all the process variables that affect CQAs, (iv) relating product attributes to support process needs, and (vi) representing the understanding in the design space.

Design space is visually represented as a three-dimensional model, as shown in Fig. 4.8.

Control strategy is defined as a planned set of controls, derived from current product and process understanding that assures process performance and product quality [14]. The controls can include parameters and attributes related to drug substance and drug product materials and components, facility and equipment operating conditions, in-process controls, finished product specifications, and the associated methods and frequency of monitoring and control.

A control strategy can include [14] control of input material attributes, product specification(s), controls for unit operations, in-process or real-time release testing, and a monitoring program for verification of prediction models.

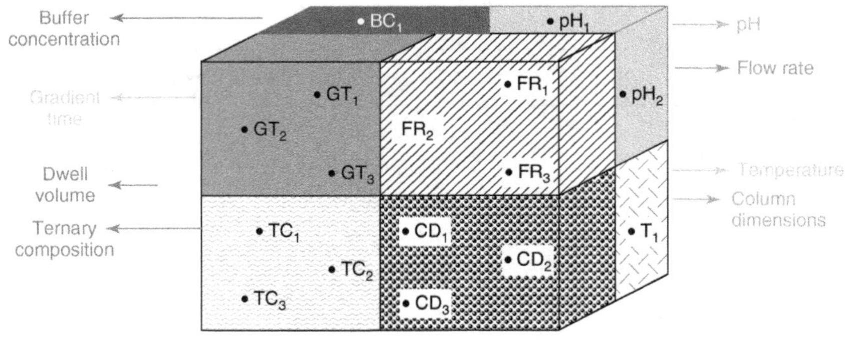

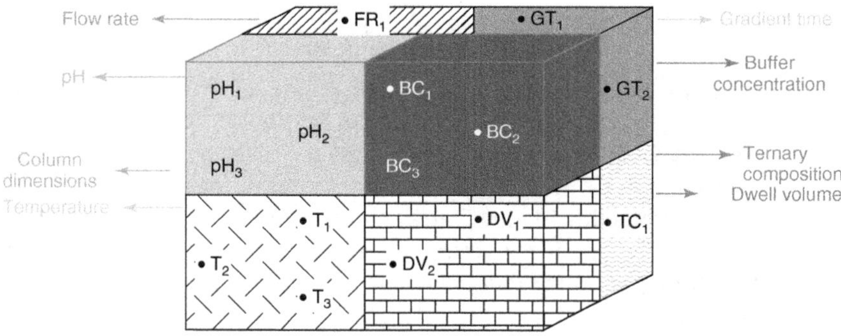

Figure 4.8 Design space model.

A holistic view of manufacturing control strategy is shown in Fig. 4.9.

A control strategy can be a complex and extensive document that contains elements that can include both conventional specifications as well as GMPs that are applied as a "baseline" approach. Some of these elements will contain both in-line and at-line testing during the manufacturing operations.

By using QbD principles, process requirements become well defined by the multidimensional design space and can be easily communicated to the project team to ensure that equipment and facility design will support PV requirements based on product and process requirements.

4.3 Risk Management

The FDA Twenty-First Century Initiative [15] placed a tremendous amount of emphasis on risk assessment and understanding. This emphasis is detailed in ICH Q9. This guidance document details the entire risk process that is represented in Fig. 4.10. Under ICH Q9, risk is defined as the combination of the

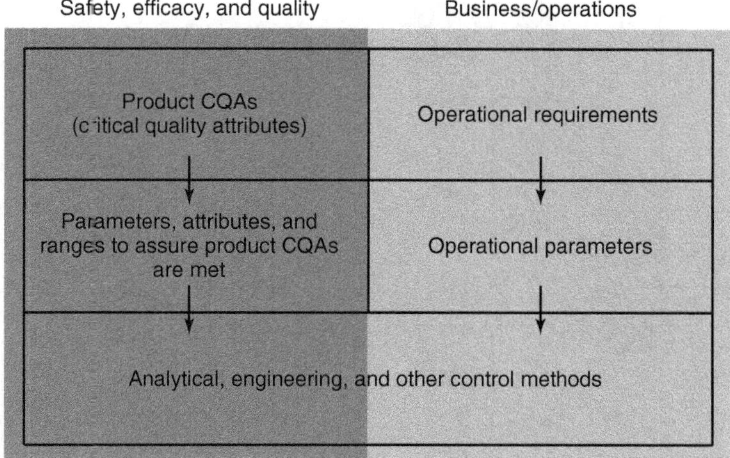

Figure 4.9 Manufacturing control strategy.

probability of occurrence of harm and the severity of that harm as they relate to drug product quality. The focus is on understanding the risks to a product during manufacturing, how to identify those risks, and what steps can be taken to mitigate or eliminate the risk to ensure product quality, safety, and efficacy [16].

The risk process has three steps: risk identification, risk analysis, and risk evaluation. Identification focuses on what might go wrong during the manufacturing process and requires there be a clear identification of the risk in question. Analysis assesses the likelihood of the risk occurrence and evaluates the impact to the product. The evaluation then assigns levels of severity and defines actions that can be implemented. Figure 4.11 provides a sample risk matrix.

ICH Q9 recognizes that some risks are unavoidable. However, to reduce or control risks to the patient, consideration must be given to the following:

- Is the risk above an acceptable level?
- What can be done to reduce or eliminate the risk?
- What is the appropriate balance between benefits, risk, and resources?
- Are new risks introduced as a result of the identified risks being controlled?

The regulatory expectation is that any risk management program must provide clear, documented evidence that these questions have been answered. The risk management process must be systematic, well organized, and build clear decision making pathways that could withstand the scrutiny of a regulatory audit.

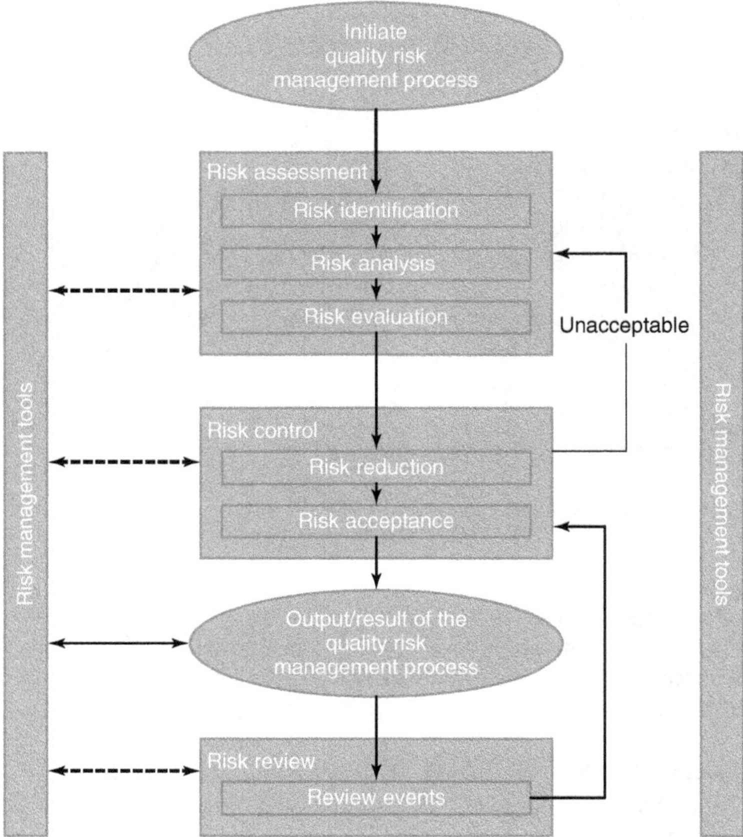

Figure 4.10 Chart from ICH Q9.

The following are two examples of risk assessment tools that are often used in industry and that are also recognized in ICH Q9. Both examples are for a production bioreactor (Figs. 4.12 and 4.13).

Evaluating risk involves placing a weight/value on each risk using severity and probability scores for each. This evaluation will identify the hazard/risk, the CQA that is affected, a score leading to a ranking of low (dark grey), medium (light grey), or high (black). Figure 4.13 provides an example of this type of risk evaluation tool.

To conduct effective risk analysis and provide the supporting data for validation, there are a number of recognized tools that can be used. These include control charts, design of experiments (DOEs), histograms, Pareto charts, and process capability analysis.

The impact of risk management on facility validation can be directly linked to the decisions made during facility design. To address process risks, it may

Process parameter in production bioreactor	Quality attributes						Process attributes			Risk mitigation
	Aggregate	afucosylation	Galactosylation	Deamidation	HCP	DNA	Product yield	Viability at harvest	Turbidity at harvest	
Inoculum viable cell concentr										DOE
Inoculum viability										Linkage studies
Inoculum in vitro cell age										EOPC study
N-1 Bioreactor pH										Linkage studies
N-1 Bioreactor temperature										Linkage studies
Osmolality										DOE
Antifoam concentration										Not required
Nutrient concentration in medium										DOE
Medium storage temperature										Medium hold studies
Medium hold time before filtration										Medium hold studies
Medium filtration										Medium hold studies
Medium age										Medium hold studies
Timing of feed addition										Not required
Volume of feed addition										DOE
Component concentration in feed										DOE
Timing of glucose feed addition										
Amount of glucose feed										DOE-indirect
Dissolved oxygen										DOE
Dissolved carbon dioxide										DOE
Temperature										DOE
pH										DOE
Culture duration (days)										DOE
Remenant glucose concentration										DOE-indirect

Figure 4.11 Sample risk matrix.

become necessary to implement a particular technology or specific vendor's equipment in the process design. This decision would then influence the development of protocols, the testing methods that are implemented, the training of operators, and even the overall configuration of the facility. That is why it is so important to address risk early; the impacts are often "trickle down" and can have a negative impact on the final completion efforts of qualification and validation activities.

4.4 Qualification/Verification

To understand the role that qualification plays in facility validation, it is necessary to look at a historical timeline showing the evolution of commissioning and qualification activities as recognized by global regulators (Fig. 4.14) [17]. Qualification was defined as the series of activities that were undertaken to demonstrate that facility, utility, and equipment systems were suitable for their intended use and would perform properly based on a predetermined set of

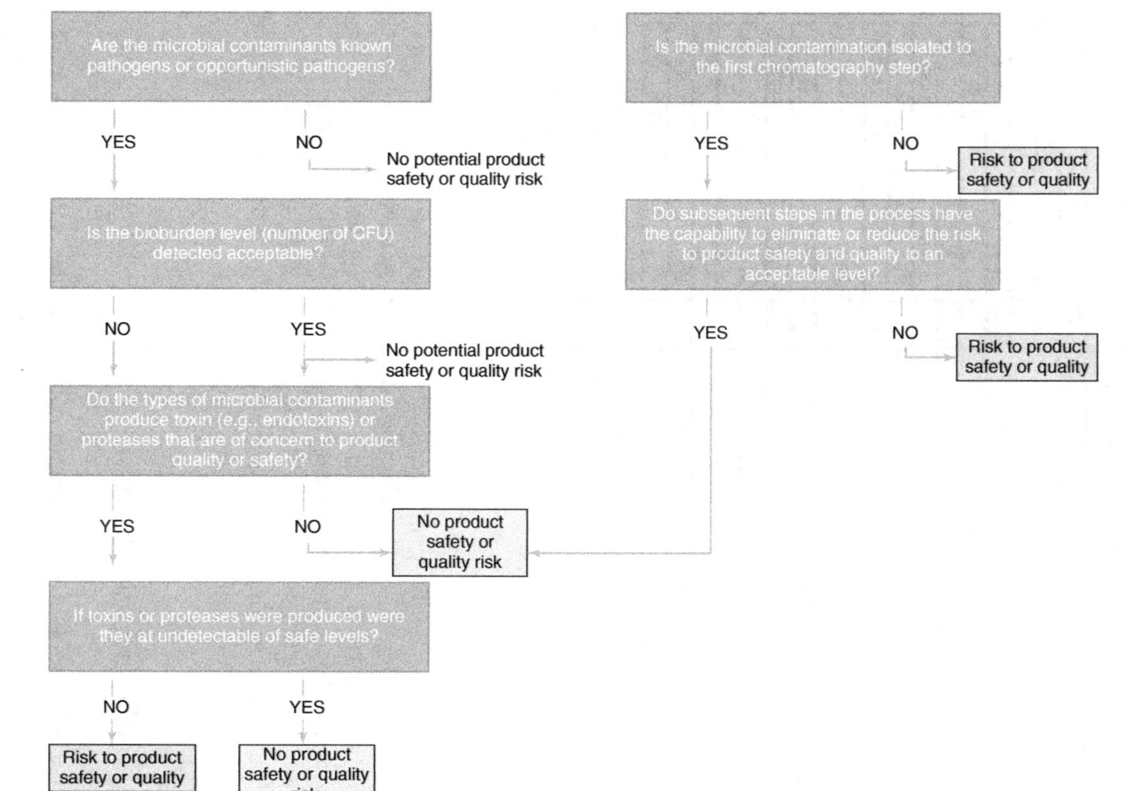

Figure 4.12 Risk assessment decision tree.

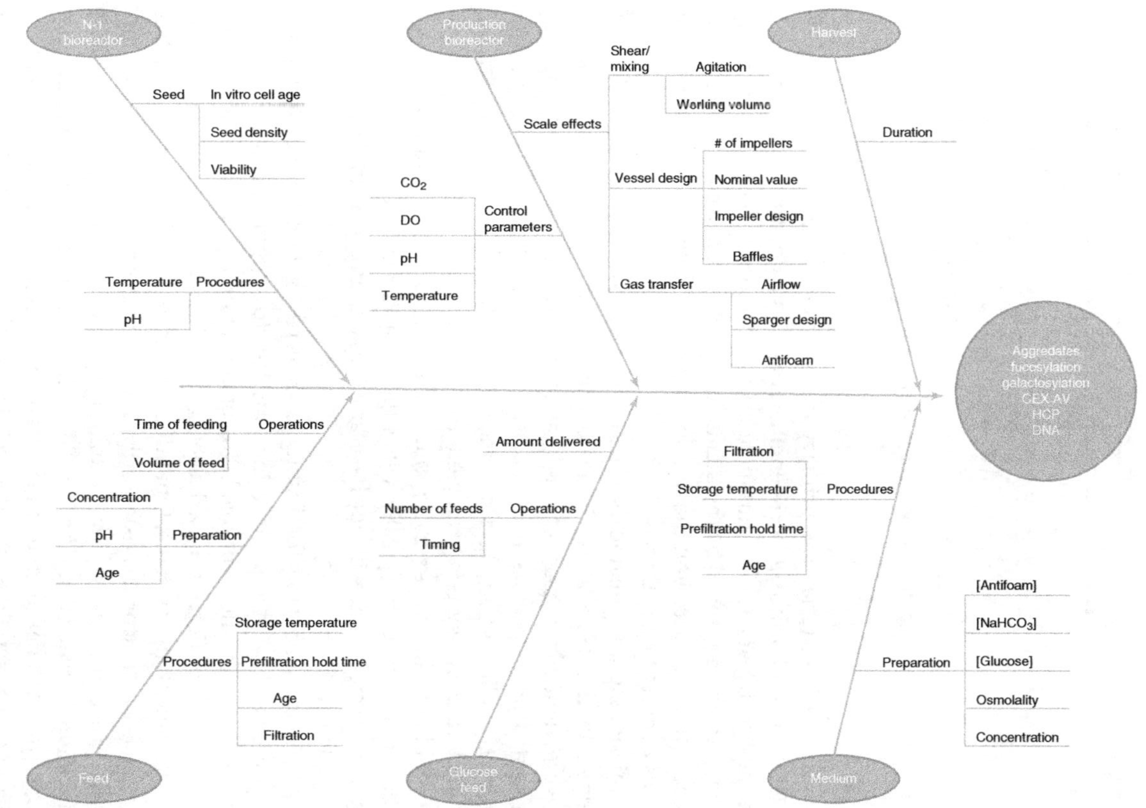

Figure 4.13 Ishikawa diagram of CPPs for a production bioreactor.

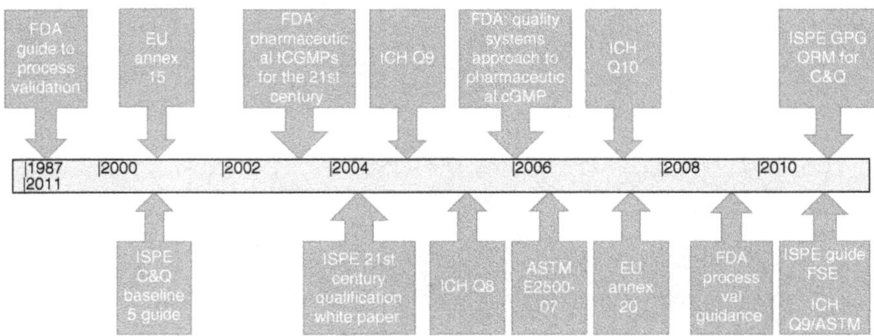

Figure 4.14 Evolution of commissioning and qualification.

operational (OQ) criteria. The primary focus was on quality control. Installation (IQ) and OQ protocols were engineering-based documents that defined specific design, IQ, and OQ requirements. These protocols were (and still today are) reviewed and approved by the quality unit and are focused on good documentation practices based on the development of a "check list" approach with approvals of execution activities.

Commissioning and qualification activities formed the foundation for PV as defined in the 1987 FDA Guide to Process Validation. System boundaries were defined as part of the Validation Master Plan. Each system was assigned an impact assessment level based on their potential impact to product quality, safety, and/or efficacy. The three levels of impact are [18] direct impact—system expected to have a direct impact on product quality; indirect impact—system not expected to have a direct impact on product quality but typically will support a direct impact system; and nonimpact—system that will not have any impact, direct or indirect, on the product quality.

ASTM (American Society for Testing & Materials) E 2500 [19] now provides the new standard for the qualification practices to support a risk-based validation system. In this new approach, the qualification activities are now based on a process of verification-based science and risk assessment to assure that cGMP equipment and systems. Quantification includes whether the activities are fit for use, can perform satisfactorily, and will they operate in a drug manufacturing, processing, packaging, or holding system in accordance with cGMP.

Verification is viewed in the ASTM E 2500 standard as a documented, life-cycle approach based on defined acceptance criteria, not a single event.

The foundation activities and life-cycle approach of an E 2500-based system are shown in Fig. 4.15 [19].

There are 10 recognized principles for executing a risk-based qualification process using the ASTM E 2500 guidance [20]. They are

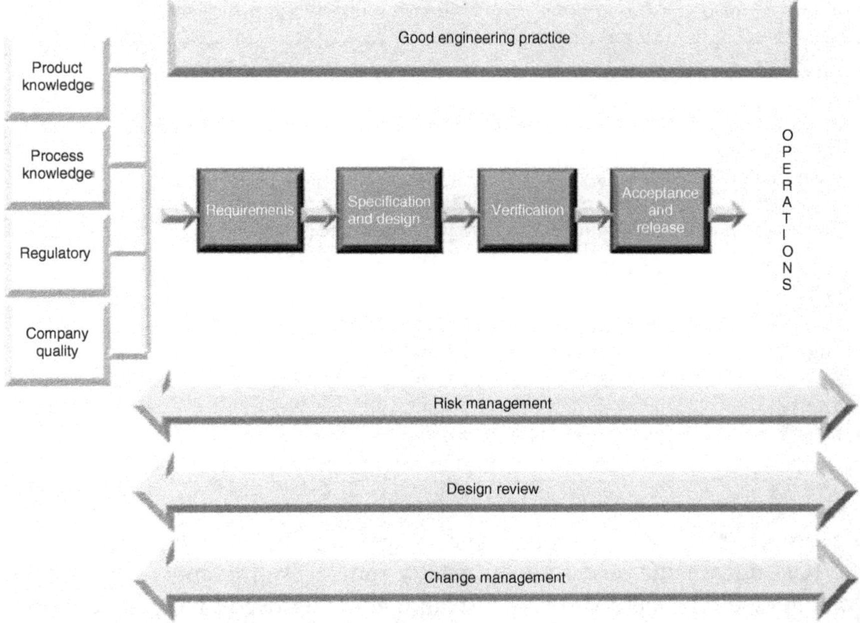

Figure 4.15 Good engineering practice.

1) Focus on attributes that affect product quality,
2) User requirements specifications (URSs) are key to acceptability; IQ and OQ support the PV effort,
3) Risk assessments and process knowledge must be used to identify the critical elements,
4) Only critical features/functions need to be qualified,
5) All activities must contribute value,
6) Risk-based asset delivery; not "cookbook" requirements,
7) Value-added documents based on technical merit,
8) Use of supplier documentation is encouraged,
9) Test planning; test once,
10) Foster innovation.

There is value to looking at the differences in the traditional versus risk-based approach to qualification (Table 4.2). Today, both these approaches are still used and recognized by regulators. Migration toward the ASTM E 2500-based system will support the current PV guidance much better.

The verification process as defined in ASTM E 2500 provides some very distinct advantages. These can be summed up as follows. Verification is a life-cycle approach; not a "check" at the end of the qualification process. It

Table 4.2 Differences in Traditional Approach Versus Risk-Based Approach

For the traditional approach • (Product) user requirements not formally documented • Protocols developed from "templates" • IQ/OQ protocols "preapproved" • Commissioning not leveraged • Engineering and "validation" personnel often distinct • Emphasis on documents and not on system performance
For the risk-based approach • Process requirements documented, approved • Risk assessments determine critical aspects of design • Engineering testing ("commissioning") verification • All documents with technical merit used as evidence of fitness for use • Emphasis on meeting process requirements

is based on product and process knowledge, regulatory guidance defined in the ICH documents, and internal quality standards. The approach must be documented through a verification plan that is approved by the quality unit. Acceptance criteria must be predetermined. There must be documentation that the equipment/system is fit for intended use [factory acceptance test (FAT) and site acceptance test (SAT)]. Acceptance and release must be documented and approved by subject matter experts (SMEs). SMEs are designated based on their combined educational and professional work experience and area of expertise.

The document types that support a risk-based verification approach include a requirements document that defines the product and process quality requirements, project verification plan that defines verification strategy and roles/responsibilities and risk assessments of the product quality/patient safety, design review summary report that confirms design review(s), verification test plan that documents the testing plan and all the corresponding check sheets, and system acceptance and release reports that confirms the system to be fit for intended use.

This verification process flow is shown in Fig. 4.16 [21].

4.5 Process Validation

In January of 2011, the FDA issued the first revision to the PV guidance since 1987. This document made significant changes to the regulatory expectations for validation and is based on the current risk-based approach that has been implemented in other guidance documents.

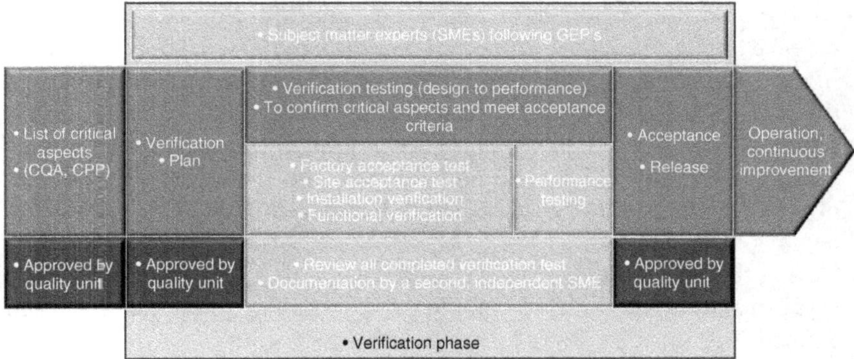

Figure 4.15 Verification process flow.

The most familiar definition of validation comes from the 1987 document.

> Establishing documented evidence which provides a high degree of assurance that a specific process will consistently produce a product meeting its pre-determined specifications and quality attributes. (22)

Under the new PV guidance, this definition takes on a different approach:

> The collection and evaluation of data, from the design stage through production, which establishes scientific evidence that a process is capable of consistently delivering quality products. (23)

The established prerequisites for PV to occur can be divided into two general categories: process and facility elements. For the process, the requirements are as follows: equipment SOP (Standard Operating Procedure), approved process development reports and master batch records, and trained manufacturing personnel.

For the facility, these requirements are as follows: cGMP production and support systems in place; facility, utility, and production equipment qualified; preventative maintenance and calibration programs in place; cleaning, sanitization, and sterilization methods defined; and quality system in place.

The new PV paradigm has a three-stage process for validation. This process implements a risk-based approach to validation utilizing the concepts that are defined in ICH Q8 and Q9 and focuses on compliance around the Q7A guidance supported by the ASTM E 2500 methodology. The three stages are stage 1: process design, stage 2: process qualification, and stage 3: continued process verification.

Stage 1: process design. This stage defines the commercial process and focuses on development and scale-up experience. Also, the design space is created. In this stage, the process is defined based on the knowledge gained through the process development and scale-up activities. Studies undertaken at laboratory, pilot, and commercial scale will be the source of significant amounts of data. These data will be used to develop an understanding of the sources of variability in the process unit operations and their potential impact on the quality of the product.

During this stage, the regulatory expectation is that variability in the process can and will be managed commensurate with the risk(s) that have been defined. The use of DOE and process modeling are some of the methods that will define these criteria. The CPPs will be evaluated at the baseline.

Stage 2: process qualification. In this stage, the focus is on the confirmation that the process can reproduce product at the commercial scale. The facility, utility systems, and equipment must be qualified. This will be a qualification executed in accordance with the ASTM E 2500 requirements. The qualification of the process will have been completed and confirmation of the design space verified. This will require a series of tests that will support process confirmation.

To qualify the process there must be a qualification of the facility. This is accomplished through the IQ and OQ activities previously discussed. It also requires that all raw materials and components used in the manufacture of the drug substance/product have been approved. All written control procedures such as SOPs and batch records must be complete and approved. Finally, all personnel must have completed any required training.

Stage 3: continued process verification. In this stage, the focus is on process verification; does the process produce a consistent product that meets the design space requirements? The stage-3 activities are heavily focused on process monitoring, data collection, and the implementation of any process changes or improvements. The process will be in a constant state of monitoring during operation. To that end, the implementation of tools such as process analytical technology (PAT) will become even more prominent. Out-of-specification and out-of-tolerance conditions will be identified in real time and this feedback will be used to confirm the design space of the product.

Using this approach and methodology results in a significant amount of regulatory flexibility. On the basis of the science-based process knowledge, analysis of process risk, and defined design space boundaries, organizations may now be able to move away from the concept of "revalidation" and also no longer need to manufacture and hold for evaluation, three conformance batches of product before commercial release.

4.6 List of Abbreviations

cGMP	Current good manufacturing practice
CPPs	Critical process parameters
CQAs	Critical quality attributes
FAT	Factory acceptance test
FDA	Food and Drug Administration
GMP	Good manufacturing practice
ICH	International Committee on Harmonization
IQ	Installation
ISPE	International Society of Pharmaceutical Engineering
OQ	Operational
PAT	Process analytical technology
PV	Process validation
QbD	Quality by design
QTPP	Quality target product profile
RTRT	Teal-time release testing
SAT	Site acceptance test
SMEs	Subject matter experts
URSs	User requirements specifications

	Regulatory Documents Sited	Web Site
1.	Title 21, *Code of Federal Regulations*, Good Manufacturing Practice Regulations	http://ecfr.gpoaccess.gov/cgi/t/text/text-idx?c=ecfr&rgn=div5&view=text&node=21:2.0.1.1.10&idno=21
2.	"Guidance for Industry, Sterile Drug Products Produced by Aseptic Processing—Current Good Manufacturing Practice," US FDA, September 2004	http://www.fda.gov/ohrms/dockets/ac/05/briefing/2005-4136b1_04_Sterile%20Drug%20Products.pdf
3.	"Guidance for Industry, Quality Systems Approach to Pharmaceutical cGMP Regulations," FDA, September 2006	http://www.geinstruments.com/library/industry-links-and-regulations/pharmaceutical-biopharmaceutical.html
4.	"Guidance for Industry, Q7A Good Manufacturing Practice for Active Pharmaceutical Ingredients," FDA, August 2001	http://www.fda.gov/downloads/regulatoryinformation/guidances/ucm129098.pdf

	Regulatory Documents Sited	Web Site
5.	EudraLex, "The Rules Governing Medicinal Products in the European Union," Annex 1, *Manufacture of Sterile Products*	http://freedownload.is/pdf/eudralex-the-rules-governing-medicinal-products-in-the-european-union-2071969.html
6.	ICH Q7, "Good Manufacturing Practice Guide for Active Pharmaceutical Ingredients," November 2000	http://www.ich.org/products/guidelines/quality/quality-single/article/good-manufacturing-practice-guide-for-active-pharmaceutical-ingredients.html
7.	*ISPE Baseline Pharmaceutical Engineering Guide*, Volume 6 "Biopharmaceutical Manufacturing Facilities," June 2004	http://www.ispe.org/
8.	"Guidance for Industry, Q7A Good Manufacturing Practice," Section 4.13	http://www.fda.gov/downloads/regulatoryinformation/guidances/ucm129098.pdf
9.	FDA form 356h, "Application to Market a New Drug, Biologic, or an Antibiotic Drug for Human Use," Item 15	http://www.fda.gov/downloads/AboutFDA/ReportsManualsForms/Forms/UCM082348.pdf
10.	"Guidance for Industry, Q7A Good Manufacturing Practice," Section 4.1	http://www.fda.gov/downloads/regulatoryinformation/guidances/ucm129098.pdf
11.	*NIH Guidelines for Research Involving Recombinant DNA Molecules*, Appendix K, October 2011	http://oba.od.nih.gov/oba/rac/guidelines/APPENDIX_K.htm
12.	ISPE Baseline Pharmaceutical Engineering Guide, Volume 6 "Biopharmaceutical Manufacturing Facilities," Bioprocess design, computer-aided, June 2004	http://www.ispe.org/
13.	"Guidance for Industry, Q8 (R2) Pharmaceutical Development, ICH," November 2009	http://www.fda.gov/downloads/Drugs/GuidanceComplianceRegulatoryInformation/Guidances/ucm073507.pdf
14.	PQLI Guide, Part 1, "Product Realization using Quality by Design (QbD): concepts and Principles," ISPE, Protein aggregation and precipitation, measurement and control	http://www.ispe.org/
15.	"Pharmaceutical cGMPs for the 21st Century—A Risk-Based Approach," US FDA, February 2003	http://www.fda.gov/drugs/developmentapprovalprocess/manufacturing/questionsandanswersoncurrentgoodmanufacturingpracticescgmpfordrugs/ucm071836

	Regulatory Documents Sited	Web Site
16.	"Guidance for Industry, Q9 Quality Risk Management," US FDA, ICH, June 2006	http://www.fda.gov/downloads/Drugs/GuidanceComplianceRegulatoryInformation/Guidances/ucm073511.pdf
17.	International Society of Pharmaceutical Engineering (ISPE) Training, March, 2012	http://www.ispe.org/
18.	*Pharmaceutical Engineering Baseline Guides*, Volume 5, "Commissioning and Qualification," ISPE, March, 2001	http://www.ispe.org/
19.	ASTM Standard for Specification, Design and Verification of Pharmaceutical and Biopharmaceutical Manufacturing Systems and Equipment, ASTM, July 2007	http://www.astm.org/
20.	"Risk Based Qualifications for the 21st Century," ISPE White Paper, March 2005	http://www.ispe.org/
21.	*Good Practice Guide, Applied Risk Management for Commissioning and Qualification*, ISPE, October 2011	http://www.ispe.org/
22.	"Guidance for Industry, General Principles of Process Validation," US FDA May 1987	http://www.fda.gov/

References

1 *Title 21,* Code of Federal Regulations, *Good Manufacturing Practice Regulations.*
2 *Guidance for industry, sterile drug products produced by aseptic processing—current good manufacturing practice.* US FDA; September 2004.
3 *Guidance for industry, quality systems approach to pharmaceutical cGMP regulations.* FDA; September 2006.
4 *Guidance for industry, Q7A good manufacturing practice for active pharmaceutical ingredients.* FDA; August 2001.
5 *EudraLex. The rules governing medicinal products in the European Union. Annex 1,* Manufacture of sterile products.

6 *ICH Q7. Good manufacturing practice guide for active pharmaceutical ingredients; November 2000.*
7 *ISPE baseline pharmaceutical engineering guide.* Volume 6, Biopharmaceutical manufacturing facilities; June 2004.
8 Guidance for industry, Q7A good manufacturing practice, Section 4.13.
9 FDA form 356h. Application to market a new drug, biologic, or an antibiotic drug for human use, Item 15.
10 Guidance for industry, Q7A good manufacturing practice, Section 4.1.
11 *NIH guidelines for research involving recombinant DNA molecules,* Appendix K; October 2011.
12 ISPE baseline pharmaceutical engineering guide. Volume 6, Biopharmaceutical manufacturing facilities, Bioprocess design, computer-aided; June 2004.
13 Guidance for industry, Q8 (R2) pharmaceutical development, ICH; November 2009.
14 PQLI Guide, Part 1: Product realization using quality by design (QbD): concepts and principles, ISPE, Protein aggregation and precipitation, measurement and control.
15 *Pharmaceutical cGMPs for the 21st century—a risk-based approach'.* US FDA; February 2003.
16 *Guidance for industry, Q9 quality risk management'.* US FDA, ICH; June 2006.
17 International Society of Pharmaceutical Engineering (ISPE) Training; March 2012.
18 *Pharmaceutical engineering baseline guides. Volume 5,* Commissioning and qualification'. ISPE; March 2001.
19 ASTM standard for specification. Design and verification of pharmaceutical and biopharmaceutical manufacturing systems and equipment. ASTM; July 2007.
20 Risk based qualifications for the 21st century. ISPE White Paper; March 2005.
21 *Good practice guide, applied risk management for commissioning and qualification.* ISPE; October 2011.
22 *Guidance for industry, general principles of process validation.* US FDA; May 1987.
23 *Guidance for industry, process validation: general principles and practices.* US FDA; January 2011.

Chapter 5

Closed Systems in Bioprocessing

Jeffery Odum

NNE, Durham, North Carolina, USA

5.1 Introduction

The concept of a closed processing system makes common sense. What may be surprising is how closed systems are defined and referenced within the body of regulatory guidance documents and how companies differ in their implementation of a "closed manufacturing" concept. What is not surprising is the impact that closed systems have on facility design.

5.2 Definition of Closed Systems

A search of the current literature within the industry will result in very little direct reference to defining a closed system. For example, there is no reference to closed process systems and their definition in 21 Code of Federal Regulation (CFR)—Part 211. However, there are some reference materials that begin to give definition to the concept of closed systems.

Commission Directive 2003/94/EC of the EMEA [1] introduces the term *closed system* in reference to equipment and system design for production. *ICH Q7* [2] also references the use of "closed or contained systems" in section IV (Buildings and Facilities) and section V (Process Equipment). However, neither of these guidance documents provides a definition for a closed system.

The *ISPE Baseline Pharmaceutical Engineering Guide® for Biopharmaceutical Manufacturing Facilities* [3] does provide a definition of a closed process or system:

Process Architecture in Biomanufacturing Facility Design, edited by Jeffery Odum and Michael C. Flickinger.

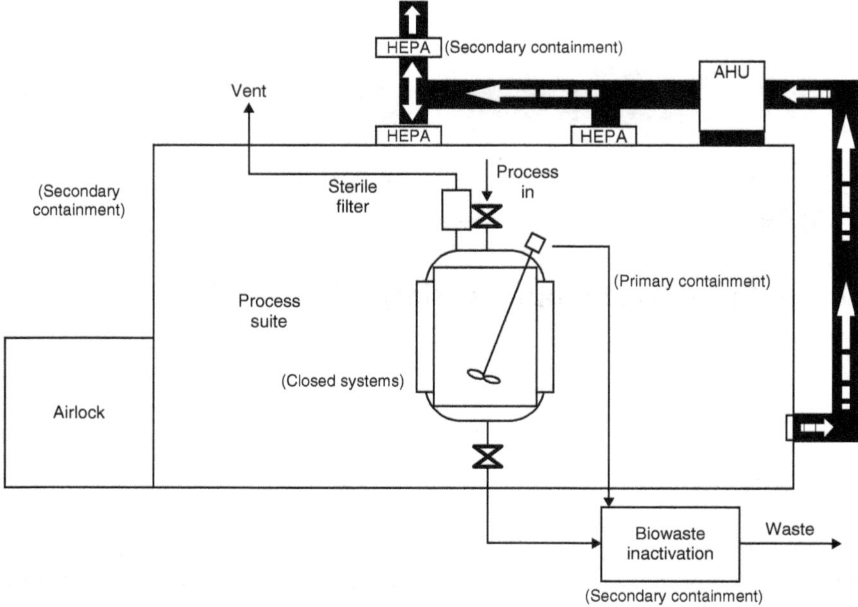

Figure 5.1 Primary and secondary containment boundaries are defined by elements of the facility design, and include the use of closed systems to provide both product protection and environmental containment.

> *A process step (or system) that utilizes processing equipment in which the product is not exposed to the immediate room environment.*

The Guide goes on to state that:

> *It is the Manufacturer's responsibility to define and prove closure for a process step.*

Closed systems (Fig. 5.1) are systems that prevent escape of the product and prevent entry of contaminants from the external environment into the product. To achieve this closed systems typically

- utilize processing equipment and components that do not expose product to the external environment;
- allow materials to enter or leave the system via predetermined control points;
- can be validated to prevent unwarranted material exchange between the process and the environment.

5.3 Closed System Design

The basic elements of containment design in the manufacture of a biological product provide a simple graphic representation of the foundation of closed systems and how they impact facility design. The implementation of primary containment elements (equipment and systems) are the primary means of both product and personnel protection.

It is critical to recognize that "locally protected" systems are not closed. Locally protected systems are those that do not meet the definition of closed but which provide greater protection than an open process or system. This would include those systems where product protection is achieved via some form of localized environmental protection such as a laminar flow curtain or biological safety cabinet/isolator system. It is also important to recognize that biotech processes are different; some lend themselves better to a closed system than others. Less refined processes requiring frequent in-process intervention are more difficult to close. Earlier stage, smaller scale operations are typically more "open" in their design approach due to an often limited availability of small closed equipment and because it is easier to control the external environment at smaller scale.

The American Society of Mechanical Engineers (ASME) provides a comprehensive set of guidelines applicable to closed bioprocessing systems and equipment [4]. The 2016 edition of the BioProcess Equipment Standard is the recognized industry guide for equipment design. The critical elements of the design standards covered in the guideline centers around sterility and cleanability. Each component is required to be cleaned and sterilized and designed to be operated in a clean and sterile manner. The accepted practices defined in the guidelines meet the intent of the GMPs for equipment design and construction [5]. These include accepted practices for cleanability, materials of construction, connections and fittings, design for system drainability, seals and gaskets, and containment.

There are many different design approaches that are employed within the industry to meet the closed system design criteria. One example that provides a clear view of the complexities and options available to engineers involves the design of systems for collecting in-process samples while maintaining aseptic conditions.

Figure 5.2 illustrates one approach for a closed system design for in-process sampling, where the product transfer occurs via a series of automated and manual valves and the sample port is cleaned and sterilized using the Sterilize-in-Place (SIP) system. In this scenario, the sample assembly is cleaned prior to the product transfer and again upon completion of the transfer. There is a specific sequence of operation that must be followed for the cleaning and sampling process. Additionally, there are a number of parts and components that are required to complete the entire design.

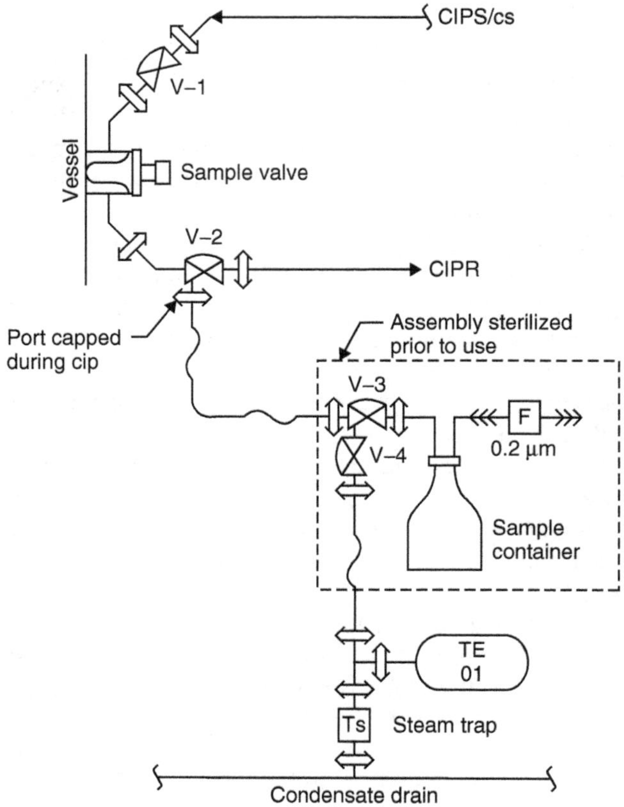

Figure 5.2 Closed system design example for in-process sampling. [Drawing courtesy of CRB Consulting Engineers, Kansas City, Missouri]

Another approach to obtaining in-process samples in a closed system design is the implementation of closed disposable sampling. The NOVAseptic NovaSeptum system is a presterilized component that has been used to accomplish the same sampling as the previously described process. In this system (Fig. 5.3), components are connected to sample bags via a closed cannula and septum. The number of samples that can be taken is limited by the configuration of the bag holder, but the system has proven successful for many sampling functions. The surface of the septum will also be sterilized during the SIP cycle.

In selecting a sampling approach, there are also a number of variables that will impact the overall cost including the number of samples required, disposal cost, cleaning frequency, and the level of automation incorporated into the entire process.

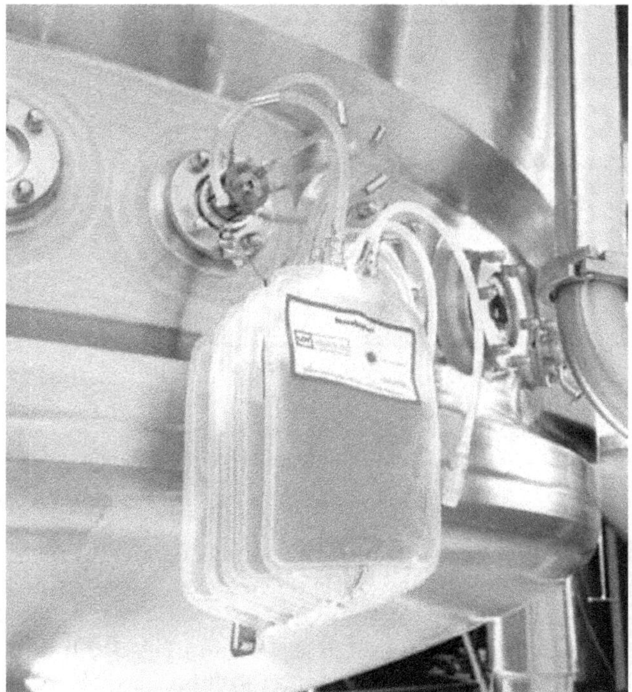

Figure 5.3 The NOVA septic NovaSeptum system with a presterilized component. [Photograph courtesy of NOVA Septic-Millipore, Gothenburg, Sweden]

5.4 Impact on Facility Design

If a system design is truly closed, it opens up a wide array of facility options and provides a very flexible approach to compliance from a regulatory perspective. This is clearly shown in the following quote from ICH Q7A:

> *Where the equipment itself (e.g., closed or contained systems) provides adequate protection of the material, such equipment can be located outdoors.*

While no one expects to see this dramatic implementation in any facility design, it does provide architects and engineers with many options from which to approach facility and process system integration, especially as firms seek to find ways to reduce not only capital costs, but annual operating and maintenance costs in order to improve cost-of-goods figures and reduce the impact of validation.

Costs for biomanufacturing space are very expensive. Some facilities can average between $700 and $2000 per square foot and greater [6,7]. It is always

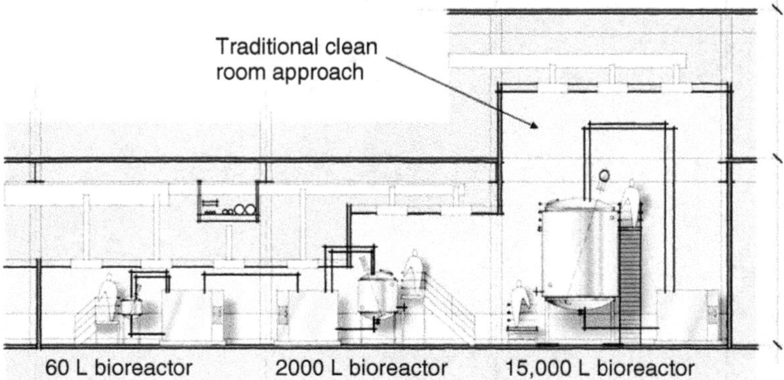

Figure 5.4 Traditional approach to clean room design; large areas of classified space where closed systems were often not implemented. [Graphic courtesy of Flour Corporation, Greenville, South Carolina]

one of the major challenges that design teams face to find creative ways to reduce facility cost by reducing the amount of classified space that is defined in the design documents.

The traditional approach to clean room design is shown in Fig. 5.4. In this approach, facility cost factors include equipment and room finishes, Heating, Ventilation, and Air Conditioning (HVAC) equipment, routine maintenance of clean room space, cleaning and gowning materials, and increased access control requirements due to open system operations.

By choosing to implement a predominately closed system approach, the same layout would look dramatically different in Fig. 5.5. In this layout, the actual amount of classified space is dramatically reduced. Critical issues such as sampling and filter change-out must still be addressed, but overall construction, operation, and maintenance costs can be significantly less.

The *ISPE Baseline Guide*® addresses the options for implementing a closed process design as a means of developing manufacturing areas designated as *controlled, nonclassified* (CNC) space. It is recognized in the industry that a large majority of process operations occur in areas that could be considered CNC space, regardless of whether they are claimed as such by the manufacturer. As discussed in the *ISPE Baseline Guide*, CNC space can resemble a wide variety of configurations and appearances [8]. But, the key attribute of CNC space is the fact that it begins with closed system design.

Figure 5.6 provides four different layout considerations for closed systems in CNC space [9]. In the first case, the layout approach and classification philosophy that are described represent what is currently the primary approach taken by the industry; all process equipment is located in a classified space that is validated and under protocol control.

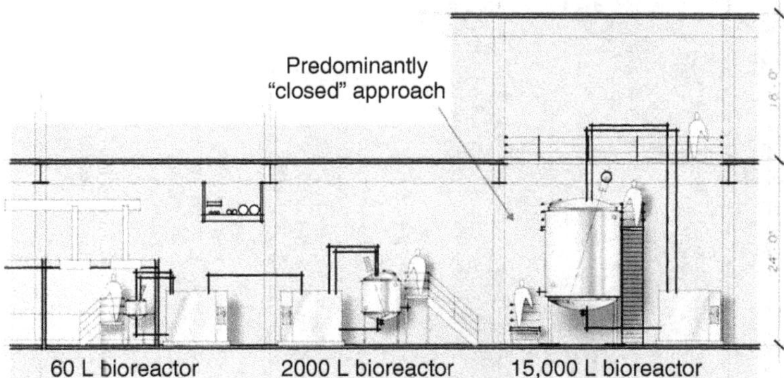

Figure 5.5 Implementing a predominately closed system approach: the layout of Fig. 5.4 looks different, as the actual amount of classified space is dramatically reduced. [Graphic courtesy of Flour Corporation, Greenville, South Carolina]

The second case resembles the approach often referred to as *gray space* implementation, where some of the mechanical and "maintenance intense" components of process equipment are located in a nonclassified space. Moving to the third case, a less-often used approach of placing only sample port locations inside a clean vestibule space is illustrated. Figure 5.7a and b provides a graphic and photo of one such installation.

The final case is the true implementation of complete closed system production. As shown, the amount of actual classified clean room space is dramatically reduced, and logically, so will the costs of operation and maintenance, gowning, and validation for this space be reduced as well.

5.5 Impact on Operations

It is considered good engineering practice to design the main process stream as closed as possible during operation. This will substantially reduce the risk of product contamination from personnel or the facility environment. It will also provide protection of personnel from potential exposure to viruses or other pathogens that may be present.

There are products and processes that are better suited to closed system design. It is much easier to design closed systems for well-characterized products and well-defined processes that are robust. However, in many cases, such as clinical or process development or smaller scale operations, it may be much easier and cost effective to place the entire process into a classified environment (Case 1) and allow for more open process operations.

Upstream process operations such as seed inoculum and cell culture unit operations are typically designed as axenic operations, where the culture is

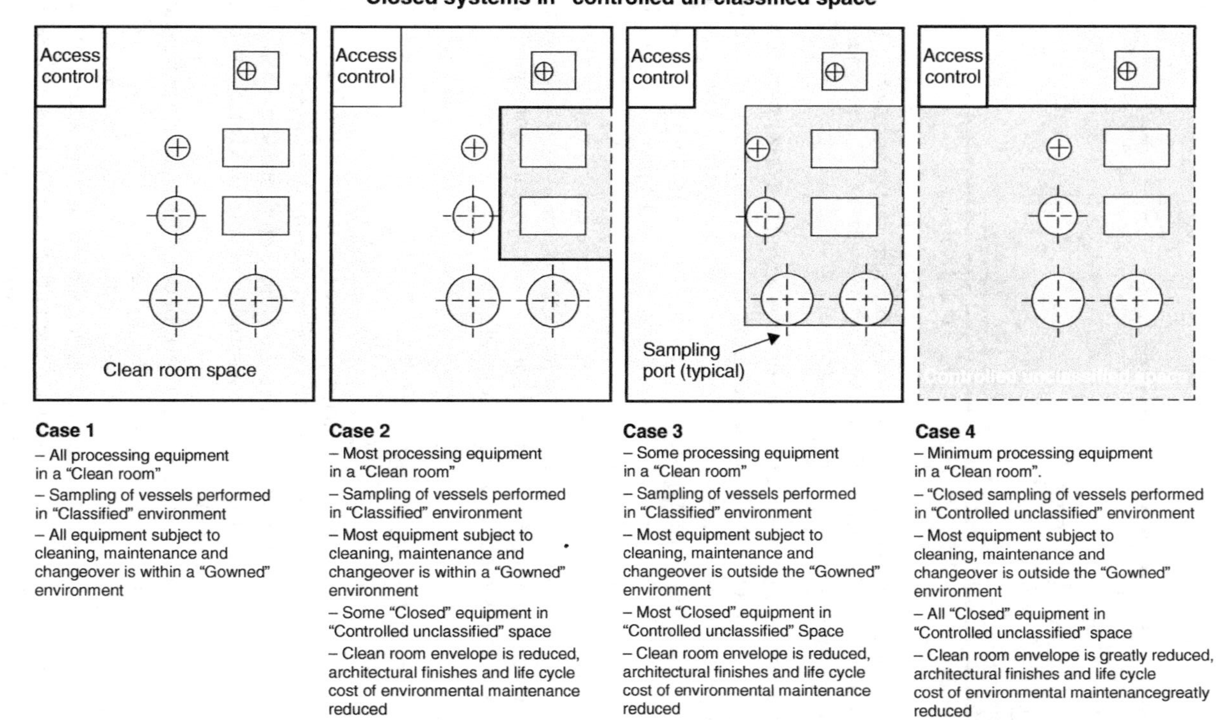

Figure 5.6 The evolution of classified space design, from large open areas (Case 1) to current implementation of closed system design philosophy as defined by ICH Q7. [Graphic courtesy of ISPE, from the "Biopharmaceutical Facilities Baseline Guide, Volume 6."]

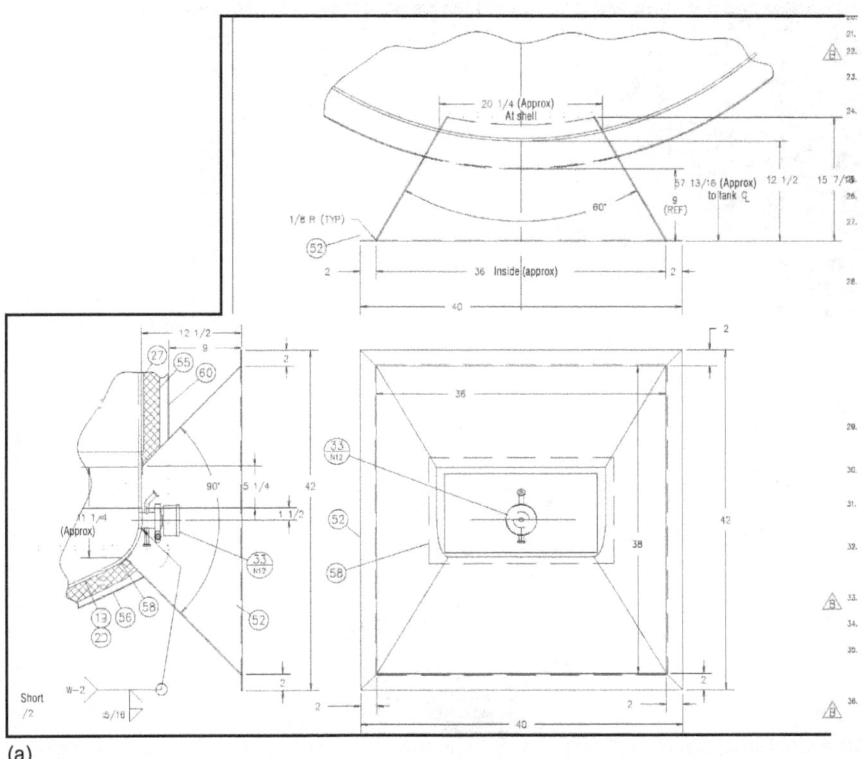

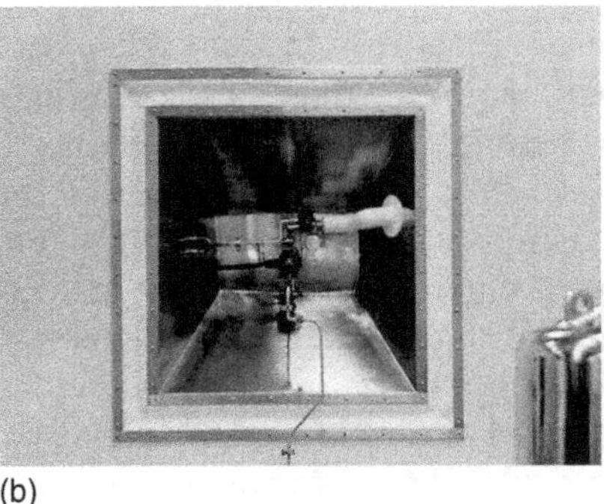

Figure 5.7 Sample port locations inside a clean vestibule space sketch (a) and photograph (b). The creation of the vestibule allowed for sample collection inside a classified clean room space while reducing overall area of clean manufacturing space. [Sketch and photograph courtesy of Biogen, Research Triangle Park, North Carolina]

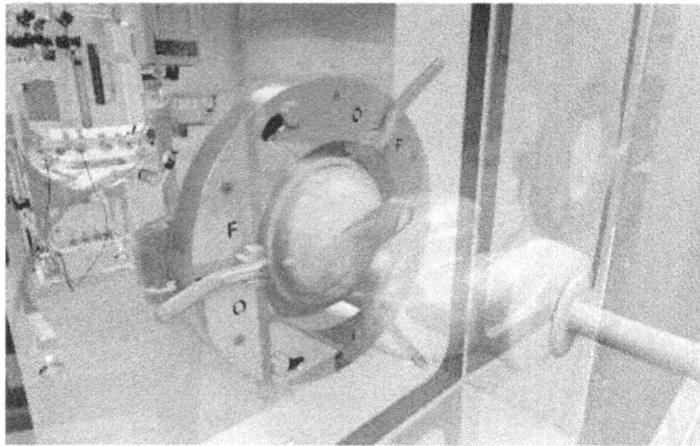

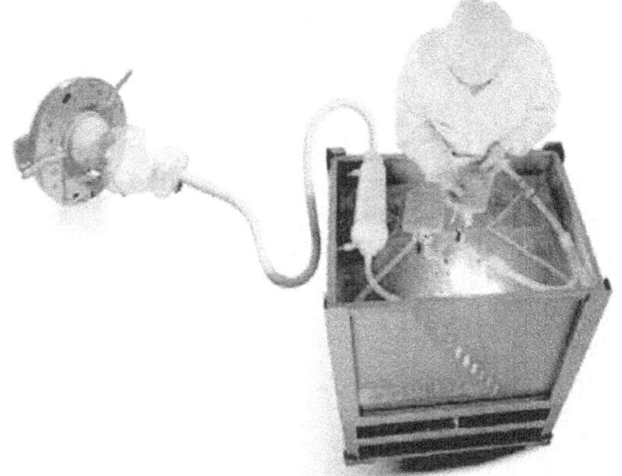

Figure 5.8 The RAFT (rapid aseptic fluid transfer) system, based on Biosafe Technology by Stedim, allows aseptic zone-to-zone transfer of fluids via closed transfer system and is used by many companies. [Photograph courtesy of Sartorius Stedim Biotech, Copenhagen]

free from living organisms other than the host cell. This, by its nature, implies a closed system design approach. Bulk and final filling operations are also designed to control contamination and bioburden, both in terms of product contamination and risk. Product contamination at this final stage of the process can have a disastrous impact economically to the manufacturer.

There are some process unit operations that are difficult to keep fully closed. Some forms of in-process sampling, column fractionation, and solids additions from bulk containers are examples. It is also important to understand that closed systems will be open at certain times. This will occur for maintenance and setup. For example, filter housings are open system components during setup, but are closed during operation. The loss of the closed state due to these routine activities does not negate closure as a key component of the facility design. It is necessary to have validated procedures for cleaning, assembly, and sterility testing to ensure that the closed state has been reinstituted. A pressure hold test to prove system integrity will also be used to validate closure of the system prior to operation.

As many companies continue to implement single-use or disposable technology into their manufacturing operations, the question of system closure must also be addressed. Many process unit operations have successfully been designed and validated for system closure. These include

- media, buffer, fluid transfer, and storage
- product transfer, holding, and shipping
- bioreactors
- distribution and filling manifolds
- sampling systems.

One example of where an enabling technology based on single-use design is being implemented is with the transfer of fluids under aseptic conditions. The rapid aseptic fluid transfer (RAFT) system, based on Biosafe Technology by Stedim allows aseptic zone-to-zone transfer of fluids via closed transfer system and is used by many companies (Fig. 5.8).

This type of component–system integration is becoming more common within the biopharmaceutical manufacturing industry.

5.6 Summary

Closed systems are critical for meeting GMP requirements. International regulatory bodies and industry standard institutions recognize the importance of this fundamental practice. While it is not necessary that all production systems be completely designed around the closed system concept, good engineering design practice encourages closed systems to protect the product, the workers, and the environment. Closed system design is a key to risk minimization, both in terms of facility capital cost and in operational expenditures.

References

1 European Medicines Agency. Medicinal products and veterinary use: good manufacturing practice. Annex 2, Volume 4, Chapter 5; 2003 and 2008. p 3.
2 International Conference on Harmonization. Guidance for industry: Q7 good manufacturing practice guidance for active pharmaceutical ingredients; 1998.
3 International Society for Pharmaceutical Engineering Baseline Pharmaceutical Engineering Guide for New and Renovated Facilities. Volume 7, *Biopharmaceutical manufacturing facilities*. 1st ed.; 2004. p 171.
4 American Society of Mechanical Engineers. ASME bioprocessing equipment standard, BPE-2016, 2016. p 9–57.
5 21 CFR, part 211, subpart D—Equipment, 211.63 and 211.65.

6 Odum JN. *Sterile product facility design and project management.* 2nd ed. Florida: Boca Raton; CRC Press; 2004. p 14.
7 Pavlotsky R. Approximating facility costs, Cleanrooms, Volume 18, Section 6.6.5. 2004.
8 ISPE Baseline Pharmaceutical Engineering Guide for New and Renovated Facilities. Volume 6, *Biopharmaceutical manufacturing facilities.* 1st ed. Section 6.6.5. p 101.
9 ISPE Baseline Pharmaceutical Engineering Guide for New and Renovated Facilities. Volume 6, *Biopharmaceutical manufacturing facilities.* 1st ed. Section 13, Fig. 13.9. p 156.

Chapter 6

Aseptic Manufacturing Considerations for Biomanufacturing Facility Design

Jeffery Odum[1], Hartmut Schaz[2], and Larry Pressley[3]

[1] NNE, Durham, North Carolina, USA
[2] NNE, Frankfurt, Germany
[3] IPS, Morrisville, North Carolina, USA

6.1 Introduction

"Processes that are devoid of measurable (detectable) bioburden" are defined as aseptic operations [1]. Aseptic operations generally require sterilization of the environment, equipment, and process solutions to achieve the sterile state before use. While the current Guidance for Industry on Sterile Drug Products Produced by Aseptic Processing is often associated with the manufacture of sterile injectable drug product(s), it also applies to the manufacture of biologics. The reasons for this are rooted in the biology of biological therapeutics.

Because biotherapies are developed using living organisms/cell lines, there is a defined need to ensure the safety of the components that will make up the finished drug product that will be delivered to the patient. The vast majority of these products are injectables. It is a well-accepted principle that sterile drugs, free of contamination and of an acceptable level of bioburden, should be manufactured using aseptic processing [2]. The Biotech Industry has adopted the fundamental principles of aseptic manufacturing to ensure this level of safety for the patients.

In a traditional aseptic process for nonbiological products, the final drug product, its container, and closure are subjected to sterilization methods, brought together, and then subjected to terminal sterilization in the final container. This is usually not an option for biologics as the cells will not survive the sterilization process. It is therefore critical that the entire process for the manufacture of drug substance components [and active pharmaceutical ingredients (APIs)] be executed in very controlled environments, including filling and sealing of the finished dosage form.

The World Health Organization (WHO) [3] and the European Union/PIC/S [4] also follow this same approach and philosophy.

Process Architecture in Biomanufacturing Facility Design, edited by Jeffery Odum and Michael C. Flickinger.

6.2 The Relationship to Biological Products

For clarity in defining the information contained in this chapter, there will be some terms that need to be clarified and revisited.

API: Any substance or mixture of substances intended to be used in the manufacture of a drug (medicinal) product and that when used in the production of a drug becomes an active ingredient of the drug product. Such substances are intended to furnish pharmacological activity or other direct effect in the diagnosis, cure, mitigation, treatment, or prevention of disease or to affect the structure and function of the body [5].

On the basis of the abovementioned definition, the term drug substance and API are often interchanged. Drug substance is the bulk form of API that has not yet been manufactured in its final dosage form for delivery to the patient; that is referred to as drug product. So, for aseptic manufacturing applications to the manufacture of biological APIs, we must focus on ensuring that measures are taken during the manufacturing process to follow the aseptic guidance practices that will be discussed in order to minimize the risk of contamination to the bulk API before its final formulation into finished drug product.

A fundamental tenant of current good manufacturing practice (GMP) for facility design and construction is that facilities should be designed to minimize potential contamination of the drug substance/drug product during all manufacturing operations [5, Section 4.10]. The facility should also be designed to limit exposure to objectionable microbiological contaminants that may be transmitted from the operating environment or operations personnel within the facility. Under current aseptic manufacturing guidance, this provides a platform of facility attributes that form the foundation for good facility design that meets current GMP.

6.3 Process Attributes—Product Protection

In simple terms, a manufacturing process can be either open or closed. If the primary goal is to protect the product during the manufacturing operations, closed processing presents less of a risk to the product. This is a risk that primarily comes from the immediate manufacturing environment. So, the ultimate goal is to ensure that the product is never exposed to the environment unless the environment is constantly maintained as bioburden free. This is a very difficult and costly goal.

One of the primary contamination sources within biomanufacturing space is the employee. Recent studies conducted by Dastex [6] support the fact that even with varying levels of gowning protection for employees, airborne contamination still occurs and therefore must be properly addressed as part of an

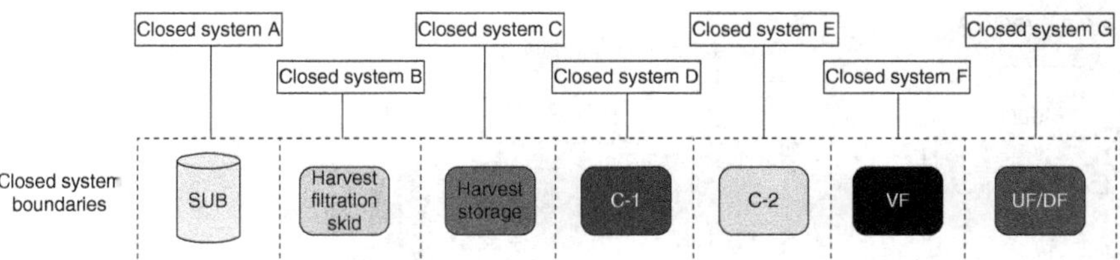

Figure 6.1 Closed system sequence.

aseptic manufacturing approach. Thus, the application of classified clean room space is introduced.

However, the reality is that mitigation of contamination risk can also be accomplished through the use of proven design techniques to achieve system closure. The American Society of Mechanical Engineers is one professional organization that has developed globally accepted standards for closed manufacturing system design that meet the requirements as stated in the GMPs [7].

When closed systems are developed to protect the product, the facility environment that houses the equipment/system no longer becomes a critical aspect of the process, thus simplifying the facility design. It also allows for a means of detection in that contamination of a closed system represents a breach of the system that can be identified and corrected as part of a regulatory investigative process.

For aseptic manufacturing operations, think of the process as a sequence of closed systems. Each unit operation is designed and run as a closed system where there will be both overlaps and interphases (Fig. 6.1).

When a process cannot be completely closed for all unit operations and/or manufacturing steps, the aseptic guidelines require that product protection still occur. This is normally accomplished by the introduction of various forms of environmental control that reduce the risk of product contamination in the event that exposed product comes in contact with external contamination sources. These include the use of clean manufacturing environments via classified work spaces, microenvironments such as laminar flow curtains and laminar flow hoods, or higher level equipment technology platforms such as restricted access barrier systems (RABS) or isolators. Each of these options not only provides a higher level of product protection but also introduces higher facility capital cost and increased annual operating costs.

6.3.1 System Closure

If you take a process system and break it down into its simplest components or unit operations, it is easy to see that there are three basic elements of the system (Fig. 6.2):

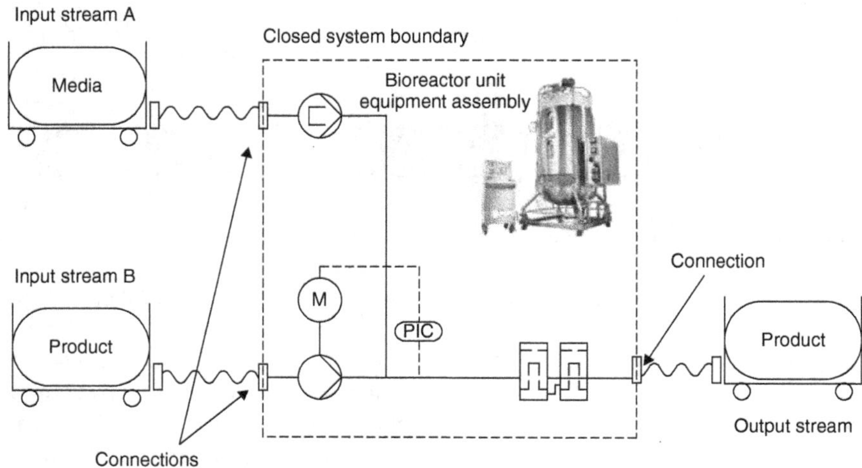

Figure 6.2 A diagram of a closed process system boundary.

- The equipment that constitutes the unit operation: vessels, bioreactor, filtration skid, and so on.
- The input streams for each piece of equipment: water, air, gases, or additions such as media or buffer materials.
- The physical connections to the equipment: piping, valves, and connectors.

These elements must be assembled and tested to verify that system closure has been accomplished and the product is not at risk of external contamination sources. The frequency that this assembly/testing activity occurs is dependent on a number of factors such as production changeover between batches/products, technology platform being implemented, and operational philosophy of the company.

In determining compliance to aseptic guidance [2] and to address issues related to risk identification defined in ICH Q9 [8], it is important that the risk of each of the three elements be reviewed and addressed via a formal risk assessment/mitigation effort. In order to accomplish this for the equipment element, the question of "where and when do we close the system" must be answered. Understanding that each unit operation may have a different answer based on numerous factors can complicate the risk analysis.

For example, if the system is implementing single-use (SU) components, the manufacturer may decide to assemble system components on their own or have a third party provide subassemblies that are then installed within the overall system. In this case, the final closure validation could have different "locations" where the closure occurs. Whether the closure technology being implemented is SU or a traditional fixed stainless steel platform, the execution of the risk analysis is critical to define the level of closure.

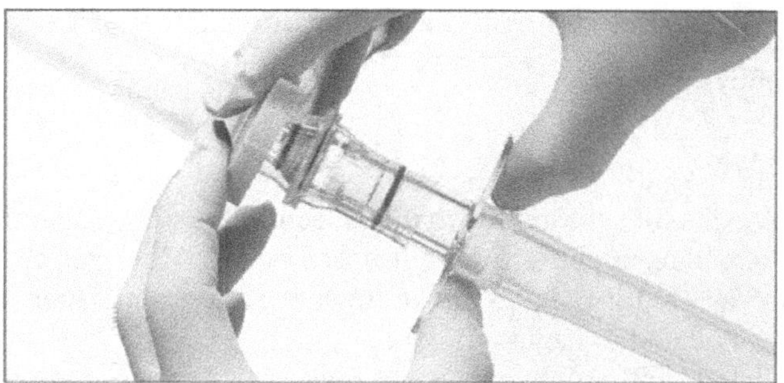

Figure 6.3 Kleenpak sterile connector courtesy of Pall Corporation.

Proof of closure for the input streams will involve different approaches, dependent on operational philosophy and the specific process parameters. For liquids and gasses, the risk mitigation approach will likely be the implementation of some form of filtration step. Quiet often, the ability to integrity test filters and their high level of reliability provides the high level of confidence that the product stream has been protected (proof of closure) as required by the aseptic guidance.

Today, closure technology provides a number of proven options around risk mitigation owing to the physical design/geometry of connectors. Stainless steel double-block valves, SU sterile connectors, and tube welders and sealers have been developed and tested across a wide range of applications and manufacturing scales. All of these components (Fig. 6.3) can be purchased with documentation certifying their compliance to current regulatory guidelines.

These types of SU components are presterilized via gamma radiation and sealed for shipment.

6.3.2 Segregation Strategy

The aseptic guidance defines expectations around product protection and how facilities should address the risk of potential contamination due to process unit operations not being closed [2, Section IV and Appendix 3]. A key consideration of the facility design must be how the facility segregation strategy is defined and implemented based on the manufacturing process definition.

The prevention of cross contamination between products and manufacturing unit operations is a mandatory requirement [9] that is easiest accomplished when there is complete separation of manufacturing operations/areas within the facility. However, this is also a costly proposition not only in capital cost but also in operational costs. So, most organizations search for solutions that provide an effective segregation strategy while also providing a more efficient and effective capital and operational cost solution.

Defining a model for segregation strategy is dependent on a number of items:

- Operational philosophy
- Product attributes
- Manufacturing technology

There is not a one-size-fits-all solution to what will lead to both a high level of risk mitigation and result in a highly efficient, licensed facility. However, the development of this segregation strategy is one of the more important aspects of regulatory compliance related to facility design.

Developing this model around an enterprise approach is a proven method that yields positive results. One such model identifies three key facility attributes recognized by the global regulatory agencies, such as process, facility, and procedural controls, and uses them as the foundation for segregation strategy development [10]. The facility space attributes include segregation of individual unit operations in some physical manner coupled with segregated HVAC systems. The process includes the physical equipment (stainless steel or single use), defined methods for system closure (validated), and engineered elements to provide system integrity during operation and change over. Procedural controls represent infrastructure support such as appropriate documentation, training, and standard operating procedures.

In facility design, an essential part of contamination prevention is the adequate separation of areas of manufacturing operation. To maintain air quality, it is important to achieve a proper airflow from areas of higher cleanliness to adjacent less clean areas. It is vital for rooms of higher air cleanliness to have a substantial positive pressure differential relative to adjacent rooms of lower air cleanliness. The aseptic guidance recommends that a positive pressure differential of at least 0.04–0.06 in. of water (approximately 10–15 PA) gauge should be maintained between adjacent rooms of differing classification (with doors closed). When the doors are open, outward airflow should be sufficient to minimize ingress of contamination, and it is critical that the time a door can remain ajar be strictly controlled. This concept of segregation is depicted in HVAC room pressure diagrams (Fig. 6.4) that are required documents in the license application package for FDA.

In the case where a facility is designed with an unclassified room adjacent to an aseptic manufacturing room, a substantial overpressure from the aseptic manufacturing room should be maintained at all times to prevent contamination risk.

6.4 Facility Design

Aseptic processes are designed to minimize exposure of sterile articles to the potential contamination hazards of the manufacturing operation [2]. Limiting

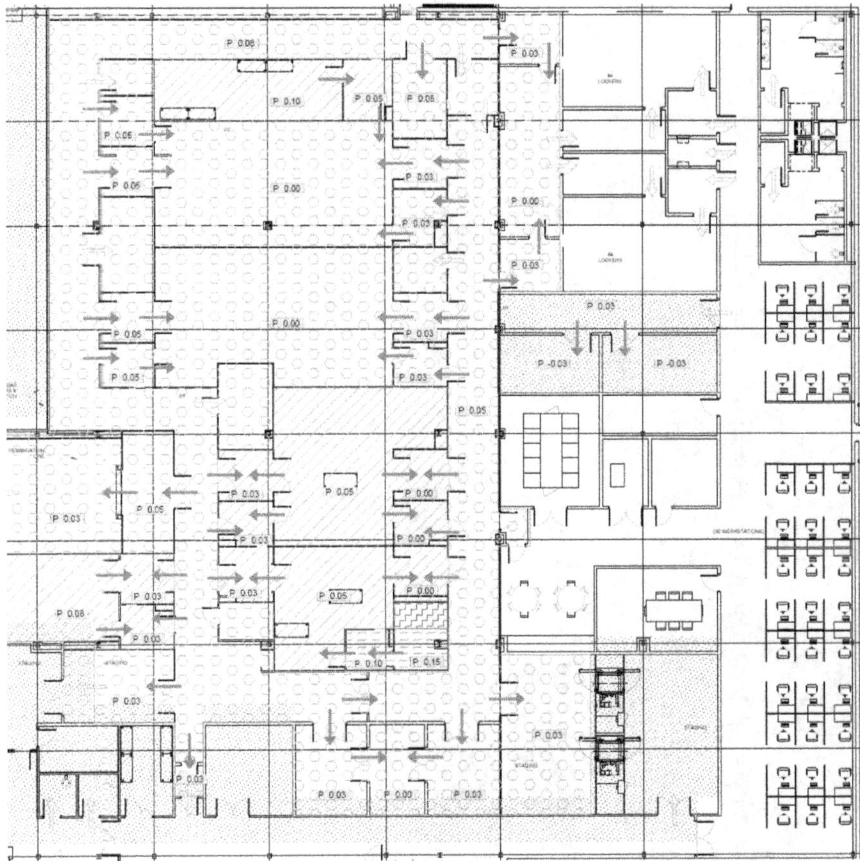

Figure 6.4 HVAC pressure diagram (courtesy of IPS).

the duration of exposure of sterile product elements, providing the highest possible environmental control, optimizing process flow, and designing equipment to prevent entrainment of lower quality air into the higher level clean areas are essential to achieving compliance to current regulatory guidance and minimizing contamination risk to the product.

It is also important that personnel and material flows be optimized to prevent unnecessary activities that could increase the potential for introducing contaminants to exposed product, container closures, or the surrounding environment. The layout of equipment within the space should provide for ergonomics that optimize comfort and movement of operators and allow access for cleaning and maintenance.

One of the operational challenges with many biomanufacturing facilities is in the level of operator support required for the manufacturing operations. The

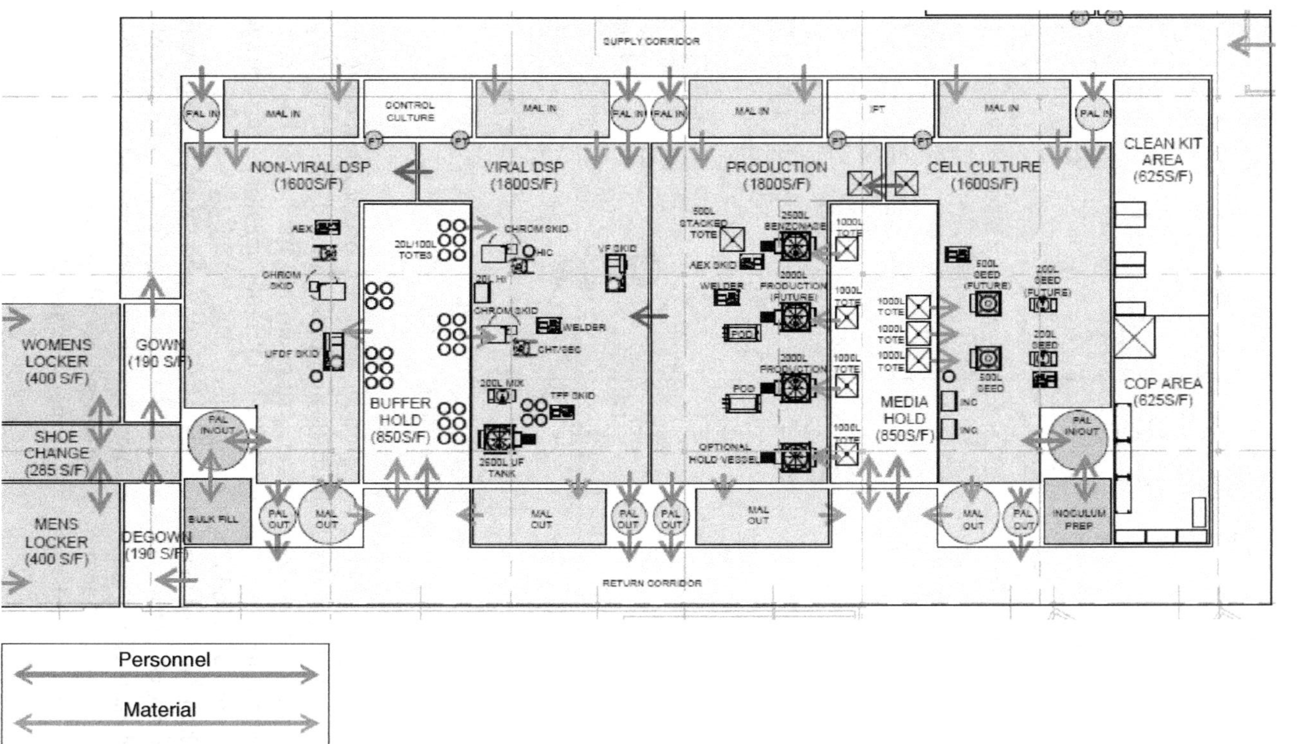

Figure 6.5 Example of flow diagrams for personnel, materials, and product as required by the FDA.

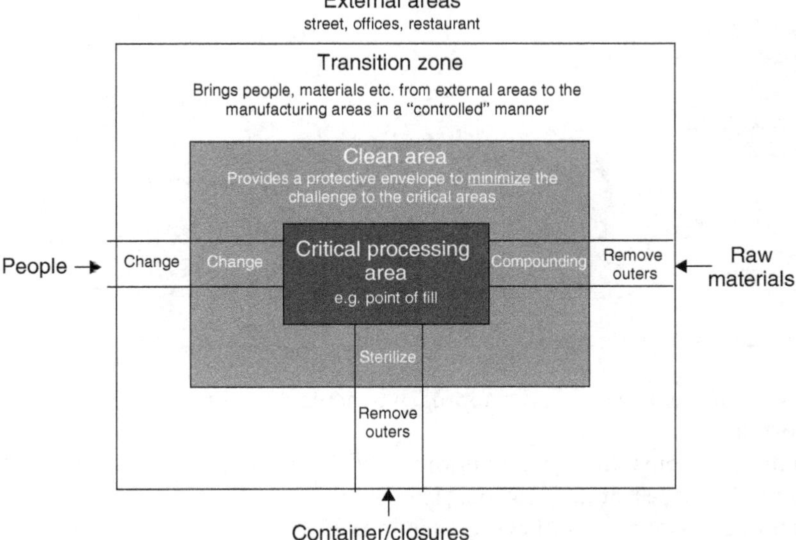

Figure 6.6 Facility segregation transitions zones.

number of personnel in an aseptic processing space should be minimized to the greatest extent possible. It is also important that the flow of personnel should be designed to limit the frequency with which entries and exits are made into and from an aseptic processing space. For the higher level classification spaces, the number of transfers into the clean room should be minimized. Flow diagrams are required supporting documents for facility licensure by the FDA (Fig. 6.5).

6.5 Critical Area

Per the aseptic guidance, a critical area is one in which the sterilized drug product, containers, and closures are exposed to environmental conditions that must be designed to maintain product sterility [2]. To maintain this implied level of product sterility, it is required that the manufacturing environment in which aseptic operations (e.g. equipment setup, product transfer, and filling) are conducted be controlled and maintained at an appropriate quality, often the highest level of environmental cleanliness.

The primary focus of environmental quality in this case is the particle content of the air. These particles are significant because they can enter a product stream as a contaminant, can contaminate the product, and can biologically impact the safety and/or efficacy of the product, thus putting the patient at risk. As previously discussed, environmental control via closed systems and classified clean room spaces are important. However, in the event that open

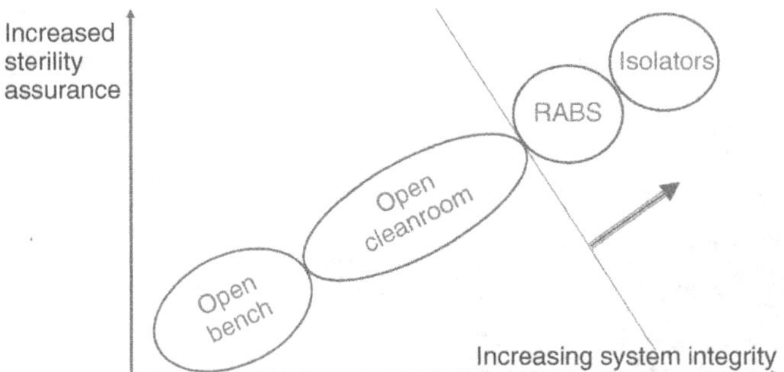

Figure 6.7 System integrity capability to ensure sterility and product protection.

systems/equipment are used, the approach to achieving this critical parameter is important.

In simplistic terms, the location of the critical area in relation to the overall facility can be represented in Fig. 6.6 [11].

Technology advancements have led to improvements in system integrity that have diminished, and in some instances, eliminated the need for a critical area as defined in the aseptic guidance. These barrier technologies include RABS and isolator system technology (Fig. 6.7). The implementation of new technologies has greatly improved the efficiency and effectiveness of environmental control around critical areas.

The general characteristics of RABS technology are as follows:

- The Grade B/ISO7 Room scale critical environment contains operators.
- Critical Grade A/ISO5 Aseptic process Core sits in Grade B/ISO7 room.
- Operators use glove ports for intervention and closed component handling from the surrounding Grade B/ISO7 room.
- Biodecontamination—generally as open room
  - Topical disinfection of surfaces
  - Room fumigation in some cases; common in Europe; validation to a 4-log reduction for viables only
- Separation between operators and Grade A/ISO5 Aseptic process core
  - Safety glass doors
  - Screens

A simple schematic of a RABS unit is shown in Fig. 6.8.

RABS characteristics include the following:

- Physical separation between operator and process via machine cladding.
- Access for sterile setup and intervention (microbiological sampling and emergency) during process via gloves.
- Transfer of format parts, stoppers, and so on into the RABS via aseptic transfer systems.

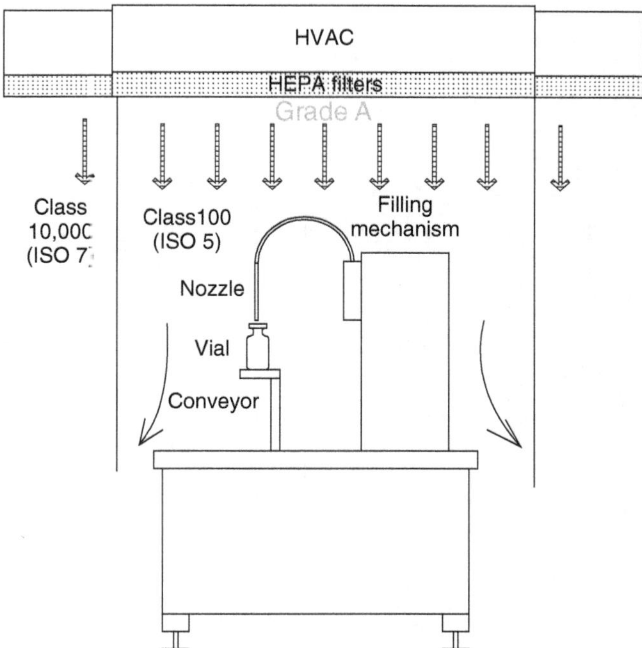

Figure 6.8 Schematic view of RABS unit (open RABS shown).

- Laminar flow units above RABS and in transition zones.
  - active RABS = machine with own LF units
  - passive RABS = laminar flow provided by site HVAC
- Open RABS
  - Outflow of air below level of product filling → active and passive RABS
- Closed RABS
  - Suction of air below level of product filling → active RABS (often for potent compounds)

The general characteristics of Isolator-based technology are as follows:

- Small contained enclosure excludes process operators.
- Critical Grade A/ISO5 Aseptic process Core sits in Grade C/ISO8 or D/ISO8 "at rest" room.
- Operators only open equipment when off-line. Human access via glove ports, and closed process loading.
- Biodecontamination…
  - Topical disinfection of surfaces inside isolator
  - or more often—Internal Isolator aerosol or VHP (vapor-phase hydrogen peroxide)

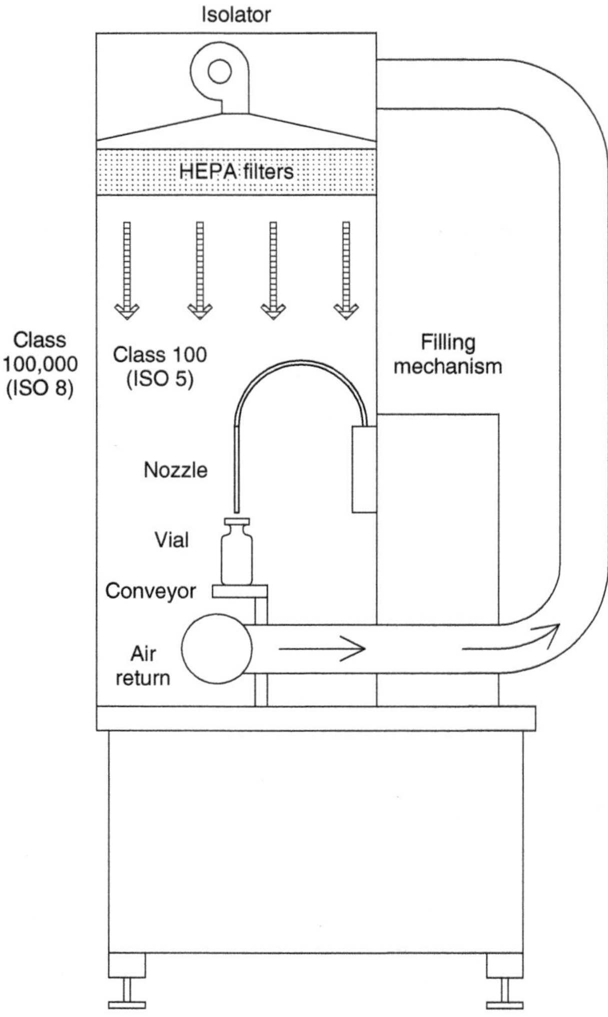

Figure 6.9 Schematic view of isolator unit.

- Separation between operators and aseptic process core
  - Isolator shell
  - Windows
  - Glove ports
  - Transfer ports sterilized change parts and components

A simple schematic of an isolator unit is shown in Fig. 6.9.
Isolator characteristics include the following:

- Barrier technology that separates the process room from the operator's area

- Highest containment possible protection of personnel over product protection
- Manipulation on the system via gloves (or robotics)
- Requires VHP cycle between batches or between multiple batches
- Transfer via RTP into the system
- Needs an isolator dedicated HVAC system or solution where the building HVAC provides conditioned air to the isolator, and the isolator is equipped with an HVAC system for air recirculation and filtration.

The implementation of either of these barrier technology platforms impacts facility design by improving the ability to provide protection of the product from environmental contamination as defined in the aseptic guidance.

References

1 Biopharmaceutical Manufacturing Facilities Baseline Guide, 2nd ed., volume 6; November 2013.
2 Guidance for Industry. Sterile Drug Products Produced by Aseptic Processing—Current Good Manufacturing Practice, US FDA; September, 2004.
3 WHO. Annex 6, Good Manufacturing Practices for Sterile Pharmaceutical Products.
4 EU Annex I. EU/PIC/S Guide to Current Good Manufacturing Practice for Medicinal Products; October 2015.
5 Guidance for Industry Q7A. Good Manufacturing Practice Guidance for Active Pharmaceutical Ingredients, Section 1.1.
6 Cleanroom Technology. Measuring Airborne Germs Produced by Humans Inside a Body Box; October 2015.
7 American Society of Mechanical Engineers. ASME Bioprocessing Equipment Standard; BPE-2014.
8 Guidance for Industry. Q9 Quality Risk Management, US FDA; June 2006, Section 4.
9 21 CFR Part 211. Current Good Manufacturing Practice for Finished Pharmaceuticals, US FDA, subparts 211.42b, 211.63, 211.100.
10 Witcher MF. Impact of facility layout on developing and validating segregation strategies in the next generation of multi-product, multi-phase biopharmaceutical manufacturing facilities. Pharm Eng Suppl 2015.
11 ISPE. Sterile Manufacturing Facilities Baseline Guide, volume 3.

Chapter 7

Facility Control of Microorganisms: Containment and Contamination

Jonathan Crane

HDR, Inc., Atlanta, Georgia, USA

7.1 Introduction

In designing a bioprocess facility producing products under current good manufacturing practices (cGMP) [1–3], it is of paramount importance to provide a design that limits the introduction of bacteria, virus, fungi (including molds), other viable microorganisms, and other environmental contaminants into the facility to reduce the possibility of product contamination. This goal is accomplished through

- physical barriers (such as surfaces including walls, ceilings, and floors; isolation through closed process equipment and systems; and isolators);
- engineering controls (such as clean critical utilities, filtration, pressure zones, and directional airflow); and
- operational procedures (such as gowning protocols; decontamination, sterilization, and cleaning protocols; and controlling the movement of personnel, equipment, and product).

The above-mentioned factors can limit the potential for contamination entering the facility, moving into process areas, and ultimately contaminating the product.

For bioprocess facilities that utilize viable pathogenic organisms as part of the manufacturing process, it is also critical to assess the hazards presented by the organism to the product, personnel, and the environment. Depending on the degree of hazard, similar principles (physical barriers, engineering controls, and operational procedures) are used to contain the organisms to the specific processes in which they are required and, in the event of an accidental release, minimize impact on personnel, the environment, the product, and on operations.

Process Architecture in Biomanufacturing Facility Design, edited by Jeffery Odum and Michael C. Flickinger.

It is important to stress that the facility design, the production process, and operational protocols (including the ability for preventative maintenance) must be integrated for a high degree of contamination control. Integration is even more important for contamination control with containment as although they both use similar principles, at time the principles (such as direction of airflow) work in opposition. Care must be taken not to compromise contamination control effectiveness with containment design, and vice versa. Although personnel harboring microorganisms may be a significant risk in the transport of organisms into and out of a bioprocess facility and may be a significant contributor to the viable particulates in an area, this article will not address personnel restrictions (health status, screening, restrictions on entry, etc.) or requirements (immunizations, training, etc.) other that discussing proper personnel flow and adequate space to allow protocols for gowning-in and gowning-out of facilities.

The risk of microbial product contamination in bioprocess facilities and the risk to product, personnel, and the environment where hazardous microorganisms are used in the process vary greatly based on

- the product and processes used in its production;
- the inherent hazards of the organism or its byproducts;
- how the organism is used and handled; and
- the scale of the work.

There is no "cookbook" method of contamination control for product protection or biocontainment for safety. The risks must be evaluated based on each unique situation and process steps required. Design and operational principles can then be applied to mitigate the risks inherent in that specific situation. Often, it is not possible to create "the ideal" facility solution. The weak points in a facility solution should be evaluated and the remaining risks should be mitigated through enhanced operational or safety procedures.

7.2 Design Principles for Controlling Microorganisms

Inadvertent microorganism entry or exit from a classified or contained zone can occur in basically four ways: (i) transport by being physically carried by personnel, equipment, packaging, materials, etc.; (ii) transport of airborne microorganisms through air supplied to the zone, air exhausted or returned from the zone, or air moving from one zone to the next; (iii) transport out of the facility through waste streams such as water drainage; and (iv) spills that may not be contained within a zone. The designer has many tools to use in controlling the flow of microorganisms within facilities for product protection, personnel protection, and protection of the environment. They fall into five general categories:

- planning concepts
- physical barriers
- engineering systems
- isolation and containment equipment
- facility design to support operational protocols.

7.2.1 Planning Concepts

Careful thinking through the conceptual planning of a bioprocess facility and applying the following planning concepts can help in the development of a facility that will increase the ability for microbial control in the facility.

Clearly delineate zones of classification and/or containment to identify area requirements. Identify classification and containment boundaries clearly, particularly where they overlap. Identify appropriate buffer spaces (airlocks or anterooms), if required, at the entry to and exit from each zone to control the transport of airborne microorganisms between adjacent zones and allow for gowning-in on entry and removal of protective garb on exit. In a bioprocess facility that utilizes hazardous microorganisms, these buffer zones must control the transport of microorganisms in both directions. Work to minimize spaces in the classified zone or contained zone to the spaces that are required to be in those zones.

Create layers of zones with escalating cleanliness and/or containment, similar to the layers of an onion, to effectively control the transport of microorganisms within the facility. Start layering from nonclassified spaces at the entrances to the building. Create layers that reduce the penetration of environmental contaminates into the facility. Continue layering at the entrance to each process area and layer zone by zone within the process area.

Provide separate personnel, material, equipment, waste, and product flows where combined flows create the risk of cross-contamination or mix-up. Evaluate the risk of cross-contamination between personnel flow, material flow, clean and used equipment flow, waste flow, and product flow. Utilize separate airlocks and anterooms where risk of cross-contamination exists. For facilities handling pathogenic organisms, specifically separate airlocks where a potential for the presence of live organisms exists from airlocks where no live organisms are present including separate personnel entry and exit airlocks. One-way flows can be significant in reducing the potential for cross-contamination within the facility.

7.2.2 Physical Barriers

Physical barriers create separation between rooms and zones. Physical barriers comprise the walls, floors, and ceilings that define the space. In a bioprocess facility, the surfaces of these barriers should be easily decontaminated and finishes should be specified to withstand constant cleaning and decontamination.

There may be multiple layers of barriers, for example, a basement, the floor of the process space, a suspended ceiling, and the roof structure above. The barrier(s) closest to the process space (floor and ceiling, in this case) should prevent the transport of air from areas in the basement or above the ceiling into any zone of the process facility and vice versa in facilities requiring containment. A general rule-of-thumb is that the higher the need for cleanliness or containment, the more tightly these barriers should perform.

Openings (doors, pass-through cabinets, pass-through autoclaves, etc.) must occur in the barriers for the movement of personnel, equipment, materials, waste, and product. These are generally the weak points for transport of microorganisms into or out of classified or containment zones.

For large-scale containment operations, physical barriers may include dikes to contain potential large-scale spills of hazardous materials to limit the area of contamination.

7.2.3 Engineering Systems

Engineering system design for heating, ventilation, and air-conditioning (HVAC) and plumbing systems is a critical factor in mitigating the entry or exit of air and waterborne microorganisms into and out of the facility and between zones.

Air movement is used in bioprocess facilities for temperature control, to reduce airborne contaminants through filtration, and to reduce the transport of contaminants. Filtration of the air entering the facility to reduce organisms entering from the exterior environment, within the facility to enhance clean room performance, and in some cases where biocontainment is required, keeping organisms from exiting the facility with the exhaust air is a significant component of engineering design. High efficiency particulate air (HEPA) filters are routinely used to eliminate microorganisms from classified bioprocess production areas and to contain highly pathogenic microorganisms from exiting the facility through exhaust air.

Air change rates are effective for reducing particulate counts and airborne microorganisms when the air is recirculated through HEPA filters within a zone. Ventilation rates increase safety by exchanging and diluting contaminated air in a space with noncontaminated air; however, this dilution or replacement does not prevent personnel exposure from a burst of an aerosol containing a pathogenic organism. In fact, it has been shown [4] that increasing ventilation rates from 6 to 30 air changes per hour has minimal effect on aerosol concentration of microorganisms in the first few minutes after release. Higher ventilation rates are effective in reducing long-term exposure to chemicals and reducing long-term sensitization to potentially allergenic materials that may be produced in bioprocess facilities.

Zone pressure differentials create movement of air from a zone of higher pressure to a zone of lower pressure to keep contaminates from moving into a facility to maintain clean conditions or to keep contaminates from moving out of a facility to maintain containment. This air movement is a critical component of airborne microorganism control; however, it is important to understand that opening a door between zones reduces this pressure differential and flow significantly. Opening the door allows movement of air (including airborne microorganisms if they are present) between zones to occur with the air currents generated by the door opening or the movement of personnel or materials through the door. For example, a typical 3′ by 7′ door may have an average 1/8 inch gap around the door (8 square inch total opening) with an air quantity that generates a flow of 100 ft/min through the gap when the door is closed. When the door is opened, the open area increases from 8 to 3024 square inches. With the same air quantity moving through the door, the air velocity is reduced below 1 ft/min. Aerobiology researchers, Chatigny and West state [4] that "it is impossible to traverse any doorway, or to have a single doorway between one space and another, without a substantial interchange of air." That is why buffer rooms such as airlocks or anterooms including directional airflow are a critical component of bioprocess facility design.

Anterooms, airlocks, and air-showers are spaces that create a buffer between two or more spaces. An anteroom may be as simple as a space between one zone and another to allow protocol (such as gowning) to take place. Airlocks and air-showers are more specialized anterooms. Airlocks utilize the control of entry and exit doors to eliminate air movement between the zones outside of the airlock. An air-shower typically uses air jets to physically dislodge and remove particles from personnel and equipment before entering a clean space. Air-showers are not typically used in biocontainment facilities, and in clean rooms, some studies have shown a potential to release more particulates than they control.

The direction of the airflow—outward to minimize airborne contaminants from entering the facility or inward to minimize airborne contaminants from exiting the facility—is a critical function of engineering design strategy. In addition, as noted earlier, directional airflow through only one door does little to prevent significant movement of air from one room to another. Pressure cascades, pressure bubbles, and pressure sinks represent the three major strategies for the airflow through an anteroom. For single purpose rooms, clean only or containment only, pressure cascades (Fig. 7.1) are typically utilized. Air flow in the same direction through both doorways creates a high degree of protection if each door is opened and then closed in series with an appropriate interval between each opening.

In bioprocess facilities utilizing hazardous microorganisms, particularly at containment levels of BSL-3 (biosafety level), both directional airflow into the facility through two doors to maintain containment and directional airflow out

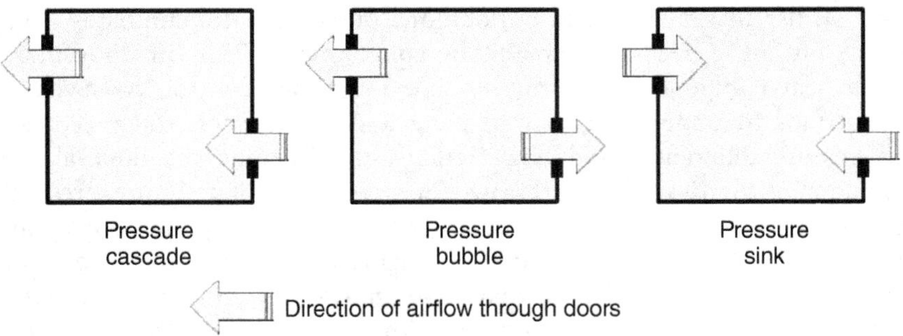

Figure 7.1 Anteroom/airlock airflow models.

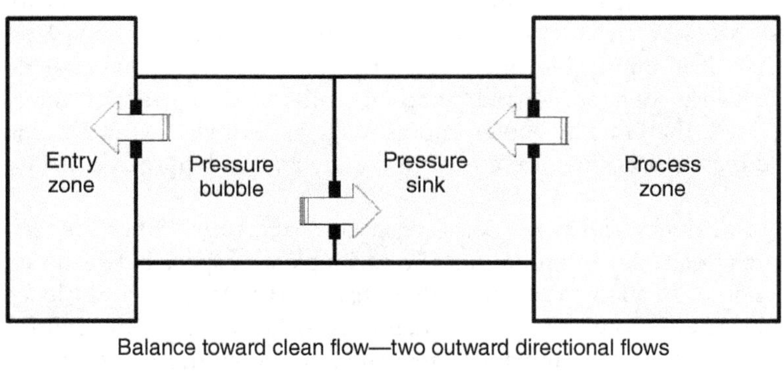

Figure 7.2 Anteroom/airlock airflow model for combining clean and contained directional airflow. This model is balanced toward clean airflow and would not meet containment best practice for pathogen use of high risk.

of the facility to maintain clean classifications may be a requirement. Multiple anterooms will likely be required to allow both to occur simultaneously (Fig. 7.2).

With two anterooms/airlocks, there are four possible conditions for airflow at the airlock doors, all three flowing out, all three flowing in, two flowing out and one flowing in, or one flowing out and two flowing in. In cases where both inward flow and outward flow are required, each compromises ideal flow to a degree. For some situations in containment, a material entry airlock, one-way entry airlock to a containment area, this balance toward clean flow may be a reasonable compromise due to the reduced risk of air moving out of the airlock into the entry zone (Fig. 7.3).

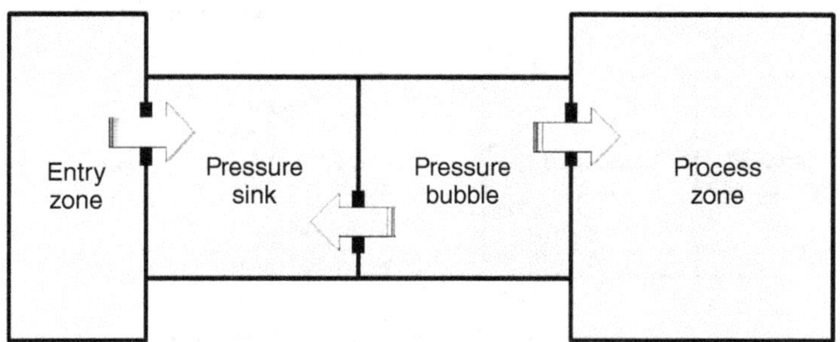

Figure 7.3 Anteroom/airlock airflow model for combining clean and contained directional airflow. This model is balanced toward containment airflow and would not meet cGMP best practice for aseptic production of biopharmaceuticals.

It also may be a reasonable compromise to balance the flow toward containment in a process facility if the production process is not aseptic and is in fully closed systems.

Figure 7.4 shows a balanced clean and contained airflow and indicates a pressure bubble in the center with pressure cascades on both sides. This does not mean to imply that this is the only method of providing a clean contained airlock. The evaluation for any specific situation should be based on how each of the anterooms might be used and how the airflow would best meet the requirements for maintaining the process space as both clean and contained. The exterior anteroom could also be widened and multiple airlocks could be placed off it to serve personnel, materials, product, one-way flows,and so on. With the correct classification and airflow, the entry corridor might be used as the initial airlock.

Although getting ideal flows for both clean and containment may seem like the waste of space and money, the cost of an added simple anteroom to get dual directional airflows both inward and outward, when compared with the shutdown of a bioprocess facility due to mold contamination or contamination from an aerosol release of a contained agent, may be a small investment to mitigate risk. This might be particularly important for facilities requiring containment of aseptic processing with the inability to terminally sterilize the final product. This would occur, for example, in a facility producing a vaccine using a live attenuated organism in the product.

HVAC systems are not the only engineering system that can transport microorganisms into and out of a facility. Microorganisms can be transported by water and other fluids or gases, vacuum systems, and in wastewater flowing out of the facility. The risk of transport through these systems should be

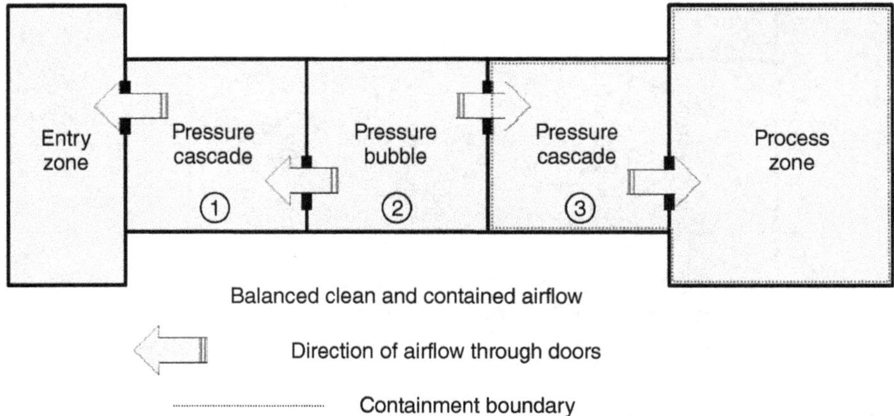

Figure 7.4 Example of anteroom/airlock airflow model for combining clean and contained directional airflow. 1. Clean airlock to support classification and provide area for donning of gown and PPE. 2. Pass-through pressure bubble to maintain directional airflow. Potentially contaminated airlock for removal of outer gowning on exit, storage of any PPE remaining in containment, and storage of supplies for emergency cleanup protocols. Provide hand washing sink in 3. Figure shows two-way personnel flow. One-way personnel flows will be different.

evaluated and filtration, backflow prevention, or local treatment should be considered to mitigate the risks. The use of separate systems only serving the classified or contained bioprocess zones is a method to reduce transport into and out of the facility.

Electrical supply stability and continuity is also critical for both contamination control and containment. Loss of power resulting in loss of control or loss of HVAC systems can result in process shutdown and the need for recleaning to maintain product quality. Shutdown and startup of HVAC systems often result in air imbalances that can draw contaminants into the process areas. Having a reliable emergency generator for power backup, including HVAC supply and exhaust system, and an uninterrupted power supply system for critical controls will keep systems running through any power brownouts or outages greatly reducing this potential risk. Backup power for process containment requirements and support systems is recommended as large-scale operations may be difficult to rapidly shut down in the event of loss of power.

7.2.4 Containment and Isolation Equipment

Total control of contaminants within a room is difficult as personnel will transport microorganisms into and out of the space. In addition, larger spaces are more difficult to control than smaller spaces and in biocontainment, the facility itself is considered secondary containment. Isolation by means of laminar flow cabinets, biosafety cabinets, positive or negative pressure isolators, or other primary containment devices, specifically designed to provide product protection or in the case of containment, both personnel and product protection add a significant additional layer of protection from the movement of microorganisms. This double layer of protection can decrease the impact of many adverse events by several orders of magnitude.

7.2.5 Design to Support Operational Protocols

Contaminant control is accomplished by a balance of facility design, primary protection equipment, and operational protocols. Well-planned facilities can reduce the level of operational protocols; conversely, poorly planned and design facilities may require significantly greater protocols to reduce the transport of microorganisms either into or out of the facility.

Develop protocols and the design concurrently based on the planned user operational specifications. Begin by looking at how the facility design will be driven by the process flows. Then look beyond process and examine how the process areas may connect with shipping and receiving, personnel ingress and egress, and flows to and from other process areas and operations such as warehouse, upstream, downstream, media and reagent prep, washing and sterilization, and fill and finish. Connect flows in sequence where possible to eliminate movement in and out of classified or contained zones.

Identify areas of high risk of microbial release, for example, plan areas where unpacking and breaking down cardboard boxes in locations that minimize the potential for aerosols released by these activities from being transported into the facility by air flow or movement of personnel. Create spaces for cleaning protocols for materials or equipment entering the bioprocess zones. Design spaces to be easily cleaned by planned protocols (including ceiling heights). Look at and plan for steps for dressing-in and dressing-out of each zone in containment, look at where hand washing should logically occur when leaving contained areas, identify potential points of cross-contamination within the facility and eliminate where possible, understand maintenance within the bioprocess areas, and design to reduce impact to get the right balance for your risks. At that point, the greatest ability to balance cost versus risk will be present. No design will be perfect and without risk, examination of the design and the institution of protocols and careful adherence that specifically addresses the risks associated with that design will significantly reduce the impact of adverse events.

7.3 Controlling Viable Environmental Particulates

Viable particulate (molds, other fungi, bacteria, etc.) control is an increasingly important and difficult aspect of bioprocess facility design. In 2012, the United

States Pharmacopeia (USP) updated USP 1116, "Microbiological Control and Monitoring of Aseptic Processing Environments" [5]. This document includes discussions on the classification of a clean room based on particulate count limits; microbiological evaluation programs for controlled environments; critical factors in design and implementation of a microbiological evaluation program; development of a sampling plan; establishment of microbiological alert and action levels; methodologies and instrumentation used for microbiological sampling; media and diluents used; identification of microbial isolates; and operational evaluation via media fills. The 2012 update of USP 1116 also defines new methodology for analysis of recovered contaminants.

The new contamination recovery rate (CRR) methodology defined by USP 1116 looks at the percentage of plates that show any microbial recovery irrespective of the number of colony forming units (CFU) on the plates. With this new analysis methodology, the sensitivity of the analysis of environmental monitoring has been significantly enhanced [6].

As molds are a significant concern in bioprocess facilities, they will be used in this article as the example for microorganisms in the environment. The danger of mold contamination in pharmaceutical products has been brought into focus by the failure of sterility control at the New England Compounding Center, also in 2012, which led to at least 48 deaths and an additional 720 persistent fungal infections from mold contamination in an injectable product [7,8]. This outbreak of preventable disease and the Congressional hearings that followed have led to an increased focus on mold contamination control in bioprocess environments by regulatory authorities.

The combination of increased sensitivity of environmental monitoring with the increased focus on viable contaminates raises the bar for process, facility design, and operations to ensure a sterile product. The full impact of this heightened awareness on bioprocess facility design is yet to be seen. Process and facility design features combined with operational practices to reduce the presence and impact of molds has the added benefit in the reduction of other environmental microorganisms and particulates. Other contaminates often have their origin from different sources than molds.

Microorganisms, particularly molds, are ubiquitous in the outdoor environment. They are present not just in warm and humid climates, but in cold and dry climates as well. In cold climates, molds may become dormant during the winter; however, it is certain that with spring and increased humidity, molds will flourish. Molds do not come in one size or variety. In fact, there are hundreds of varieties of mold, many with unique properties and environmental conditions required for growth. Contamination control for molds in a bioprocess facility includes both design and operational practices to reduce the potential for the transport of mold from outside the facility into classified areas and design and operational practices to minimize the potential growth of mold within the facility including growth in processes areas. Both approaches must be applied to keeping mold counts in process areas within the limits set by cGMP. These two concepts are applied in concert and are critical to the successful control of unwanted microorganisms:

1) Minimizing transport of organisms into the facility
2) Reducing opportunities for growth of organisms within the facility

7.4 Reducing the Transport of Mold into the Bioprocess Facility

As stated earlier, molds are ubiquitous in the exterior environment and are found in significant quantities in most indoor environments [9,10]. There are limited measures to control molds in the outdoor air; however, there are precautions that can be taken to reduce the quantities of molds that occur at air intakes and building entrances. Take steps during design or operations to reduce the growth of molds directly adjacent to the facility. Many molds require moisture and nutrients to promote growth. Eliminate sources of moisture and standing pools of water through good drainage of the roof and at the ground. Areas that remain constantly wet should be remediated. Promptly deal with leaks. Limit planting and areas where dead leaves can gather. Inspect areas for signs of mold growth. Good drainage and good housekeeping around the building, particularly near entrances and air intakes, is a good starting point to minimize the potential for a mold bloom to occur close to a weak point in the building envelope allowing the transport of significant quantities of spores into the facility.

7.4.1 Environmental Zoning

A detailed discussion of classified environmental zones is provided in Chapter 1. Careful planning and development of the environmental zoning concept for a bioprocess facility is critical to proper contamination control for product quality. Not counting the uncontrolled and not-classified space in a facility outside of the process area, based on EU criteria, there are typically five grades of process environmental zoning: Controlled Not-classified, Grade D Classified, Grade C Classified, Grade B Classified, and Grade A Classified. The level of facility nonviable and viable particulate control increases from not-classified to Grade A. All may be important to consider when maximizing contamination control in bioprocess areas.

Spaces that are not controlled or classified represent space not considered part of the process zones. However, the thought put into the contamination control of these spaces even though they are not part of the process zones is also important. Minimizing the quantity of contamination in these areas will minimize the potential for contamination to be transported into process zones.

As personnel move through entry doors into a facility, air and any microorganisms contained in the air or on their clothing are also transported into the facility. Creating an entry airlock, capturing the air and exhausting the air from the airlock, and minimizing the opportunity for both doors to be opened at the same time can reduce outside contaminants from entering further into the facility. In addition, providing space for storage of coats and outwear in this outer airlock will reduce the transport of molds and other particulate further into the facility.

Careful design of loading docks and other spaces used for taking materials into and out of the facility is also important in reducing contamination coming into the facility. Providing a closed vestibule at the loading dock that will allow loading doors to be shut when internal doors require opening creates both a staging area and limits air pressure variations within the building due to wind currents outside of the building. Creating an exhausted space for receiving, unpacking, and breaking down any cardboard cartons can reduce the entry of contaminants as cardboard cartons can harbor mold and vermin. Opening and breaking down these boxes are activities that are high risk for creating aerosols of contaminants.

When nonproduction spaces, such as offices and break rooms, are present in the facility design operate them to minimize the potential for mold growth within these areas. If moving or construction occurs in these areas, use temporary barriers to segregate these areas from the production zones to eliminate transport of the aerosols generated by these activities. A better strategy is to fully segregate nonproduction activities from areas of production.

Controlled not-classified zones have a low degree of contamination control. Incoming air typically is not HEPA-filtered, and as a result, some levels of mold may be directly transported into the space through the HVAC system, particularly during seasons when higher concentrations of molds are present in the outside air. Typically, these zones are filtered with 85% filters. For facilities where contamination control a particular issue, consider moving to HEPA filtration to reduce mold contamination in this zone when this zone directly relates to zones of higher classification and when airflow for both containment and contamination control is required.

Grade D Classified and Grade C Classified zones are clean rooms with specified environmental performance to meet the classifications (*see* Chapter 1). Grade C zones are typically used for upstream and downstream process space with closed systems for products that can be terminally sterilized. Grade C zones are also used as glassware washing and preparation space for materials to be sterilized into Grade B and Grade A aseptic processing zones. Grade D zones are typically support spaces for Grade C zones. As clean rooms, it is important to note that airlock entry to each zone and between zones is a critical component of maintaining classified environmental conditions.

Grade B Classified and Grade A Classified zones are clean rooms with higher levels of specified environmental performance to meet the classifications. These are used for aseptic upstream and downstream process operations such as seed stock and inoculum manipulation, aseptic processing with systems that are not fully closed, and fill and finish operations. Often Grade A zones are provided by isolating areas within a Grade B zone or through the use of a biological safety cabinets within a grade B zone. Airlock entry to these zones is a critical component of maintaining classified environmental conditions.

Creating a layering of zones and spaces from lower to higher level of cleanliness is a good strategy for maximizing contamination control through layered gowning and cascading airflow. For facilities requiring both clean and contained space, better planning for clean space allows the containment zoning to be applied with less risk of compromising clean conditions. Depending on the level of cleanliness required and the BSL required, adding a layer or two seeking to balance clean and contained space with no compromise of either may be a good strategy to adequately achieve and maintain both conditions.

7.4.2 Filtration of Molds and Mold Spores from Incoming Air

Molds and mold spores are present in most environments in many sizes and configurations. Sizes generally range from 1.5 to 9 μm in diameter. One goal in the design of bioprocess facilities should be to eliminate all molds reasonably practicable in the process areas of the facility. This may suggest different design strategies for bioprocess facilities.

Typically, process zones with classifications of Grade D through Grade A have incoming air filtered through HEPA filters which essentially remove all incoming molds; however, other nonclassified areas (controlled nonclassified, or below) areas of the bioprocess facility are typically filtered with lower efficiency filters [generally 85–90% minimum efficiency reporting value (MERV)]. While 85–90% filters may eliminate significant percentages of mold present in the outdoor environment as it is passed into the indoor environment through the building HVAC system [11,12], some, possibly significant, level of mold or mold spores will be passed through into the space. Three significant issues may be created depending on the design and other conditions of the facility. (i) Mold present in the air in lower classified or unclassified areas can be transported into areas of higher classification. (ii) Mold or mold brought into the lower classified or unclassified areas facility may find appropriate conditions for growth in the facility creating internal sources for high quantities of mold raising the risk of transport into clean production areas. (iii) Molds passing through the filtration may find design issues in the HVAC system (excess moisture in the ductwork, poor drainage of condensate, etc.) that provide a site for mold growth therefore, again, generating high quantities of mold raising the risk of transport into clean production areas.

During the facility design, assess areas of the facility to determine whether risk exists that would make it advisable to provide supply air HEPA filtration to reduce overall mold counts within the facility (or even immediately outside of the bioprocess facility). Reducing mold counts in unclassified areas of the facility reduces the risk of transporting mold into classified areas of the facility.

7.5 Reducing Mold Sources within the Bioprocess Facility

As some level of mold will inevitably be transported into a bioprocess facility, a key factor in design and operations is to keep mold from actively growing within the facility and therefore increasing the risk of transporting mold into clean classified areas of the facility. Presence of moisture is a key factor in the growth of mold; therefore, moisture control in the facility and protocol to decontaminate sources of moisture that cannot be avoided should be a primary consideration in design and operations. Obviously, roof leaks and other unplanned sources of moisture entry should be immediately addressed and remediated; however, other sources of moisture in a bioprocess facility are unavoidable. Any gypsum board walls, ceilings, and other material subject to moisture infiltration that become wet and damaged should be removed and replaced. Other areas of concern within bioprocess facilities include

- *Sinks.* Sinks, particularly sinks with closed cabinets where leaks may remain undetected for significant periods of time. Where sinks are provided, consider open sinks with easy access below for cleaning and disinfection. Equipment processing sinks, such as boiling sinks or wash stations, are sometimes closed and insulated resulting in an ideal growing environment for mold below the sink when leakage occurs.
- *Environmental Rooms.* Cold rooms, freezers, warm rooms, and other environmental rooms often require condensate drainage from their mechanical systems and may, depending on environmental conditions within the facility, have condensate issues related to their surface temperatures. Utilizing "classified" environmental rooms that remotely mount mechanical equipment, dehumidifying the air, and HEPA filter entering air can significantly reduce the risk of moisture in the facility to promote mold growth. In addition, the space between the walls of the environmental room and the adjacent construction can provide ideal growth conditions for mold if moisture is present. Air movement into and out of this space is often uncontrolled.
- *Autoclaves and Glassware Processing Equipment.* This equipment often introduces water, steam, and other sources of moisture into the facilities that can create conditions for mold growth. In addition, the service spaces for this equipment are difficult to clean and decontaminate and also have open

drains for waste and condensate. Ideally these service spaces are located outside of both contained and classified areas.

- *Mechanical Spaces and Service Rooms.* Bioprocess facilities have strict environmental conditions and process requirements that require significant HVAC and plumbing equipment for proper operation. The spaces that house this equipment have significant sources of moisture and have ideal environments for mold growth. Provide good housekeeping in these areas to reduce the potential for mold growth and eliminate openings and transport paths between theses areas and the bioprocess areas.
- *Difficult to Clean Spaces or Surfaces.* Production spaces with high ceilings, platforms with small openings, crack and crevices, equipment that is too crowded, piping without appropriate standoffs all can create areas in a facility that cannot be adequately cleaned. Lack of cleaning, combined with moisture sources, will contribute to the growth of viable microorganisms within a production space.

Where sources of moisture are unavoidable, institute operational protocols to ensure they do not become sources of mold growth within the facility.

7.5.1 Cleaning and Decontamination

It is inevitable that some amount of contamination will enter all spaces in the bioprocess facility. USP 1116 recognizes that minimal contamination recovery will occur even in Grade A Classified zones. Recognizing that viable microorganisms will be present in every space, cleaning procedures are critical to ensure that the contamination remains in control and within acceptable limits. Finishes and spaces should be designed to allow thorough cleaning with disinfectants and sporicides. Finishes should be planned to withstand extensive cleaning with the specific products planned for use without surface failure. Spaces should allow for thorough complete cleaning without special cleaning tools. Space should be adequate to allow cleaning around all process equipment without having to relocate equipment for cleaning. Cleaning and disinfection of a process space is a difficult process. The facility design must make it as simple an effort as possible. Consideration of planning for space decontamination through gaseous or vapor phase disinfectants should be given during design.

7.6 Biocontainment: An Overlay to Process Design

The other side of microbial control in bioprocess facilities relates to containment in facilities where pathogenic microorganisms are used as part of the production process. It is important to first state that the primary design issue in bioprocess facilities is design to support product quality control. Product quality control must not be compromised by facility design compromise. Design to provide the appropriate measures for biocontainment should be an overlay to process design for product quality control. In that respect, creating appropriate containment must follow design for product quality control and be complementary in supporting that primary goal. This does not mean to imply that there should be compromise made in the measures taken for biocontainment. Both appropriate product quality control and appropriate biocontainment must be provided.

Biocontainment is a balance of primary containment, engineering controls, operational protocols, and personnel protective equipment (PPE). Unlike the design of many other biocontainment facilities, bioprocess containment requires containment with adherence to good manufacturing practices, some of the requirements of which are contrary to typical containment design and operational practices.

While drug regulatory agencies understand and easily regulate the contamination control aspects of bioprocess facility design and operations for product quality, they are generally unfamiliar with biocontainment principles and practices. Conversely, biocontainment regulatory authorities bring little understanding of the requirements of product quality to their oversight of biocontainment facility design and operations. The bioprocess operations and engineering staff must bridge these gaps by providing thoughtful analysis of both risk to product quality and risk from the use of pathogenic organisms.

Clearly documenting the rational and principles on which the containment design and operations are based can effectively communicate the issues to the drug regulatory authorities to speed any approvals required. This documentation can also help support the current expectations of biocontainment facilities that require stringent adherence to the principles of containment, particularly when facilities use regulated pathogenic organisms.

An assessment of risk at each stage of the process must be an integral part of the planning and design process. Risks, and therefore design and operational solutions, may vary greatly for each part of the process. For example, in a facility producing inactivated virus vaccines, risk in dealing with the live virus before inactivation will be significantly higher than after inactivation. Risk of large-scale live virus production will be significantly different that laboratory-scale handling of the same virus in the quality control laboratory. Risk of fill and finish will be different for a live attenuated virus vaccine than for an inactivated virus vaccine. There is no "cookbook" approach to design for containment; understanding the risk and providing the appropriate design and operational response is critical to creating a facility that mitigates the risk to the appropriate level. The overall goal must be to design a facility that maximizes protection of products and related processes while controlling unwanted microorganisms and containing process biohazards.

7.7 The Biocontainment Regulatory Environment

Biocontainment facility design is complicated by the fact that the risk of using a pathogenic organism and therefore the design response varies by the organism, the specific strain of the organism, how it is used, the scale of use, route of transmission, whether vaccines are available for personnel, location of use, and other factors. As a general principle, pathogenic organisms are kept and manipulated in primary containment (sealed containers, biosafety cabinets, fermentation vessels, etc.). The exception to this rule is when pathogenic organisms are used in large animals where primary containment is impractical. Therefore, the typical biocontainment facility is designed for secondary containment. This means that the containment aspects of the facility only come into play in the event of an accidental release of the organism from primary containment. In a facility with proper use of primary containment and strict adherence to practices and protocols, there should be no viable pathogenic organisms present in the air or on surfaces outside of primary containment; however, as accidental release may be difficult to detect, it should be operated as if the agents were present. Experience has taught us that human error and mechanical failures occur, possibly at greater levels than recognized. It is critical that the secondary containment provided is adequate when adverse events occur.

The regulatory environment for biocontainment varies significantly per country. It is important to understand both the guidance and the regulatory environment in the specific country where the bioprocess facility is to be built. Owing to highly publicized breaches of containment protocols in numerous facilities combined with increasing concerns related to the potential use of pathogenic organisms for bioterrorism, the regulation of use of pathogenic organisms and the facility requirements to handle them has changed significantly over the last decade. Regulatory requirements have been increasing and one should expect additional oversight and regulation in the future.

In the United States, the only true regulations (except for a few localities) related to the use of pathogenic organisms relate to the "Select Agent Program". This program regulates a specific list of agents where there is deemed to be a risk of use for bioterrorism. From the Select Agent website [13],

> "The Federal Select Agent Program is jointly comprised of the Centers for Disease Control and Prevention Division of Select Agents and Toxins and the Animal and Plant Health Inspection Services Agricultural Select Agent Program. The Federal Select Agent Program oversees the possession, use and transfer of biological select agents and toxins, which have the potential to pose a severe threat to public, animal or plant health or to animal or plant products. The Program greatly enhances the nation's oversight of the safety and security of select agents by:

- Developing, implementing, and enforcing the Select Agent Regulations
- Maintaining a national database
- Inspecting entities that possess, use, or transfer select agents
- Ensuring that all individuals who work with these agents undergo a security risk assessment performed by the Federal Bureau of Investigation/Criminal Justice Information Service
- Providing guidance to regulated entities on achieving compliance to the regulations through the development of guidance documents, conducting workshops and webinars
- Investigation of any incidents in which non-compliance may have occurred."

Additional information including a list of the select agents and toxins can be found on the website: http://www.selectagents.gov/.

7.7.1 Laboratory-Scale Use and Use in Animal Models of Disease

The most common use of pathogenic organisms is done at laboratory scale. This would include use of pathogenic organisms in animal models. In the bioprocess arena, this scale would be found in seed manipulation areas, quality control laboratories, and research and development facilities. The US Government through the Centers for Disease Control and Prevention (CDC) and the National Institutes of Health (NIH) provides guidance for the use of other hazardous microorganisms. This guidance is provided through "Biosafety in Microbiological and Biomedical Laboratories" (BMBL) [14], which identifies the agents, risks of various uses and establishes four basic levels of containment for laboratories, BSL-1, BSL-2, BSL-3, and BSL-4, and use in animals ABSL-1, ABSL-2, ABSL-3, and ABSL-4. BSL-1 is for use with less hazardous organisms and BSL-4 is for use with the most hazardous organisms. Adherence to the BMBL as determined by the Select Agent Program is mandatory for use of pathogenic organisms or their byproducts named as select agents and toxins. For all other pathogenic organisms, the application and use of the BMBL to guide design and operation is not a requirement unless the funding for facility construction is provided by a US Government agency that requires its use. In other cases, the use of the BMBL for facility design and operation is voluntary and is normally based on corporate or institutional policy.

It is important to understand that the BMBL takes a risk-based approach, not a "cookbook" approach, based on the anticipated organisms to be used and how the organisms will be used. A "cookbook" approach to biocontainment defines the specific requirements to be used at each level of containment. Necessarily, this approach must be proscriptive and must look at each level and define the requirements for containment at each level based on the worst case possible.

This creates an additional burden of requirements on most facilities, as most facilities do not operate at the worst case scenario.

According to Dr John Richardson [15], former Director of Biosafety at the CDC and a coeditor of the first edition of BMBL, there was significant concern that adding undue burden to laboratories, particularly clinical laboratories handling diagnostic samples might result in the loss of significant laboratory capacity impacting public health. For that reason, a risk-based approach was taken that gives guidance and allows each facility to be designed to match the risk contained specifically at that facility. From firsthand knowledge as an editor of the fourth edition of the BMBL, this intent to not require unnecessary containment features at each BSL was continued. The features described at each level represent the recommended containment when all aspects described for the level can be applied as a whole. By assessing the risk of the work and the specific circumstances of the facility, additional enhancements (such as exhaust filtration or effluent treatment) may be found advisable. This approach allows the right facility to be developed to support the work without overdesign. The difficulty in this approach is that it takes expertise to apply, particularly with complex containment issues, not specifically addressed in the BMBL, such as large-scale use.

There are other considerations that make full use of the guidelines, a recommended practice. Compliance with BMBL provides a level of safety for both workers and the environment; thus, it provides a good baseline for establishing good corporate practices related to biocontainment. In addition, as mentioned above, regulatory authorities fully familiar with good manufacturing practices may not be as familiar with biocontainment design practices. Questions often arise as to interpretations of containment principles; clear documentation of compliance with BMBL may speed the regulatory approval process. An accidental release of a biohazard within a facility will require shutdown and decontamination of the facility. This may require that manufacturing operations stop for a significant period of time and have significant costs for the cleanup and the loss of production. Risk assessment and design to mitigate that risk can reduce this potential liability. Lastly, issues with biocontainment facilities have received significant attention in the press over the last 5 years. Negative publicity related to a facility, particularly, publicity questioning safety, will impact community relations, the corporate image, and potentially the ability to conduct business operations within the community.

7.7.2 Large-Scale Use of Pathogens

The BMBL does not address biocontainment at large scale. Large scale is generally defined as work with over 10 L of a culture containing pathogenic organisms. There is other guidance referenced by the BMBL that can be used for guidance when large-scale use of pathogens is planned [16,17].

In Canada, a wider regulatory process has been developed. New biocontainment guidance, the Canadian Biosafety Standards and Guidelines [18], was published in 2013 that consolidated all Canadian federal agency biocontainment requirements into this single document. A regulatory process to review and verify that specific facility types meet the regulatory requirements was also created. For bioprocess facilities, this new guidance contains standards for large-scale use of microorganisms and toxins of biological origin. This guidance, taken as a whole, can be helpful in planning large-scale facilities in countries where the guidance for large scale is not as robust.

There are also specific pathogenic organisms that may have additional containment guidance based on unique requirements and issues. An example would be for facilities manufacturing inactivated polio vaccine (IPV). Polio has been eradicated from North America and Europe, and the World Health Organization (WHO) has been working to eradicate it in the few parts of the world where the disease has been found. The manufacture of IPV requires the large-scale use of wild-type poliovirus, which if released, could cause disease in the increasingly naive populations in North America and Europe where many of the vaccine plants are found. For that reason, the WHO has developed specific guidelines [19] for facilities manufacturing IPV. These guidelines are being put into place as full eradication becomes near.

7.7.3 Animal and Plant Pathogens

Generally, biosafety and biocontainment is thought of as related to organisms causing human disease; however, there are many plant and animal pathogens that can cause disease in crops or livestock with severe economic impact. In the United States, the importation and use of certain plant and animal pathogens are regulated by the US Department of Agriculture (USDA) and a number are classified under the select agent act [13]. The USDA Agricultural Research Service has developed guidance [20] for the safe handling of these pathogens.

The USDA guidance also includes requirements for the design of facilities that are identified as BSL-3 Ag containment facilities. This is a very specific facility designed to allow the safe use of large animals (cows, horses, etc.) infected with high risk, highly virulent animal pathogens where primary containment is not practicable. Examples might include livestock infected with an organism such as Foot and Mouth Disease Virus or Chickens infected with highly pathogenic strains of Newcastle Disease Virus. There has been demonstrated risk of environmental consequence from bioprocess facilities producing vaccines from these organisms. The evidence suggests [21] that the 2007 Foot and Mouth Disease outbreak in Surrey, UK, was from leaking effluent lines coming from a vaccine production facility.

It is unlikely that bioprocess facilities would require BSL-3 Ag containment; however, features in the USDA guidance [20] may be useful for bioprocess facilities handling large-scale high risk organisms where process systems are not fully closed.

7.7.4 Genetically Modified Organisms (GMO) and Synthesized Organisms

Since the advent of modern molecular biology techniques and equipment to modify the genes of organism to create specific characteristics in the organism (less or more virulence, the ability to produce specific proteins or chemicals for therapeutic use, etc.), there has been a rapid increase in the use of genetically modified organisms (GMOs) in research and in the production of products. GMOs have been commercialized since the 1970s with many medicinal products produced through their use. The NIH first developed guidelines [16] for the use of GMOs in 1978 and the last update of this guidance was in 2013. These guidelines include requirements for large-scale bioprocess facilities.

In the United States, this guidance is not regulatory. Only institutions receiving NIH funding are required to demonstrate compliance. The use of this guidance by other organizations would be voluntary. GMOs can be significantly more or less hazardous than their natural form. It is important to perform risk assessments of their use to understand and mitigate risk to personnel, the community, and the environment.

Synthesized organisms will become an increasingly important part of bioprocess manufacturing in the future. The first report of viral synthesis came in December of 1991 [22] when scientists at the State University of New York at Stonybrook created an artificial poliovirus in a test tube. In the United States, this work, like work with GMOs, is not currently regulated except when the institution or the work receives NIH funding.

7.7.5 Toxins

Toxins are chemicals of biological origin. Some microorganisms, particularly bacteria and fungi, produce hazardous peptides, proteins (*see* Chapter 1), and other small biological molecules capable of causing disease from contact, adsorption, or ingestion. There is guidance for safe handling of toxins from CDC [14], Canada [18], and the Center for Chemical Process Safety [23], including large-scale guidance in the Canadian Biosafety Standards and Guidelines [18]. Some toxins are regulated as select agents in the United States. Botulism toxin would be an example due to its threat as a weapon-of-mass-destruction.

Many of the severe disease effects from bacterial infection are from toxins produced by the bacteria during an infection. A few examples of the many diseases that are impacted by bacterial toxins include botulism (from *Clostridium botulinum*), cholera (from *Vibrio cholerae*), diphtheria (from *Corynebacterium diphtheriae*), and necrotizing fasciitis (from *Streptococcus pyogenes*). The effects of these toxins can range from diarrhea in the case of cholera, to potential rapid death in the case of botulism, and to severe necrosis of tissue in the case of the "flesh-eating" strain of Strep infection.

Many vaccines for bacterial disease contain toxoids (inactivated toxins), derived from toxins as a component of the vaccine. For these processes, large-scale volumes of toxins are produced, purified, and inactivated as part of the production process. Risk assessment of the toxin and process used for production should be performed to add process and facility features to mitigate the risks. The Canadian Biosafety Standards and Guidelines require the same facility features for the large-scale use of toxins as for the large-scale use of the pathogenic microorganisms that produce them.

7.7.6 Allergens and Biologically Active Products

Biological material that are not classified as pathogenic or toxic may still provoke an immune reaction in individuals, and in some individuals, these reactions may be life-threatening. Therapeutic proteins and other biological products have provoked immune responses in patients to whom they have been given. The FDA has developed guidance [24] for assessment of patient safety related to the use of therapeutic proteins; however, there is little information available on the workplace risks in bioprocess facilities. Care should be taken when working with therapeutic proteins to reduce exposure to components biological products, particularly in the case of products in development for which safety data may not be well-characterized.

It is important to remember that bioprocess facilities are producing medicinal products with significant planned impact on body function and potentially severe side effects. Even with products produced from nonpathogenic organisms, it is important to take all necessary process and facility design and operational steps to minimize the potential for inadvertent exposure of personnel and community to these products or components.

7.7.7 Biosecurity

Biosecurity is another important consideration with the use of pathogenic organisms in a facility. When using regulated "select agents" as described above, biosecurity is a requirement of use and a biosecurity plan must be developed and followed. Even with the use of pathogenic organisms that are not regulated, a biosecurity risk assessment should be performed to determine whether any measures should be taken to ensure that pathogenic organisms are not taken from the facility for nefarious uses.

7.8 Principles of Biosafety

The BMBL lists three "principles of biosafety" (i) laboratory practices and technique, (ii) safety equipment (primary barriers and personal protective equipment), and (iii) facility design and construction (secondary barriers). These principles work in balance to create the appropriate level of safety for personnel and the environment.

7.8.1 Risk Groups

Risk groups classify agents by the hazards they pose to personnel, the community, and the environment. The risk groups may or may not correspond with the BSL for which the work with the agent is assigned as the use may dictate a higher or lower BSL based on how the agent is used. A use that often suggests a higher BSL is large-scale use in bioprocess facilities. It is important to note that the risks of disease identified are for normal healthy adults. The risk of disease from the agents may be significantly higher for persons with compromised immune systems or other high risk conditions. Table 7.1 is from the BMBL [14] and identifies how the definitions of two institutions for risk group classification classify agents according to risk.

A complicating factor in assessing the risk of an organism is that different strains of the same organism may have significantly different risks. Vaccine strains of organisms are often attenuated strains with lower virulence and risk. Different strains may be transmitted differently; for example, Ebola Reston can be transmitted through the aerosol route but does not cause disease in humans, Ebola Zaire causes severe disease in humans but is transmitted through contact with blood or bodily fluids.

The risk of influenza virus stains varies greatly. Seasonal influenzas are generally handled at BSL-2, whereas pandemic and avian strains generally require BSL-3. Genetically reconstituted strains of the 1818 pandemic influenza are select agents handled at BSL-3 or BSL-4 due to their potential for creating a highly lethal pandemic. Risk group assignment is a starting point for understanding the risk of the organism. How the organism is used is equally important.

7.8.2 Biosafety Levels

The CDC/NIH BMBL states:

Table 7.1 Risk Groups for Pathogenic Organisms

Risk Group Classification	NIH Guidelines for Research Involving Recombinant DNA Molecules 2002[a]	World Health Organization Laboratory Biosafety Manual 3rd Edition 2004[b]
Risk group 1	Agents not associated with disease in healthy adult humans	(No or low individual and community risk) A microorganism unlikely to cause human or animal disease
Risk group 2	Agents associated with human disease that is rarely serious and for which preventive or therapeutic interventions are *often* available	(Moderate individual risk; low community risk) A pathogen that can cause human or animal disease but is unlikely to be a serious hazard to laboratory workers, the community, livestock, or the environment. Laboratory exposures may cause serious infection, but effective treatment and preventive measures are available and the risk of spread of infection is limited
Risk group 3	Agents associated with serious or lethal human disease for which preventive or therapeutic interventions may be available (high individual risk but low community risk)	(High individual risk; low community risk) A pathogen that usually causes serious human or animal disease but does not ordinarily spread from one infected individual to another. Effective treatment and preventive measures are available
Risk group 4	Agents likely to cause serious or lethal human disease for which preventive or therapeutic interventions are not usually available (high individual risk and high community risk)	(High individual and community risk) A pathogen that usually causes serious human or animal disease and can be readily transmitted from one individual to another, directly or indirectly. Effective treatment and preventive measures are not usually available[b]

a) Department of Health and Human Services, National Institutes of Health, NIH Guidelines for Research Involving Recombinant DNA (NIH Guidelines), November 2002.

b) World Health Organization Laboratory Biosafety Manual 3rd Edition 2004, Geneva Switzerland.

"Four BSLs (Biosafety Levels) are described in Section 4, which consist of combinations of laboratory practices and techniques, safety equipment, and laboratory facilities. Each combination is specifically appropriate for the operations performed, the documented or suspected routes of transmission of the infectious agents, and the laboratory function or activity. Often an increased volume or a high concentration of agent may require additional containment practices.

Biosafety Level 1 practices, safety equipment, and facility design and construction are appropriate for undergraduate and secondary educational training and teaching laboratories, and for other laboratories in which work is done with defined and characterized strains of viable microorganisms not known to consistently cause disease in healthy adult humans.

Biosafety Level 2 practices, equipment, and facility design and construction are applicable to clinical, diagnostic, teaching, and other laboratories in which work is done with the broad spectrum of indigenous moderate-risk agents that are present in the community and associated with human disease of varying severity.

Biosafety Level 3 practices, safety equipment, and facility design and construction are applicable to clinical, diagnostic, teaching, research, or production facilities in which work is done with indigenous or exotic agents with a potential for respiratory transmission, and which may cause serious and potentially lethal infection.

Biosafety Level 4 practices, safety equipment, and facility design and construction are applicable for work with dangerous and exotic agents that pose a high individual risk of life-threatening disease, which may be transmitted via the aerosol route and for which there is no available vaccine or therapy."

The BMBL provides good guidance for laboratory-scale activities; however, many bioprocess operations use significantly higher quantities or higher concentrations of pathogenic organisms. Large-scale operations and facility design features are not specifically covered in BMBL. Bioprocess facilities have been designed at all four BSLs; however, BSL-4 bioprocess facilities are rare and typically use processes of significantly smaller scale than found at BSL-2 or BSL-3.

7.9 Principles of Biocontainment Facility Design

Biocontainment facilities use a number of basic principles to create a redundant layering of protection and containment for minimizing the potential for the exposure of personnel in the containment area, personnel outside of the containment area, the community, and the environment.

7.9.1 Risk Assessment

Biocontainment facility design should be based on a risk assessment of the planned work. Biosafety levels and guidance are based on the risk posed by the organism and the processes in which the organisms are used. For example, the risk with an organism such as *Mycobacterium tuberculosis* which is easily aerosolized, infectious through the aerosol route (by breathing in the organisms), and has a low infectious dose (less than 10 organisms may cause disease) is significantly different than an organism such as the rabies virus which is not generally transmitted through an aerosol route.

Handling a small sample of each organism for laboratory testing would have very different risks versus growing large concentrated quantities for production or deliberately aerosolizing the organisms for aerosol challenge. Each type of risk would require a different facility response.

Facility design for biocontainment in bioprocess facilities is already one step ahead of normal laboratories. The requirements for product quality generally dictate the following facility features for product quality: tight construction, easily cleanable surface finishes, airlocks, and space for gowning-in and PPE.

7.9.2 Primary Containment

One fundamental principle of biocontainment protocols and facility design is the primary containment of biohazards. Biosafety cabinets are used to handle smaller scale use of pathogenic organisms. Examples of use in bioprocess facilities include seed stock manipulation and initial seeding, quality control laboratory, animal work that utilizes pathogenic organisms, and smaller scale processes. Biosafety cabinets also provide Grade A Classified environments for product protection.

Primary containment in bioprocess facilities might also include closed or partially closed systems such as fermenters and bioreactors used for the growth of organisms. Containment can be improved with process equipment for live organisms that is decontaminated in place. Issues to consider in partially or fully closed systems include obtaining quality control and retention samples from the process equipment, how media and other additives used during the growth process are introduced, and how chemicals used in the process and for organism inactivation are introduced. Fully piped systems combined with heat sealed or welded closed sampling tubes and containers can reduce containment risk and product quality risk.

For systems with open sampling ports, the method of sterilizing the sampling ports can release aerosols. Aerosols may also be released during the sampling process. Appropriate PPE and decontamination procedures should be used when process equipment handling pathogenic organisms is used. Procedures should also be in place to identify and decontaminate leaks that might occur in the equipment connections. All process lines should be evaluated for the potential for discharge of live pathogenic organisms into the environment. This might include condensate return lines from steam-in-place (SIP) systems which may be a discharge point when the valves fail or are not set properly.

Other primary containment devices might include isolators, sealed centrifuges, sampling bags and other sampling containers, closed processing equipment such as microfilters, transport piping between equipment in process steps, and sealed transport containers for moving pathogenic organisms within and between laboratories.

The primary thoughts during process equipment design are product specifications and maintaining product quality. Containment is often an afterthought. It is important to thoroughly evaluate the risks of live organism discharge from the equipment, either accidentally or during normal operations.

7.9.3 Secondary Containment

As noted earlier, pathogenic organisms should routinely be contained within a primary containment system. Facilities are therefore secondary containment. Their design should prevent the release of organisms into other areas of the facility and the environment or in the event of release of pathogenic organisms outside of primary containment. Architectural and engineering barriers form this secondary containment which prevents the release of pathogenic organisms.

Architectural barriers consist of the arrangement of space, including airlocks. A series of spaces may be provided creating multiple layers that must be breached before loss of containment. Often the arrangement and design of bioprocess facilities inherently create multiple layers, improving the opportunity for containment. The walls, floors, and ceilings form the physical architectural barrier.

The general requirements of the architectural barrier for containment are the ability to seal the barriers and the ability to clean and decontaminate the barriers without damage. Finish durability is particularly important in bioprocess facilities. The finish must be able to withstand the frequent cleaning with harsh decontaminants that are used for environmental microbial control. The ability for surfaces to be sealed, cleaned, and decontaminated which is inherent in the design of bioprocess facilities is generally fully adequate for biocontainment purposes.

Openings, such as doors, are the weak points in containment barriers. Airborne, pathogenic organisms inadvertently released into the spaces can be transported through the doorways out of the facility. The discussion of directional airflow in the first part of this article provides guidance on aerosol control of pathogens at doors. Space for the removal of potentially contaminated gowns, booties, hair protection, and gloves; the provision of sinks for hand washing after glove removal; and space for holding potentially contaminated materials for autoclaving are critical components of the space for personnel movement out of a biocontainment facility. In some cases, the ability for personnel to fully shower out may be warranted.

Engineering barriers also contribute to the secondary containment. As noted in the discussion of airlocks, directional airflow into the facility helps prevent the transport of pathogenic organisms released into the air from exiting the facility. In general, one door with inward directional airflow does not prevent significant mixing of air between spaces. Air currents generated by the act of opening the door, movement through the door, temperature differentials, etc. can create movement of aerosols from the space being contained. An airlock or anteroom, with both doors having directional inward airflow, creates a buffer and reduces the potential for aerosols containing microorganisms to leave the containment zone.

Pressure differentials between spaces are required to create directional airflow. The pressure in spaces requiring containment should be lower than the pressure in surrounding spaces to create directional inward airflow, not only through the doorways but also through any gaps or pathways in the walls such as unsealed electrical conduits. In bioprocess facilities, care should be taken to ensure air infiltrating through these pathways is the same quality as air introduced into the room through the HVAC system to minimize the introduction of contaminants into the space.

Engineering barriers would also include the HVAC supply, return, and exhaust systems that are required to maintain the appropriate temperature and humidity conditions within the space and to allow the creation of directional airflow. This would include filtration of the airflow to provide appropriate levels of product protection as well as removal of contaminates.

7.9.4 Impact of Scale and Process

As described earlier, BSLs and guidance are based on the risk posed by the organism and the processes in which the organisms are used. Again, larger and more concentrated volumes of organisms used in production have a higher risk than smaller samples used for laboratory testing. There is very different risk in a 10-L container of *Mycobacterium tuberculosis* than in a sputum sample containing live tuberculosis bacterium which is being sent to a laboratory for analysis even though with the 10-L rule-of-thumb, each would be considered laboratory scale. The engineering barriers should reflect the increased risk.

Owing to the large quantities and high concentrations of pathogenic organisms found in areas with large-scale fermenters, HEPA filtration of exhaust air from these spaces should be considered based on the risk of release outside of primary containment, the organism used, potential for re-entrainment into the facility, and the impact on areas close to the facility.

The potential for effluent discharge into the drainage system should also be evaluated. Considerations might include the following:

- Avoidance of floor drains in process areas. If they are required, spill protection should be provided.
- The provision of floor dikes to keep large-scale spills from migrating out of the immediate containment area.
- Evaluation of potential discharge of organisms in the SIP and clean-in-place (CIP) processes, including initial condensate from tanks and transfer lines.
- Heat or chemical treatment of all effluents before release to the sanitary sewer system as an additive engineering barrier creating redundancy against environmental contamination.

Other process services, including drainage for pure water and water-for-injection systems, condensate drainage for clean steam, vacuum system, and other drainage or venting from the facility should be evaluated for filtration or treatment of waste and against backflow. Some process equipment, particularly vessels used for bacterial fermentation, may require exhaust venting as part of the fermentation process. As the venting is coming directly from the fermenter containing high quantities of live organisms, filtration on these vent lines should be provided. In addition for bacterial fermentation with toxins as a product, filtration or heat inactivation to prevent the venting of toxins should be evaluated.

Autoclaves or other methods of decontamination should be provided in the facility. To minimize the potential for pathogenic organisms leaving the facility, it is recommended that pass-through autoclaves be provided at the process zone containment barrier.

An advantage of bioprocess facilities is that the processes are defined before design and construction work. This allows a specific risk assessment to be made to determine the facility-specific containment design requirements. The change control process typically found in bioprocess operations limits the rate of change from the defined process and containment risks. Facilities that require serial campaigns of multiple products and modular process facilities that are designed allow rapid process change add a challenge to designing appropriate biocontainment. In these cases, the facility must be planned to provide appropriate biocontainment for the full range of intended uses.

7.10 Design for the Entire Process

It cannot be emphasized enough that it is not adequate to design a balance between the requirements for quality requirements and containment; each must be designed to adequately meet its needs. When designing for both the containment of hazardous microorganisms and the control of unwanted environmental contaminants, consideration of the entire process is important to reduce all risks. Facilities that can link the process components but also provide appropriate segregation of the process can be an important tool in meeting both goals. Facility design can reduce the risk of moving pathogenic organisms from one step of the process to the next and can also reduce the transport of unwanted microorganisms into the facility. For this to occur, the entire process must be considered during design.

7.10.1 Upstream Process Facilities

For processes involving microorganisms that are grown as a part of the product production process, seed stocks, of the specific strains of the organisms, are used to start the growth process. Ideally, seeds stocks requiring a specific level of biocontainment are stored within the contained zone of the facility. Seed organisms are taken to an aseptic processing area to inoculate the culture media for the initiation of the growth process. This is often performed within a biosafety cabinet in an aseptically zoned room. Generally, this would be a laboratory-scale process.

Once the process has been inoculated with live organisms (whether bacterial fermentation or viral cell culture), the organisms are grown to the scale required to produce the quantity of material necessary to produce the product. The scale of production may range from smaller scale vessels of 20–200 L to larger vessels of 2000–5000 L and above. This production scale would obviously be large scale. Overall scale is less obvious in some cases where cell culture may be at a small scale but the cumulative activity creates a large-scale process. For example, influenza viruses for vaccines are often grown in fertilized eggs rather than larger culture vessels. Even though each egg is below the 10-L rule-of-thumb, the quantity of eggs used overall and therefore the quantity of virus is large, making this a large-scale process.

Once the organism is grown in sufficient quantity, it is harvested. In most cases, the live pathogenic organisms are inactivated or filtered out of the material requiring further processing in downstream operations. This would not be the case in vaccines using live attenuated organisms as a component of the vaccine. These organisms would be present throughout the production process.

7.10.2 Downstream Process Facilities

Once harvested, the material is usually subject to further steps of processing for final use in the product. This may include inactivation; extraction of components, including toxins; purification; or pooling of batches into larger quantities. Containment may still be required during the inactivation process until quality control tests come back demonstrating full inactivation. In addition, nonlive components, such as toxins, may still require containment and may also be quantities that would be considered a large-scale risk until the toxins are inactivated through processing. Some products, such as live virus vaccines, contain live organisms and require large-scale containment throughout processing.

7.10.3 Fill and Finish Facilities

Facilities used for fill and finish activities such as vial filling, sealing, and capping may still require containment for products containing live organisms such as live virus vaccines. These fill and finish operations would typically contain product tanks of a size to be considered large-scale operations.

7.10.4 Quality Control Laboratory Facilities

Quality control laboratories are a critical part of bioprocess facilities. These laboratories will be handling samples pulled throughout the process and therefore would be handling the same organisms or toxins that are utilized in the process. Generally, laboratory-scale containment would be required based on the risk of the organism used. In some cases, animal models are used as part of the quality testing process. Containment requirements for animal use may be required if the quality testing includes viable pathogenic organisms.

7.10.5 Cross-contamination "Live" to "Nonlive"

One principle of cGMP is the segregation of process steps to eliminate the potential for cross-contamination. This is particularly important in facilities utilizing live pathogenic organisms. When live pathogenic organisms are grown for use in products that do not contain the live organisms in the final product such as inactivated or component vaccines, there is potential risk that live organisms may inadvertently be introduced into the final product. Specific attention must be made in these cases to the flow of materials, personnel, equipment, product, and waste to ensure adequate separation exists to eliminate the potential of cross-contamination from live to nonlive processing areas. Segregated one-way flows should be strongly considered as part of the design of the facility.

7.11 Conclusion

Appropriate design for contamination control to provide product quality in bioprocess facilities is becoming an increasing challenge. When biocontainment is an overlay to contamination control, the challenges are multiplied. There are well-grounded, proven design principles for the design of facilities to provide contamination control and containment. The design and planning principles outlined in this article can be applied to fixed or modular facilities, single- or multiple-product facilities, CIP or single-use equipment, fixed or mobile equipment, or any variation thereof.

Control of microorganisms within a bioprocess facility, whether for product protection or for the containment of hazardous microorganisms that are required as part of the process, requires a balanced combination of facility design measures, equipment, and operational protocols. For maximum effectiveness, this balance should be created during the design process by developing the design and documenting the operational protocols concurrently. Reducing the potential for breach of containment as well as contamination by environmental microorganisms requires a proactive process to ensure that the facility is maintained and operated within the initial design and operational protocols. Application of design and operational principles reducing the potential for unwanted migration or transport of microorganisms within a bioprocess facility can significantly reduce the potential of and the impact from adverse events.

References

1 CFR 21, Part 211 – Current Good Manufacturing Practice for Finished Pharmaceuticals.
2 US Food & Drug Administration. Guidance on Sterile Drug Products Produced by Aseptic Processing, September, 2004.
3 European Commission, Enterprise and Industry Directorate- General, The Rules Governing Medicinal Products in the European Union, Volume 4, EU Guidelines to Good Manufacturing Practice Medicinal Products for Human and Veterinary Use, Annex 1, Manufacture of Sterile Medicinal Products, Brussels, 25 November 2008 (rev.)
4 Chatigny M, West D. Laboratory ventilation rates: theoretical and practical considerations. Proceedings of the Symposium on Laboratory Ventilation for Hazard Control; 1976; Frederick, MD.
5 United States Pharmacopeia, USP 1116, Microbiological Control and Monitoring of Aseptic Processing Environments, USP 35, vol. **1**, 2012a, p 697–707, 2012.
6 Sutton S. Accuracy of plate counts. *J Validat Technol* 2012;**18**:79–83.
7 FDA, MedWatch The FDA Safety Information and Adverse Event Reporting Program, New England Compounding Center (NECC) Potentially Contaminated Medication: Fungal Meningitis Outbreak, Updated 10/06/2012, http://www.fda.gov/Safety/MedWatch/SafetyInformation/SafetyAlertsforHumanMedicalProducts/ucm322849.htm. Accessed 2014 Aug 7.
8 The Commonwealth of Massachusetts, Executive Office of Health and Human Services, Massachusetts Department of Public Health, Board of Registration in Pharmacy Report, New England Compounding Center (NECC), Preliminary Investigation Findings, October 23, 2012.

9 Hurst C, editor. *Manual of environmental microbiology*. Washington, DC: ASM Press; 1997.

10 Sandle T. A review of cleanroom micro-flora: types, trends, and patterns. *PDA J Pharm Sci Technol* 2011;**65**(4):392–403.

11 American Society of Heating, Refrigerating, and Air Conditioning Engineers. Method of Testing General Ventilation Air-Cleaning Devices for Removal Efficiency by Particle Size, ASHRAE Standard 52.2., 2000.

12 Camfil-Farr, Mold Filter Recommendations, Filtration of Mold and Mold Spores, http://www.filterair.info/articles/article.cfm/ArticleID/ABEFC51E-F66D-4A4A-BE15A665464FC549/Page/1. Accessed 2013 Sep 12.

13 CDC/APHIS National Select Agent Registry, http://www.selectagents.gov/. Accessed 2013 Oct 12.

14 U.S. Department of Health and Human Services. Biosafety in microbiological and biomedical laboratories (BMBL). 5th ed. Washington, DC: Government Printing Office; 2007 http://www.cdc.gov/biosafety/publications/bmbl5/. Accessed 2014 Aug 27.

15 Discussions between Jon Crane and Dr. John Richardson, Atlanta, GA, 1989.

16 Department of Health and Human Services, National Institutes of Health, NIH Guidelines for Research Involving Recombinant or Synthetic Nucleic Acid Molecules (NIH Guidelines), November 2013.

17 Cipriano M. Large-scale production of microorganisms. In: Fleming D, Hunt D, editors. Biological safety: principles and practices. 4th ed. Washington, DC: ASM Press; 2006. p 561–579.

18 Public Health Agency of Canada, Canadian Biosafety Standards and Guidelines for Facilities Handling Human and Terrestrial Animal Pathogens, Prions and Biological Toxins, 1st ed., 2013.

19 World Health Organization, Annex 2, Guidelines for the Safe Production and Quality Control of Inactivated Poliomyelitis Vaccine Manufactured from Wild Polioviruses (Addendum, 2003, to the Recommendations for the Production and Quality Control of Poliomyelitis Vaccine (Inactivated)), WHO Technical Report Series, No. 926, 2004.

20 United States Department of Agriculture, ARS-ERS-NASS-NIFA Manual, ARS Facilities Design Standards, ARS-242.1, 1 May, 2012.

21 Spratt B, Chairman, Independent Review of the Safety of UK Facilities Handling Foot-and-Mouth Disease Virus, Presented to the Secretary of State for Environment, Food and Rural Affairs and the Chief Veterinary Officer, August 2007.

22 Cello J, Paul A, Wimmer E. Chemical synthesis of poliovirus cDNA: generation of infectious virus in the absence of natural template. *Science* 2002;**297**(5583):1016–1018.

23 AIChE Center for Chemical Process Safety. Guidelines for process safety in bioprocess manufacturing facilities. Hoboken, NJ: John Wiley & Sons, Inc.; 2010.

24 Food and Drug Administration, Center for Drug Evaluation and Research (CDER), Center for Biologics Evaluation and Research (CBER), Guidance for Industry, Immunogenicity Assessment for Therapeutic Protein Products, Draft Guidance, February 2013.

Further Reading

Bioprocess Manufacturing Facilities; John Wiley & Sons, Inc., 2011.

Block S. Disinfection, sterilization, and preservation. 5th ed. Philadelphia, PA: Lippincott, Williams and Wilkins; 2000.

Crane J, Richmond J. Design of biomedical laboratory facilities. In: Fleming D, Hunt D, editors. Biological safety: principles and practices. 4th ed. Washington, DC: ASM Press; 2006. p 273–294.

Dillon H, Heinsohn P, Miller J, editors. Field guide for the determination of biological contaminants in environmental samples. Fairfax, VA: American Industrial Hygiene Association; 1996.

Friedman R. Aseptic processing contamination case studies and the pharmaceutical quality system. *PDA J Pharm Sci Technol* 2005;**59**(2):118–126.

Halkjær-Knudsen V. Design considerations for large-scale production of biologicals: GMP and containment synergies. In: Richmond J, editor. Anthology of biosafety VIII: evolving issues in biosafety. Mundelein, IL: ABSA; 2005. p 39–67.

Halkjær-Knudsen V. Designing a facility with both good manufacturing practice (GMP) and biosafety in mind: synergies and conflicts. *Appl Biosaf* 2007;**12**(1):7–16 ABSA.

International Society for Pharmaceutical Engineering. Baseline pharmaceutical engineering guide for sterile manufacturing facilities. 2nd ed. Tampa, FL; 2011.

Macher J, editor. Bioaerosols: assessment and control. Cincinnati, OH: American Conference of Governmental Industrial Hygienists; 1999. ISBN: 1-882417-29-1.

Chapter 8

Process-Based Laboratory Design

Henriette Schubert and Flemming K. Nielsen

NNE A/S, Gentofte, Denmark

8.1 Introduction

The new pharma reality is characterized by fast changeability and unknown future demands. Often, the laboratories are established before the processes, facility use, and final technologies are fully uncovered.

Hence, there is a trend toward flexible, modular, and changeable laboratory facilities that are adaptable to new research needs or process knowledge.

Planning and designing biopharmaceutical laboratories for research, development, and quality control (QC) require an approach where understanding the project-specific challenges and business drivers is essential, to ensure compliant and futureproof laboratory operations.

One of the key messages in this chapter is the importance of thoroughly analyzing operations before actual design initiation, with the purpose of establishing a factual and informed basis for management decision early in the project life cycle. In addition, establishing a solid basis for design before putting pen on paper and by that ensuring project success and delivering superior laboratory facilities for the biopharmaceutical industry fit the purpose (Fig. 8.1).

8.2 Areas of Application/Scope

The purpose of this chapter is to provide guidance and a methodology that is based on a process and operational driven design approach. The methodology can be used for QC laboratory projects and other types of laboratories such as R&D (research and development) laboratory facilities, *in vivo* animal research, trial facilities, facilities for development, or small-scale manufacturing of tailored therapies.

Process Architecture in Biomanufacturing Facility Design, edited by Jeffery Odum and Michael C. Flickinger.

Figure 8.1 R&D Bioreactor/Fermentation laboratory, customized design of vertical shafts with supplies and tables. [NNE.]

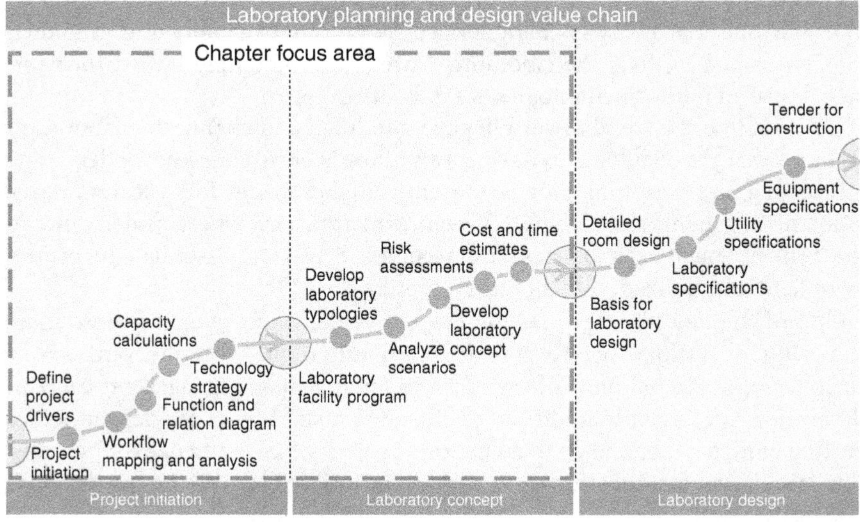

Figure 8.2 Laboratory planning and design value chain. [NNE.]

The approach and methodology described in this chapter is applicable to initial planning, project initiation, and the earliest concept design phases. However, the methodology can also be applied to manage and evaluate design development associated with laboratory in other phases, for example basic design (see Fig. 8.2 for an overview of laboratory planning and design value chain).

The focus of this chapter is on the guidance to create basis for design development and management decision rather than comprising a detailed design and construction guideline. The methodology and principles described in this chapter can be used globally regardless of the local requirements where a detailed design guideline is largely driven by country-specific requirements.

8.3 Translation of Process Elements into Laboratory Architecture

The relationship between the laboratory product and processes and how it is handled in operations should be the key drivers for laboratory design.

Hence, comprehensive understanding of laboratory work processes, workflow, and operations is the basis for translating process into planning, concept, and design of laboratories fit for purpose (Figs. 8.6, 8.9, 8.10 and 8.12).

What does this mean to the consultants and their effort of creating basis for decision and design?

To understand laboratory processes and workflows, consultants need to listen, learn, and conduct comprehensive analyses and mapping before accelerating into initial concept and design development. The following sections in this chapter provide guidance and give examples on how to conduct mapping and analyze user and customer deriving input.

The users and customers are the laboratory experts, and a major part of the consultants' effort is to drive and facilitate a process that uncovers laboratory needs and requirements for the specific project.

In addition, consultants should work beyond their background (single) discipline and make use of a holistic mindset to manage functional and regulatory requirements.

Translating process into architecture requires a structured way of working, where the use of a seamless cross-disciplinary and holistic approach are key elements. It is not about the architect, the engineer, or the laboratory planner working on their own and passing on single disciplinary outcome between them—everybody really needs to work closely together, including close cooperation and interaction with customers and users in focus. To succeed, it is key to have the right competencies and experience present both from the consultants and the customer side.

The consultants' side involvement may include the following:

- Laboratory planners
- Architects
- Mechanical engineers
- Logistic engineers
- Safety, health, and environmental engineers
- Quality and compliance specialists

- Heating, ventilation, and air conditioning (HVAC) and plumbing engineers
- Organizational change management (OCM) consultants

The customer side involvement may include the following:

- Laboratory managers
- Senior laboratory staff
- Scientists and researchers
- Laboratory technicians
- Representatives from the compliance group (QA and others)
- Working environment representatives

The project team, meaning consultants, and user team should work as one team. The key steps of the work processes should be broken into defined activities to facilitate a structured and documented approach allowing the provision of basis for management decision and laboratory design in subsequent phases.

8.4 Key Steps in Planning Approach and Methodology

8.4.1 Laboratory Planning Process

As mentioned in Section 8.2, this chapter focuses on the initial phases of planning; in this section, we will shortly explain the different phases of planning to ensure understanding of the terms used, again with the primary focus on the initial phases.

When initiating planning and design of laboratories, it is important to focus on the flow of events and ensure data availability. The natural flow would typically be:

In this chapter, we have chosen to divide the design process into the above illustrated phases but will focus on project initiation and conceptual design.

Bear in mind that the naming and content of the design phases may differ from company to company, but in general, the overall content most likely is the same.

8.4.1.1 Project Initiation (Analyze Data)

The project initiation phase is all about analyzing and understanding customer needs and objectives with the purpose of establishing a solid baseline for project execution. To do this, it is important to understand the current As-is situation and the desired To-be situation.

Typically, key subjects for programming will include the following:

- Mapping of project/program vision, targets, success criteria, and focus areas
- Workflow mapping (As-is and To-be situations)
- Technology strategies
- Capacity calculations (samples, analysis, and persons)
- Space, area, and adjacency needs
- Mapping of overall functional needs
- High-level technical needs on functional level
- Project risk assessment (product and project-related risk)

8.4.1.2 Conceptual Design (Develop Concepts)

In the conceptual design phase, the focus is to develop different conceptual scenarios to clarify and illustrate the best possible conceptual solution for the desired future state, as mapped out during project initiation. An important part of conceptual design is not just to develop solutions but to analyze and develop recommendations that allow the customer to choose the optimal direction to ensure fulfillment of project success criteria.

The documentation and data from the project initiation phase form the basis for the conceptual design and the development of the following:

- Overall functional description or user requirement brief
- Laboratory concepts (overall facility and layout concepts)
- Mechanical concepts (utilities and HVAC)
- IT and automation concepts [building management system (BMS) and facility management system (FMS)]
- Good manufacturing practice (GMP) and cross-contamination strategies
- SHE (safety, health, and environment) strategy (including hazardous materials and, e.g. fire/explosion evaluation)
- Area need calculations
- Cost estimates, total project cost (TPC)

8.4.1.3 Basic Design and Detailed Design (Develop Solutions)

As mentioned in the introduction, this chapter focuses on project initiation and the earliest concept design phases of laboratory design; hence, we will not go into the details regarding the basic and detailed design phases but only touch upon the primary focus areas for the two phases:

Basic design: the overall target of the basic design is to convert the strategies and concepts from conceptual design into tangible layouts and feasible technical solutions. The solutions should provide proof of concept and with the use of design rationales, pros and cons on proposed design details guarantee a design that can fulfill the desired operational capabilities of the laboratory.

Detailed design: in detailed design, involvement from users and customer should be limited to an absolute minimum, and the solutions decided upon during basic design should be converted into technical documentation ready for tender to contractors and construction. Furthermore, plans for qualification and validation activities should be finalized.

8.4.2 Creating an Informed Basis for Design

The importance and effort of carefully analyzing and understanding laboratory processes and operations before initiation of design is often underestimated.

It is likely that a substantial effort invested in the early phases will ensure an informed basis for decision and solid basis for design in the consecutive phases.

Typically, the initial programming and planning will impact around 80% of the total project investment; this is why a frontloading of the project activities most likely will benefit the project going forward with better overview of cost and quality (see Figs. 8.3 and 8.4 for typical cost impact spread). Furthermore, experience emphasizes the fact that changes to the project are most effective when they occur in the early stages of programming and design development [1].

It is most likely that using a methodology based on an interactive, dialogue-based process between customers and consultants via meetings,

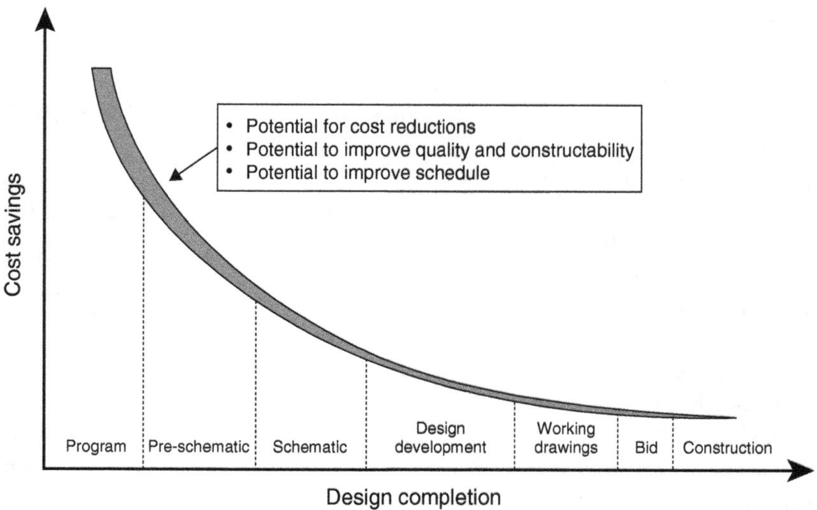

Figure 8.3 Cost savings versus design completion. [ISPE Good Practice Guide: Quality Laboratory Facilities (ISBN 9781936379422).]

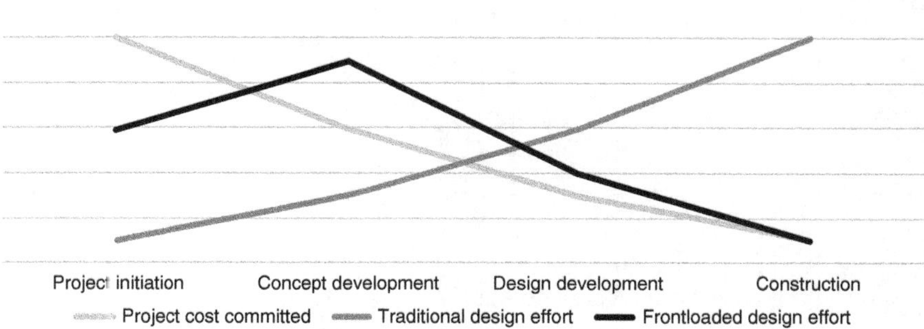

Figure 8.4 Traditional versus Frontloaded design method. [NNE.]

reviews, and workshops will ensure the best possible alignment of project stakeholders as described in this section.

8.4.2.1 Mapping of Design Drivers and Project Targets

Ensuring alignment on project mission, drivers and success criteria are the most critical project initiation activities. During the initial project mobilization, it is important to start the dialog with key project stakeholders to align project expectations and ensure that the entire team has a common understanding of the project's mission, objective, strategies, key assumptions, and focus areas and targets.

Typically, the following areas would be agreed and documented:

- Project mission
- Project objectives and success criteria
- Project background
- Project assumptions
- Project focus areas

The information regarding the projects' overall strategic targets could be collected in multiple formats; one way is to develop a one-page strategy (OPS) that sums up all the abovementioned information and forms the project "guidebook," a tool that should be used to validate all major project decisions (see Fig. 8.5 for an example of an OPS).

Project strategies are usually developed in the project initiation phase of a project but should be reviewed and updated as the project progresses. Typically, it is the document with the highest ranking of importance in a project, and it should be revisited periodically to make sure that the project does not get lost in the details but rather adheres to the overall project's mission and objectives.

One of the most important parts of the project strategy is mapping of project objectives and success criteria; mapping of project objectives is normally done in a workshop or a series of workshops with participation from client senior management.

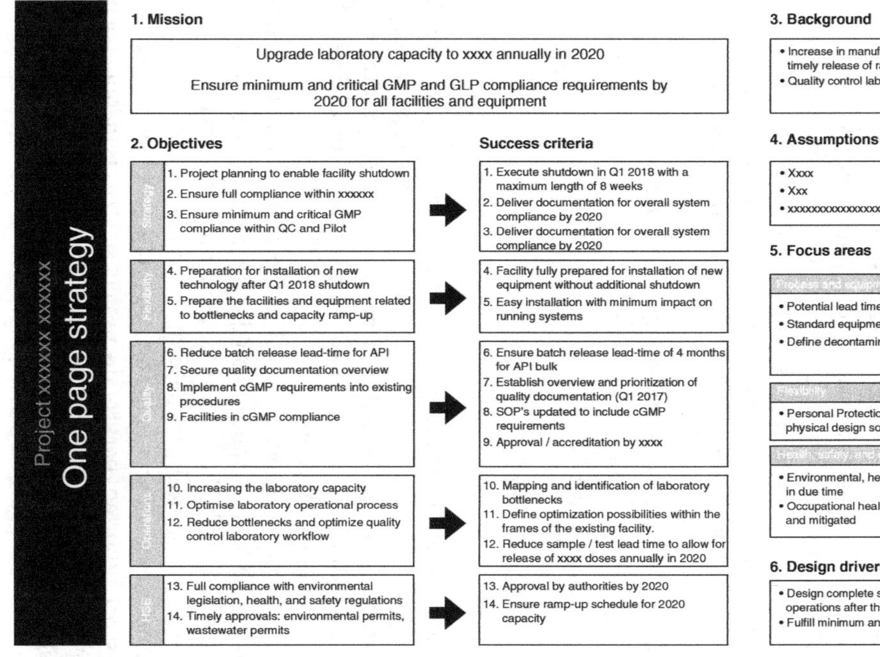

Figure 8.5 One-page strategy. [NNE.]

Depending on the project complexity and scale, the below-listed subjects should be addressed as part of the target setting process when developing the OPS.

- Capacity (e.g. number of research projects, QC samples, and staff)
- Flexibility and changeability (desired level of flexibility needed, e.g. new products, procedures, or equipment)
- Expansion (include design for the unknown or planned future expansion)
- Technology (desired level of automation and systems for data integration or management)
- Sustainability (e.g. desired level of energy optimization or certification, manual handling vs mechanical/automated/other aid)
- Safety
- Maintenance (e.g. expected facility lifetime and mechanical access)
- Compliance [e.g. good (anything) practice (GxP), genetically modified organisms (GMO), biocontainment, or containment]

When defining project targets, it is critical that all targets are linked with measurable success criteria, which allows for measurement and evaluation of the individual targets to ensure project success (Fig. 8.5).

As a part of the project assumptions, it is key to conclude the project-relevant regulations, standards, and guidelines to ensure that it is clear from the project initiation which requirements and industry standards to comply with.

8.4.2.2 Designing for the Desired Laboratory Work Culture

Laboratory design is an overall process and operations based but is equally driven by the desired laboratory work culture. The laboratory planning approach needs to cover both (culture and operations) in order to create the basis for laboratory design that can meet client business drivers and the overall project vision and mission.

A typical design driver for a research laboratory could be that the facility design should facilitate and support knowledge sharing and promote a seamless and an innovative collaboration culture in the laboratory setting. To fulfill the overall visions and mission of new or refurbished laboratory settings, product, process, and compliance objectives must be fully supported by a desired work culture.

This may materialize into diffident settings of collaborative spaces ranging from more formalized meeting rooms to breakout spaces such as informal touch points, for example the coffee area and other shared spaces for socializing and interaction.

When looking at the full workday of laboratory scientists and technicians, the actual time spent in the actual laboratory is typically a fraction of the workday activities (see example of mapping of work day activities in Fig. 8.6). Taking a holistic view of the main activities, the functionalities range from

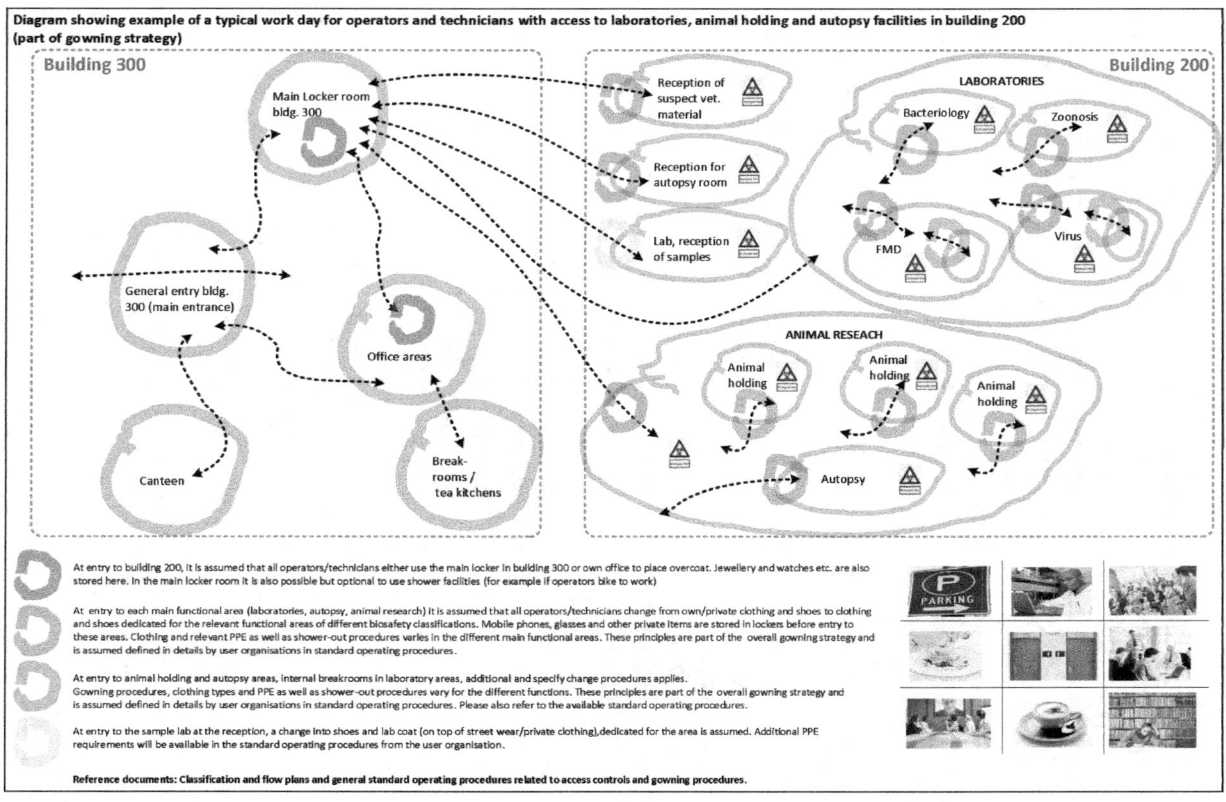

Figure 8.6 Example of typical day work flow and overall activity analysis. [NNE.]

the parking space, the locker room to the laboratory, and meeting and interaction activities including the coffee spot. All functionalities in the periphery play a role in creating a desired work culture of knowledge exchange for a laboratory and should therefore be taken into account in the project initiation.

To design for a desired work culture is a major design driver and a typical bioprocessing manufacturer vision of attracting the most skilled scientists and technicians. Innovative and supporting physical premises surrounding the laboratory operations and activities should be seen as a key part of an attractive workplace that can directly and indirectly support laboratory products and processes.

Typically, the way of working changes, and new facilities induce resistance or reservations among stakeholders. This is an often neglected element, which in the end can cause project delays and delayed start of operations or inefficient productivity in a laboratory setting. Hence, the human factor and motivation for changes is essential to address on equal terms to physical facility planning.

One way to address the process for a successful and desired work culture change may involve an organizational change process to prepare the organization for working in new ways in, for example new facilities, use of new technologies, and new ways of collaboration. OCM is typically a support provided by consultants to the management and to end-user groups and should be part of the initial planning process.

The trend for office and write-up spaces in laboratory settings is moving toward more shared and open-space write-up and office zones, mixed with study booths for work demands of concentration work. The open office space can facilitate exchange of knowledge and information but require a change management process.

The ratio between laboratory and office and write-up spaces has changed significantly over the past decade. Owing to new technologies, the ratio between direct laboratory/support and office space is currently around 40–60% and may even move further in the direction of an extended need for more collaborative and office space. One of the parameters for the ratio change is increasing automated analysis and use of equipment-integrated robotics. This type of equipment does not need constant operator attention or frequent action. Hence, some of the analysis work and robots are moving to technical room types fit for removing equipment heat and exhaust and with limited facility requirements for longer periods of operator-related work operations.

At the same time, new technologies and analysis equipment generate faster an increasing amount of data to be handled in both QC and R&D laboratory settings. This calls for specific data evaluation spaces for interaction between scientists and technicians. This may materialize into collaborative spaces that may include the use of new IT technologies for 3D simulation and viewing, holographic touch screens, and real-time and interactive global videoconferencing.

8.4.2.3 Risk Assessment (GMP, Biocontainment/High Potent Product Containment)

Laboratories handling different types of products or materials that require specialized handling with deriving facility requirements and procedures will typically be subject to risk assessment related to, for example:

- Cleanliness, sterility, cross-contamination, and product segregation (GMP requirements)—quality risk assessments
- Biohazardous products or material (GMO, biological agents)—biorisk assessments (biosafety and biosecurity)
- High potent product substances—high potent product risk assessment

Addressing risk at an early project stage is an essential element of creating an informed and fact-based basis for design and risk-based decisions. In addition, assessment of risk is directly linked to regulatory requirements and customer expectations. Scope and responsibility for planning, execution, and contribution to risk assessment activities must be clearly defined as part of the contract between customers and consultants.

GMP Risk Assessment Quality risk management supports a scientific and practical approach to decision making. It provides documented, transparent, and reproducible methods to accomplish steps of the quality risk management process based on current knowledge about assessing the probability, severity, and sometimes detectability of the risk.

Addressing risk related to GMP, the focus areas are as follows:

- Pharmaceutical products of high quality which do not pose any risk to the consumer or public

Related to laboratories, GMP risk assessment is relevant for development, QC, and in-process laboratories (IPCs).

The pharmaceutical industry and regulators typically assess and manage risk using recognized risk management tools, and it is recommend to use the same tools, examples of them being the following:

- Basic risk management facilitation methods (flowcharts, checklist, etc.)
- Failure mode effects analysis (FMEA)
- Failure mode effects and criticality analysis (FMECA)
- Fault tree analysis (FTA)
- Hazard analysis and critical control points (HACCPs)
- Hazard operability analysis (HAZOP)
- Preliminary hazard analysis (PHA)
- Risk analysis and mitigation matrix (RAMM) [2]

It is important to note that no tool or set of tools is applicable in every situation in which a quality risk management procedure is used. In addition, the relevant tool depends on the project phase and available product and process details.

One of the tools that should be considered to be used is the principles laid down in ICHQ9 [3], where the overall diagram of a typical quality risk management process is as follows:

The ICHQ9 Guideline provides principles and examples of tools of quality risk management that can be applied to all aspects of pharmaceutical quality including development, manufacturing, distribution, and the inspection and submission/review processes throughout the life cycle of drug substances and drug (medicinal) products and biological and biotechnological products (see Fig. 8.7 with a flow chart of a typical quality risk management process).

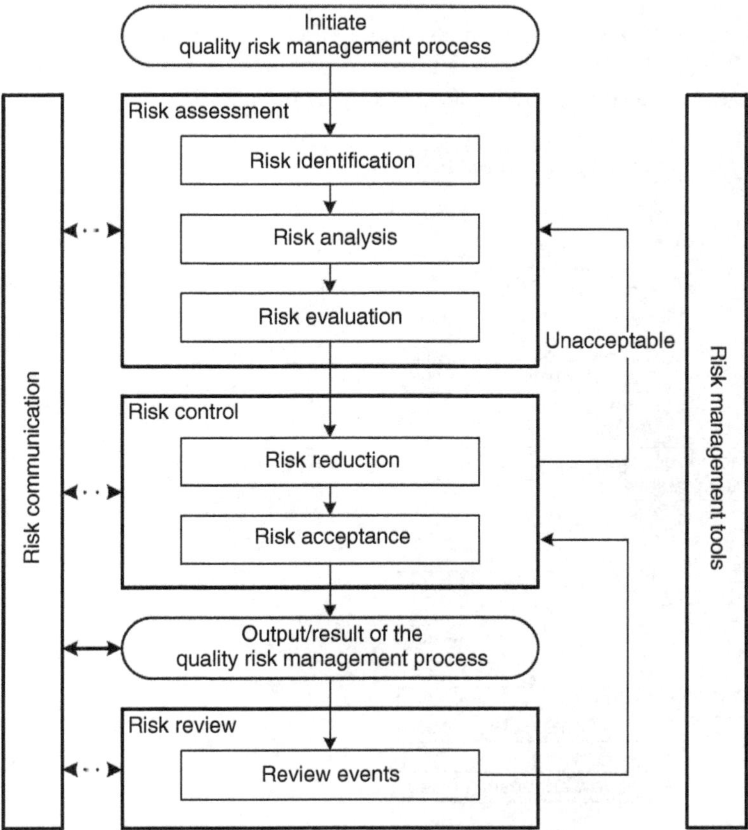

Figure 8.7 Overview of a typical quality risk management process. [Figure 8.1 from ICHQ9.]

Biorisk Assessment (Biological Agents and GMO)
The focus of biosafety is as follows:

- to minimize the risk of exposure of biological organisms (including GMOs)—to protect the operators and minimize the risk of release to the environment.

Biorisk assessment is a process of identifying the hazards and evaluating the risks associated with biological agents and toxins, taking into account the adequacy of any existing controls, and deciding whether the risks are acceptable. Biorisk covers both biosafety and biosecurity.

- Biological risk assessment is a subjective process requiring consideration of many hazardous characteristics of agents and procedures, with judgments often based on incomplete information. There is no standard approach or tool for conducting biorisk assessment. However, a number of standards and guidelines provide guidance suggesting a sequenced structure [4,5].

CWA [6] suggests a Risk Assessment Strategy focusing on Risk Characterization and Risk Evaluation, using the above shown principles with focus on risk characterization and risk evaluation (see flowchart in Fig. 8.8).

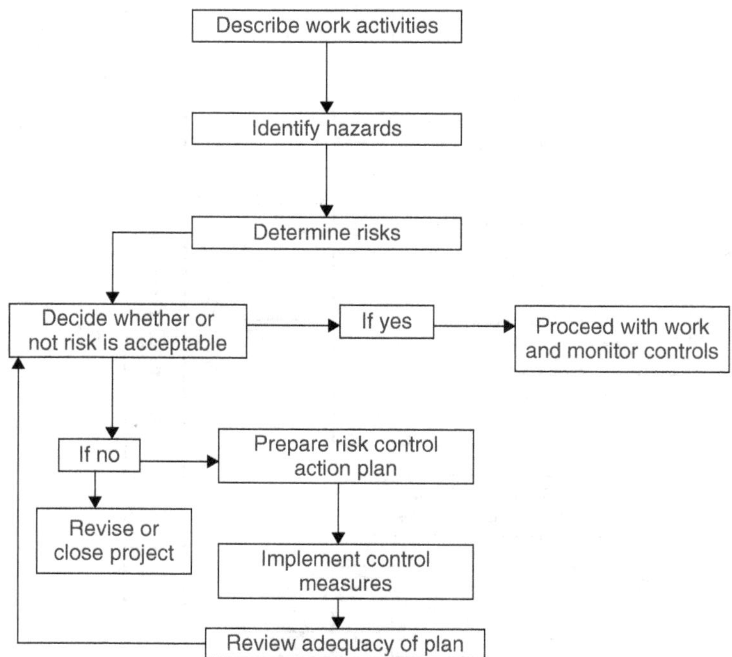

Figure 8.8 Diagram for addressing biorisk—risk characterization and risk evaluation. [CEN Workshop agreement, CWA 15793 Laboratory biorisk management.]

Examples of biorisk methodologies and tools are as follows:

- HAZOP (hazard and operability study is a systematic way to identify possible hazards in a work process)—where the HAZOP technique is qualitative and aims to stimulate the imagination of risk assessment participants to identify potential hazards and operability problems.

Biosafety levels (BSL) are defined in four levels (BSL1–4), where BSL1 is the lowest level (basic containment) and BSL4 is the highest level (maximum containment).

National regulations and international standards and guidelines list minimum requirements for engineering controls. However, it is important to notice that the lists of minimum requirements are not key lists, and it may be relevant to use requirements from different biosafety levels depending on the biological agent and project situation. Both National regulations and international standards and guidelines require biorisk assessment of the project-relevant biological organisms and GMOs.

High Potent Risk Assessment A significant proportion of new drugs under development contain high potency active pharmaceutical ingredients (HPAPIs), which leads to explosive growth in demand for their production. The cytotoxicity of HPAPIs, however, presents handling challenges and requires specialized containment.

High potent products include active product ingredients (APIs) in powder, granulate, tablet, or liquid form and may pose a risk to operators during processing in the laboratory. Hence, it is mandatory to assess the risks as required by health and safety regulations.

The focus for risk-assessment-related high potent products is as follows:

- To ensure that employees and their environment are protected from exposure of high potent products including APIs.

The risk evaluation will assess the processes and conclude if the processes comply with relevant and defined exposure level limits, which are defined as OELs (occupational exposure levels). Be aware that no common international standard exists for occupational exposure band. However, in some countries, OEL band 1–5 is used, where 1 is the lower and 5 the highest level.

One of the methods to evaluate the risk is to use calculation and simulation of the real-time operator intake. This can be conducted by ROI (real operator intake) calculation assessing the relevant process steps.

The ROI describes the amount of API that gets into the body of the operator, while being for a certain period in an area with a certain airborne drug concentration.

For further guidance related to highly potent containment, ISPE has published a number of design guides and handbooks [7,8].

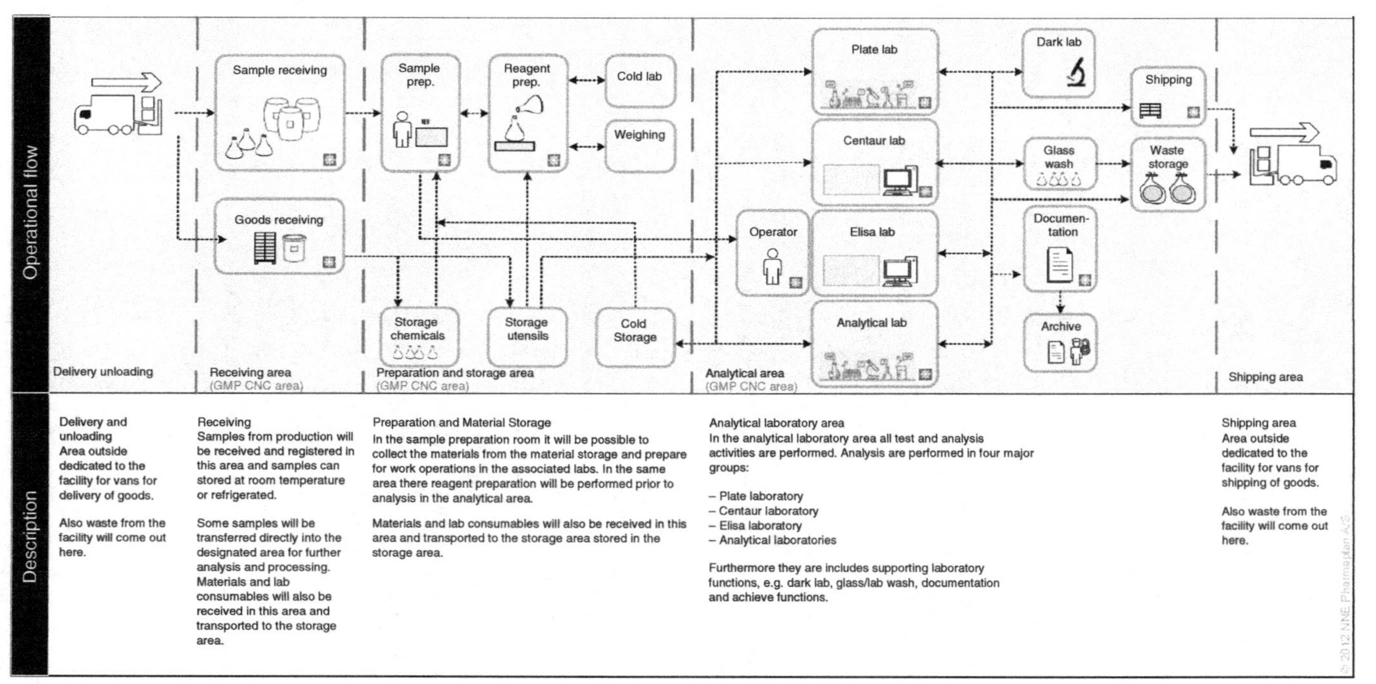

Figure 8.9 Operational flow diagram, quality control laboratory. [NNE.]

SHE Risk Assessment For assessment of SHE elements, refer to Section 8.6 (SHE laboratory design considerations).

Be aware that risk assessments should address GMP, Biorisk, high potent substances, and general SHE impact laboratory facilities, mechanical and utilities, and the work procedures and PPE (personal protection equipment) in the laboratory. Consequently, it is key for the consultant to contribute to and understand risk assessments including the rationales behind them to translate the conclusive requirements into concepts and design.

8.4.2.4 Operational Workflow Mapping and Visual Planning

Mapping of operational workflows should be part of the project initiation phase and is an essential tool to ensure full alignment between all project stakeholders. Using this powerful visual planning tool furthermore facilitates mutual understanding of the laboratory workflow and is key for us as consultants to challenge critical processes and base our assessments and recommendations on factual data (see example of an operational work flow diagram in Fig. 8.9).

Using laboratory workflow mapping does not only ensure a common picture of the main work activities and processes but is also used to address process risks related to workflows, activities, processes, or equipment at an early stage of the project.

Workflows and operations mapping may also be used to analyze not only overall but also specific complex or critical areas of laboratory operations, for example specialized operations to analyze the more detailed steps of operations and potential alternatives. The purpose of analyzing the selected areas in more detail is to ensure a common understanding between the consultants and the customer/users and be able to use the analysis and understanding directly in the concept design development (see Fig. 8.10 for an example of a detailed operational workflow analysis). In addition, the analysis can be used to address risk-related aspects and conducting an initial assessment of compliance with relevant regulations, standards, and guidelines. If the project involves, for example research animals, it would be relevant to perform a first check regarding animal research and animal welfare standards, one of them being AAALAC (Association for Assessment and Accreditation of Laboratory Animal Care International) standards.

The laboratory workflow assessment is also the ideal tool to identify optimization opportunities for lean workflow processes. Hence, the analysis should also be used to evaluate alternative settings and overall sequence of operations to comply with the project charter and defined design drivers.

In addition to the operational flow diagram (OFD) assessment, it is of great advantage to document the overall functional needs for main functionalities in the specific laboratory. This may be documented in a functional description that can support the subsequent analysis and assessments in the initial planning phase. In some cases, the customer may

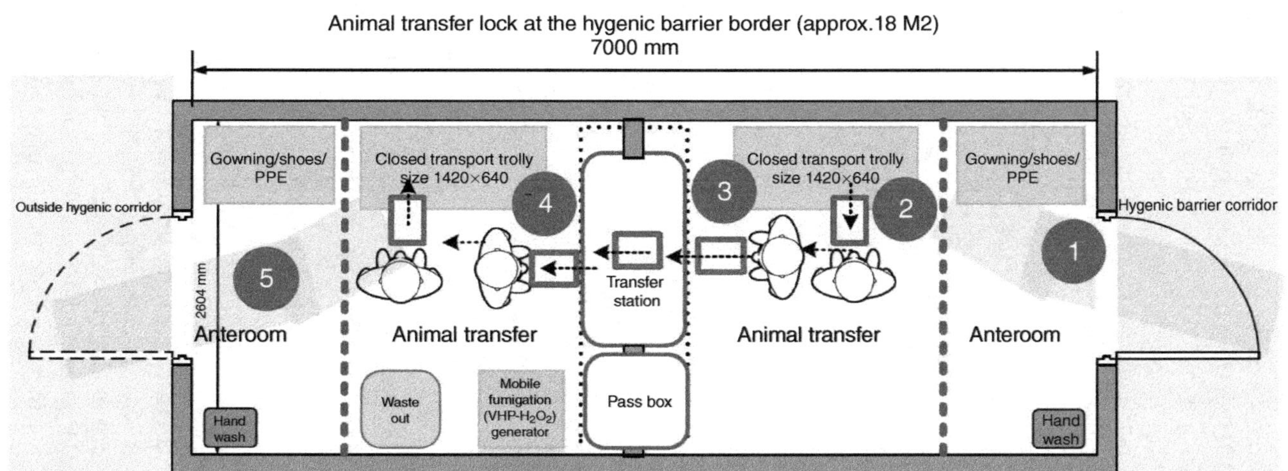

Figure 8.10 Operational flow diagram, *in vivo* drug development laboratory. [NNE.]

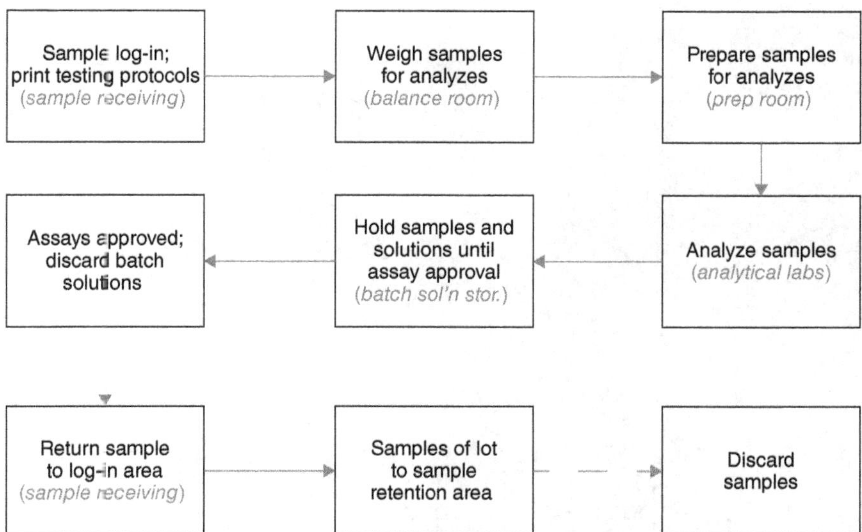

Figure 8.11 Process flow for a quality laboratory. [ISPE Good Practice Guide: Quality Laboratory Facilities (ISBN 9781936379422).]

have a detailed user requirement brief that may replace the functional description.

It is ideal to perform operational workflow mapping as a dialogue-based workshop where future operations are mapped, assessed, and discussed. In dialogue with principal investigators and leading laboratory technicians. It is essential that participants (representing or consisting of users) have deep insights into existing operations and laboratory processes.

The workshops are typically executed as a whiteboard or brown paper session, where the participants are facilitated through a process of mapping out processes, operations, key equipment, and bottlenecks using predefined icons or post-its to illustrate a typical day of operations (Fig. 8.11).

8.4.2.5 Functional Adjacency Analysis (Function/Relation)

If an operational workflow mapping has been conducted, this should form the basis for developing functional and relations analysis that is typically documented in function/relation diagrams. On the basis of the information gained from the operations workflow mapping, the most critical flows and interrelations between main functions should be mapped to create a detailed and clear overview of facility functions and how they interact and interlink. Another feature of the function/relation diagram is the possibility to clearly indicate what part of the project scope is included in, for example a GxP [relevance to Good Practice related to either Good Laboratory Practice (GLP) and/or Good Manufacturing Practice (GMP)] or biohazard or other type

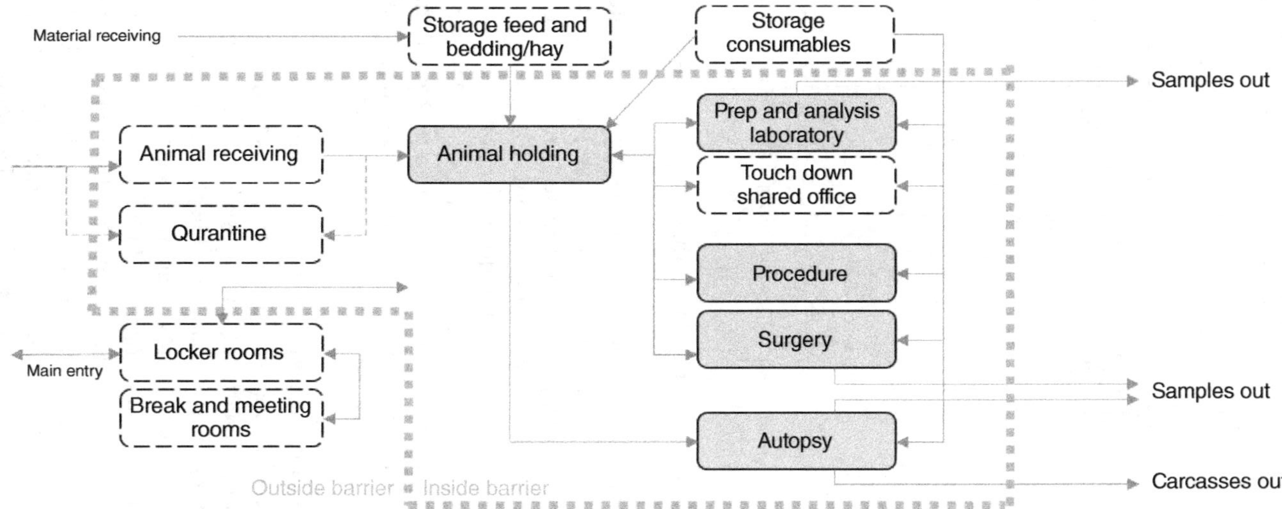

Figure 8.12 Overall functional relation diagram example, indicating main functionalities with gray boxes and support/administrative functions as white boxes. [NNE.]

of barrier classified area and what is outside classification (see an example of an overall function/relation diagram including a barrier indication in Fig. 8.12).

The function/relation analysis is used to align the consultants understanding of the laboratory setting and support functions with the user and customer requirements and expectations.

The analysis and evaluation of the functional relations is typically conducted in close cooperation between the consultants and the customer/users team and is typically conducted in a workshop setting as a subsequent activity to the workflow operations clarification.

Function/relation analysis is used for the analysis of potential conflicting functional adjacencies or crossing flows that may pose a risk to, for example GxP, biohazard, or barrier.

Finally function/relation analysis is used for space outline of main functions and overall facility organization into a diagrammatic and holistic overview to form the basis for concept development.

8.4.2.6 Laboratory Typologies as a Planning Tool

To form the basis for laboratory concepts, it is useful to conduct laboratory typologies of typical laboratory and support functions. A typology is a project defined, generic type of laboratory that may be used as a repeatable "stamp" in the laboratory planning and subsequent design. The typology defines the overall uniform characteristics of area need, access principles, equipment, and main utilities.

The purpose of preparing the typologies is to define functional "Lego" blocks of typical laboratory functional units, for example product sample reception and related analysis laboratories and support. Another example is an animal research suite including procedure room and other support functions.

First Step of Typology Development The first step of typology development should be diagrammatic assessment based on the OFDs and functional flows. Furthermore, principle indications of the internal flows of operations (processes, people, material, samples, and waste) are relevant to add to the diagrams. In relation to potential need for anteroom or airlocks, general people and material access should be considered, depending on the laboratory use and classification. When adding a number of known laboratory equipment (based on, e.g. an initial laboratory equipment list) to the diagrams and evaluation, the typology diagram indicates the main principles related to the specific laboratory typology (see an example of a first step laboratory room typology in Fig. 8.13). The principles for main equipment location in laboratory and support functions should be assessed.

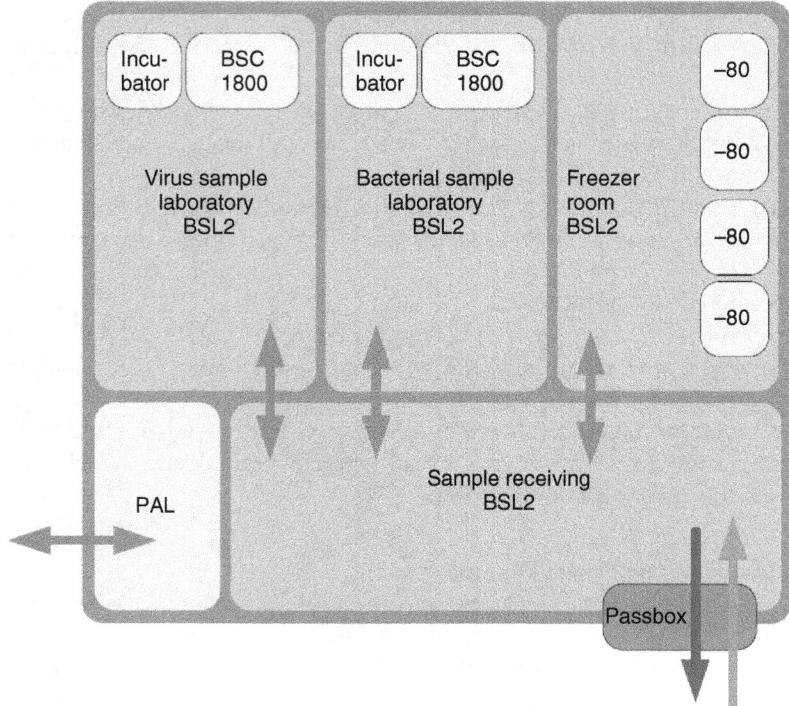

Figure 8.13 Room typology example—illustration of the first step of typology development. [NNE.]

After customer and user alignment of the principles of the diagrammatic typology, it is possible to lay out different typologies in a suggested diagrammatic patchwork of an overall facility functional relation and flow diagram, with the opportunity to discuss overall flows and functional relations.

This first step of diagrammatic assessment of laboratory typologies can support dialogue and alignment with customer and users. The strength of the diagrammatic form is a focus on the overall level of functionalities, their interaction, flow, and relations, rather than a more detailed direct concept focus on, for example door swings and exact sizing of equipment and rooms.

Second Step of Typology Development When the first typology step with customer and users is aligned, the second step of laboratory typology development has a focus on area need evaluation for the "Lego" blocks of functions, using the first step as basis. For an initial analysis of area need, the assumed laboratory work activities described in OFDs and function/relation diagrams [see Sections 8.4.2.4 and 8.4.2.5], the number of operators, the number and sizes of

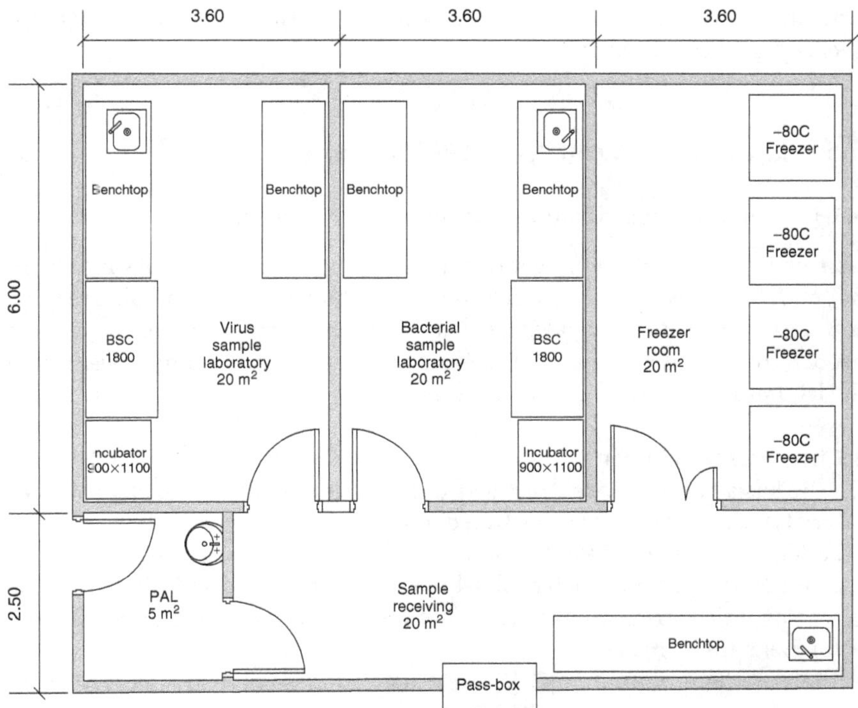

Figure 8.14 Room typology examples—illustration of the second step of typology development. [NNE.]

main laboratory equipment, and the need for laboratory bench space should be used. For the area need analysis in typologies, it is not yet essential to have a defined laboratory module as the result of the analysis evaluation will indicate an appropriate module that will be assessed further in the subsequent concept analysis (see Section 8.5.1.3).

In the more detailed laboratory typology diagram, the principle location of doors may be evaluated related to flows and location of equipment and logic work flow sequences and operations that may be sensitive to traffic or disturbance of airflows [e.g. Biosafety Cabinets (BSCs), flow hoods, and weighing equipment] (see an example of a second step laboratory room typology in Fig. 8.14). Several design manuals [one of them being (National Institute of Health) NIH Design requirement Manual [9]] provide guidance to placement and distance recommendations for BSCs and fume hoods to minimize impact on operations and equipment functions (see examples of distance guidance in Figs. 8.21 and 8.22).

The laboratory typologies provide input to the area need overview chart and create basis for overall concept development. In addition, the typologies may support the development of initial equipment lists of main equipment for overview and cost estimation.

8.5 Laboratory Concept Development

8.5.1 Planning Considerations for Laboratory Concepts

Converting data such as project drivers, workflows, and typologies into laboratory concepts will often require an understanding of, for example area distribution ratios and the tendencies within the industry—in other words, it is needed to look into the future and evaluate what will be assumed required in the laboratory space in the years to come.

8.5.1.1 Area Distribution

On the basis of experience, area distributions for a traditional analytical laboratory typically would be as illustrated in Fig. 8.15.

Looking at industry trends for QC laboratories as described in Section 8.5.1.5, the needs for general analytical laboratory areas are decreasing, and the space for office/write-up and speciality laboratories with highly automated equipment are increasing.

The paradigm shift in QC laboratories requires an increased focus on changeability within the facility design where we, as consultants, address the possibility to change function from, for example analytical laboratory to equipment or noise laboratory to ensure the desired life time of the facility. This can only be done by addressing the need for changeability within

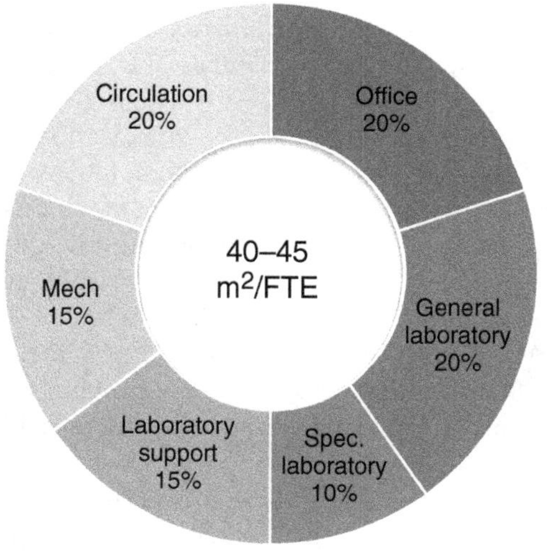

Figure 8.15 QC laboratory area distribution, gross square meter. [NNE.]

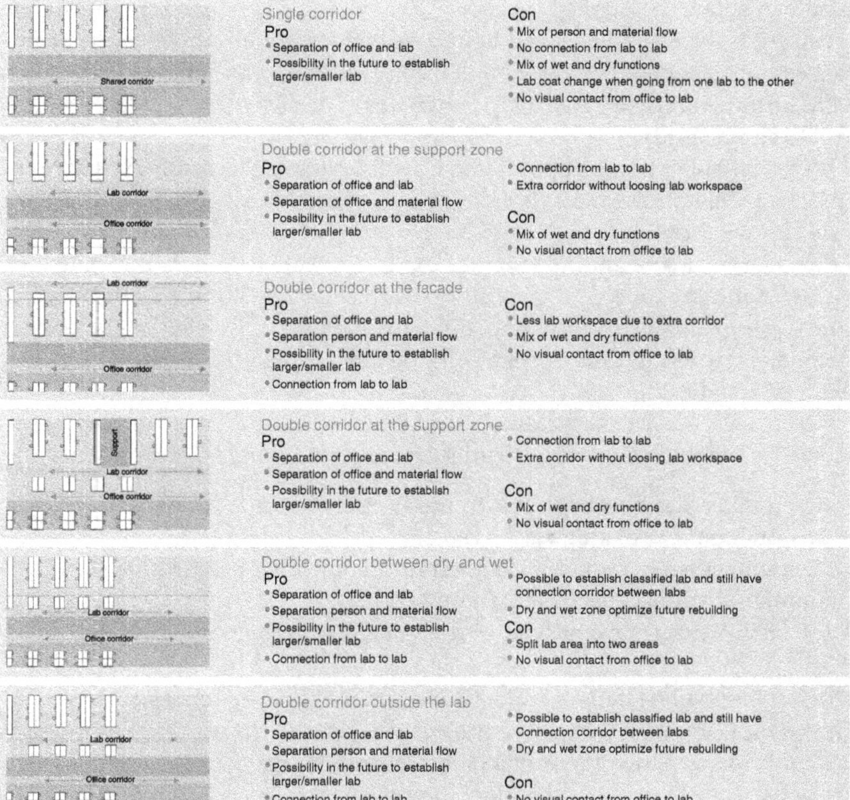

Figure 8.15 Laboratory concept evaluation example. [NNE.]

primary facility structure, main routings, and mechanical concepts early in the process.

8.5.1.2 Laboratory Concepts

Concepts for laboratories need to support the vision and mission of the individual project, taking into account what is possible within the available space, plot of land, or existing building. It is important to evaluate multiple concepts fit for purpose before concept lock (Fig. 8.16).

The abovementioned concepts are only used for illustration purpose; the overviews are not conclusive and do not cover all facility or function scenarios.

8.5.1.3 Capacity Considerations

During the early planning of laboratories, ~32 sqm (net) per FTE (full-time employee) is commonly accepted as a good "rule of thumb" [10]; this number can be used as an initial planning number, but going forward a more fact-based approach should be applied to calculate the specific area need. For R&D laboratories, it would be beneficial to look at the personnel split between principals/senior scientists, laboratory technicians, and support staff in combination with the number of expected drug candidates to be developed and the expected staff to support this.

When planning new or optimizing QC laboratories, an approach where the number of analytical samples is defining the needed future capacity is preferred. This approach requires the use of benchmark numbers for test times and/or a thorough dialogue and mapping of the individual customers time spent on each test and the time each laboratory technician spend in the laboratory and other tasks. Example of company standards indicates that ~35% of laboratory technicians working time is spent on other work than test/analysis activities (Figs. 8.17 and 8.18).

For *in vivo* R&D trial facilities, the following critical capacity parameters should be considered as part of initial capacity planning:

- the number and type of animal models planned for the facility
- the total number of animals
- the caging system required from a research and SHE perspective
- number of animals per cage/penning [11]
- the facility type: conventional, barrier, biocontainment, or SPF

8.5.1.4 Translating Strategic Project Drivers into Laboratory Concepts

The translation of the strategic project drivers and vision from the OPS is the conversion of the requirements into overall design concept elements. The project-specific drivers and vision should be used continuously during the planning process and concept development or as minimum be revisited to align the concept development with the strategy and design drivers.

In the translation process from OPS to laboratory concepts, some of the typical considerations are as follows:

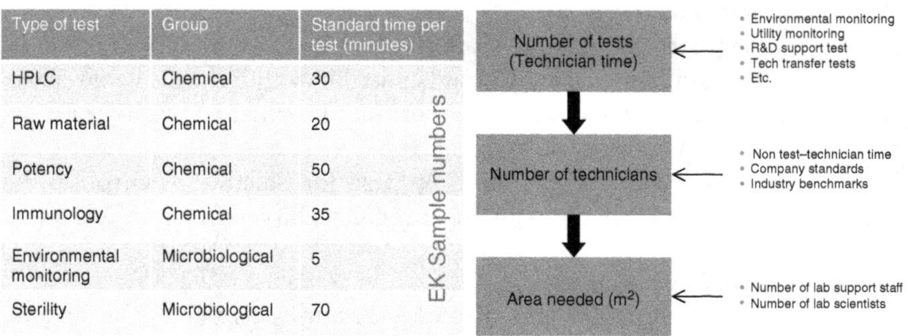

Type of test	Group	Standard time per test (minutes)
HPLC	Chemical	30
Raw material	Chemical	20
Potency	Chemical	50
Immunology	Chemical	35
Environmental monitoring	Microbiological	5
Sterility	Microbiological	70

Figure 8.17 Using standard QC test time as basis for capacity planning. [NNE.]

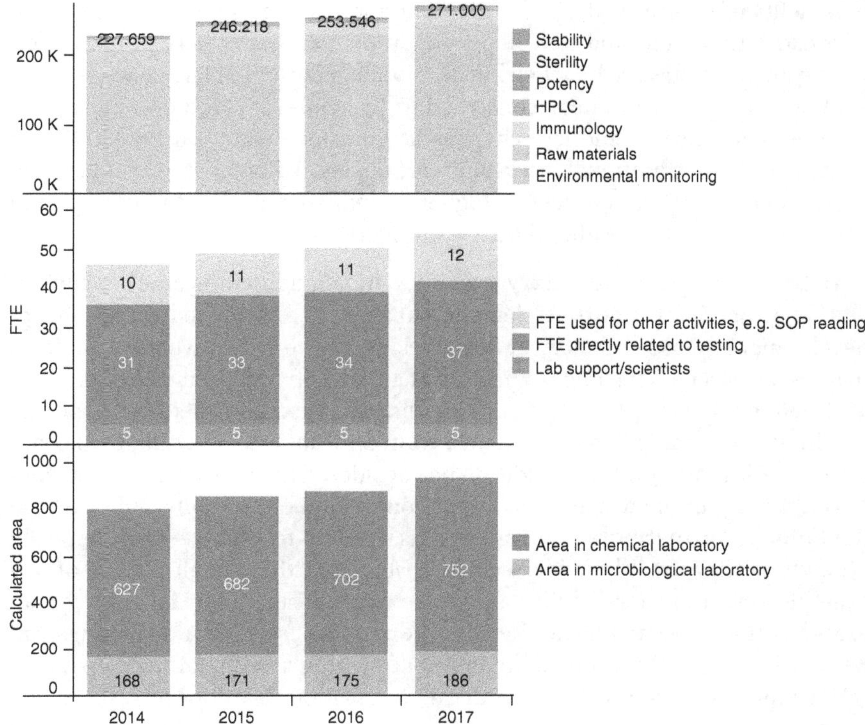

Figure 8.18 Realized capacity numbers QC laboratory (confidential customer). [NNE.]

8.5.1.5 Generic Versus Tailor-Made/Specialized Laboratory Concepts

- Evaluation of the benefits and disadvantages related to the overall project drivers such as flexibility and/or agility.
- Benefits of tailor-made/specialized laboratory concepts include optimal fit for current and known requirement situation and equipment, whereas benefits for generic concepts include a high level of adaptability for future changes and technology developments.
- Disadvantages for tailor-made concepts include less flexibility for adoption of future/unknown needs including technology development and equipment changes, whereas generic concepts may have to be compromised related to adoption of very specialized equipment.

Modular Versus Stick-Built Construction Principles

- Evaluation of the benefits and disadvantages related to modular laboratory systems versus stick-built construction, compared to project drivers such as time schedule, modularity, and/or future expansion drivers.

- Benefits of modular laboratory construction principles include less on-site construction time and less dependency on weather, where construction on-site principles are beneficial in relation to adaption of late changes.
- Disadvantages of modular construction principles include decisions need to be made earlier, and late changes have major impact on time and cost compared to traditional stick-built principles, whereas disadvantages of stick-built construction have a higher dependency on local craftsmanship competencies and weather during construction.

Mobile versus fixed laboratory inventory and equipment is another consideration, which is an important element that should be addressed in the concept development phase. The driver for considering mobile inventory systems may be a driver of flexibility and fast changeability. Here, it is key to have well-defined and measurable design drivers, for example what is the specific maximum time taken for a full change from one laboratory setting to another setting? With this, it must in addition be considered how the changeability and flexibility requirement may have impact on equipment options and potential limitations. Furthermore, the operations related to change over between different settings will be addressed to ensure that the overall drivers of, for example operational reliability are not compromised (see Fig. 8.19 for an example of a translated flexibility design driver, where a laboratory space can be transformed into an animal holding space within a defined timeframe).

Working with the translation of drivers into laboratory concepts, it is again key to use a holistic approach as the complexity is typically high, and design drivers cover all disciplines involved.

8.5.1.6 Typical Objectives for Laboratory Types (R&D, QC)

Good laboratory design should always enable the best possible functionality for the intended laboratory function; here, it is of outmost importance to understand the difference between laboratory types and functions and the related objectives (Fig. 8.20).

Typical objectives could be as follows:

- *Research*: The typical objectives are to foster creativity, and to allow for a high level of flexibility within the laboratory areas, research activities typically consist of both *in vivo* and *in vitro* activities. Creativity and innovation in research is often driven by a combination of spaces for informal interdisciplinary interaction and areas for quiet contemplation and concentration as specified in Section 8.4.2.2.

  The need for flexibility within the laboratory or animal holding space within research is primarily driven by the volatile nature of research, the irregular and unscheduled activities and the unknown change in methods, equipment, or animal models used for research. Research activities are typically not GxP regulated.

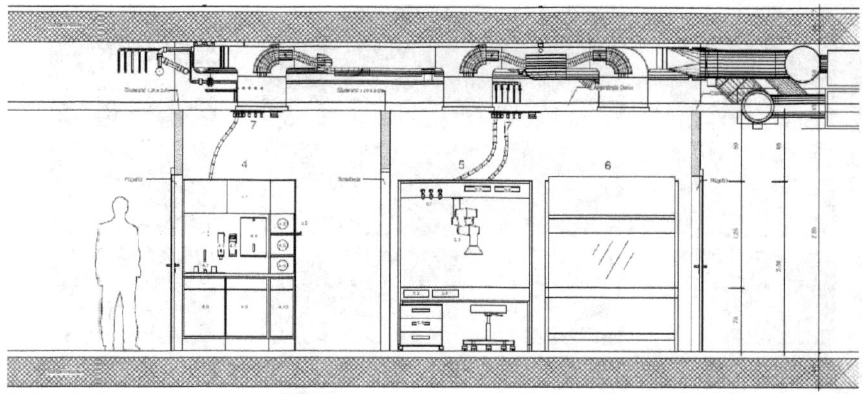

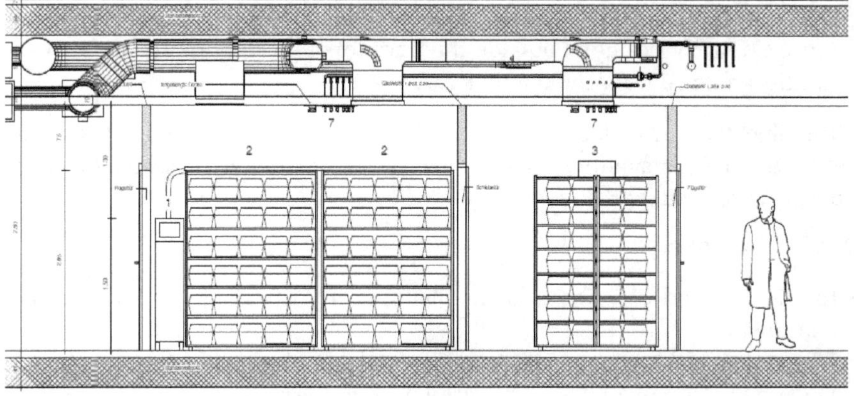

Figure 8.19 Example of a translated design driver of defined flexibility into a concept—with a requirement of changeability from laboratory setting to IVC animal holding (and vice versa) within 2–4 weeks. [NNE.]

- *Development*: Within development laboratories, we will see a need to support the irregular activities and the frequent changes in equipment at the same time we also need to support a higher degree of productivity and compliance as we are moving into the GMP-regulated area of activities. The overall objectives of development laboratories are to merge the creative space from research with the regulated constraints and productivity from GxP regulations and standards.
- *Quality control*: In QC laboratories, the dominant objective is to ensure full compliance and high productivity. The typical work operations are regular and planned owing to the close link to manufacturing throughout the entire product value chain (raw materials, intermediates, and final product release); furthermore, many tests and analyses performed are regulated within the

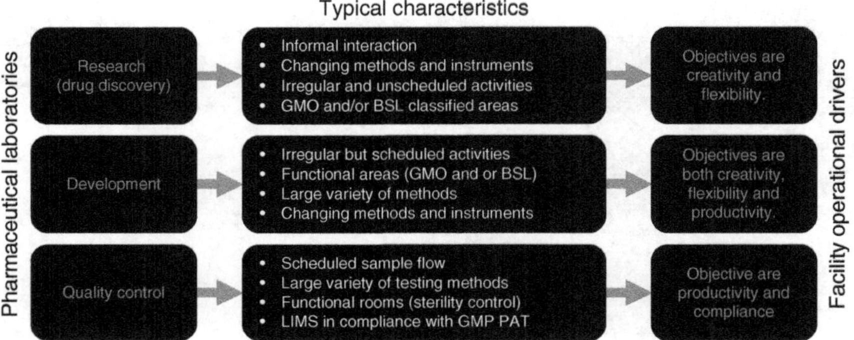

Figure 8.20 Laboratory objectives diagram. [NNE.]

individual product filings. Recent tendencies within QC show a move toward more QC activities done outside the traditional QC laboratory and at-line, in-line, and on-line at the manufacturing area.

The objective for collaborative space is different in QC laboratory settings compared to R&D facilities. The key drivers for collaborative spaces for these typically are as follows:

Quality Control:

- Interaction and collaboration for planning of critical process parameters (compliance)
- Discussion and exchange of information related to analyze and test results
- Documentation of analyses and test work processes

Research and Development:

- Interaction and collaboration for knowledge sharing and innovation
- Data handling/interpretation and understanding
- Planning of R&D activities
- Documentation of R&D results

8.5.1.7 Laboratory Planning Modules and Floor Height

Planning modules for laboratory facilities need to take into account both spaces for large equipment, fume hoods, and BSCs, ensure an efficient use of planned areas with high utilization per square meter, and at the same time secure a safe environment with space for safe operations within the laboratory.

Planning modules of 3300 mm would normally fit the purpose for most laboratories. Even 3300 mm can be a challenge to ensure a minimum distance of 1400 mm as normally recommended [12] if the laboratory operations require multiple fume hoods or BSCs (Fig. 8.21).

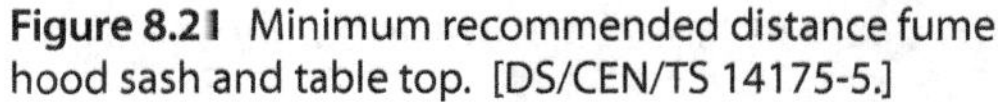

Figure 8.21 Minimum recommended distance fume hood sash and table top. [DS/CEN/TS 14175-5.]

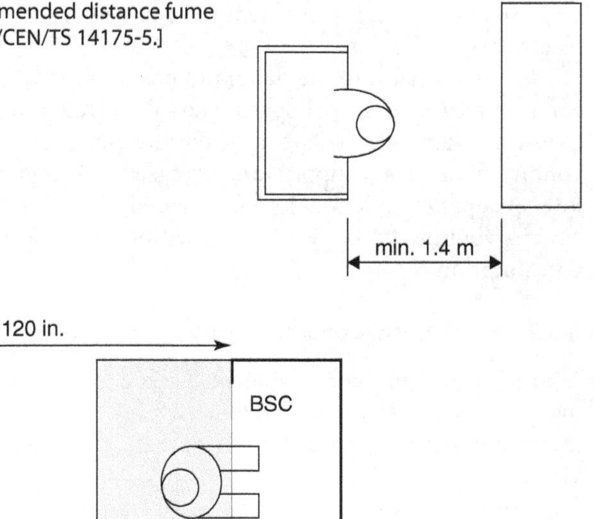

Figure 8.22 Recommendation for distance between facing BSCs. [NIH Design Requirement Manual, Appendix I: Biosafety Cabinet (BSC, Placement Requirements for new Buildings and Renovations).]

If the BSCs are facing sash to sash on opposite walls, it will require, as a minimum, a double planning module of 3300 mm and a laboratory layout that will use extensive space to be in compliance with, for example NIH Design Requirement Manual [9] recommendations and should therefor in general be considered avoided (Fig. 8.22).

An important precondition for optimal flexibility in the laboratory workspace is the room height and especially the space available for mechanical and electrical installations above-suspended ceiling or above the workspace if no suspended ceilings are installed. The mechanical space should allow for changes in existing routings and the possibility for new laboratory fit-out, for example

- Installation of new fume hoods, BSCs, or point exhausts without interference with main routings
- New utility tie-ins for new equipment
- Possibility for position of air diffusers below main ducting to allow for optimal laboratory layout without discomfort from the airflow

Recommendations for minimum total floor height are country specific and primarily driven by the size of main HVAC ducting and country-specific requirements for air change rates or air conditioning in the individual laboratory space. In general, a minimum space of 400 mm below main HVAC ducts to suspended ceiling is sufficient to ensure area for distribution of utilities and electrical distribution.

The height within the laboratory workspace, from floor to suspended ceiling or to mechanical space, should normally not be less than 2700 mm, if lower room heights are applied the possibility for discomfort from air conditioning or air supply are increased substantially. Support rooms with low occupancy rate could be lowered to 2500 mm to reduce the overall air volume to be ventilated within the room and to reduce energy consumption.

8.5.2 Mechanical Considerations

If possible, within the available facility concept, main supplies of heating, cooling, and potable water should be located in corridor zones to allow for service and maintenance, as these typically are the systems requiring most regular service.

Laboratory gasses, vacuum, clean air, and instrument air would normally be favorable to position in the mechanical space above the work area to facilitate short routings and easy access if new user points are required (see example in Figs. 8.1 and 8.23).

To ensure that the mechanical systems support the highest possible flexibility, it is beneficial to distribute primary laboratory utilities to each of the laboratory areas in the facility and then terminate them above ceiling for later access and distribution if needed; furthermore, it is normally money well spent to allow space for additional utilities in risers and main pipe racks.

As part of the mechanical considerations, it should be addressed in the programming if the new of refurbished laboratory facility should be able to

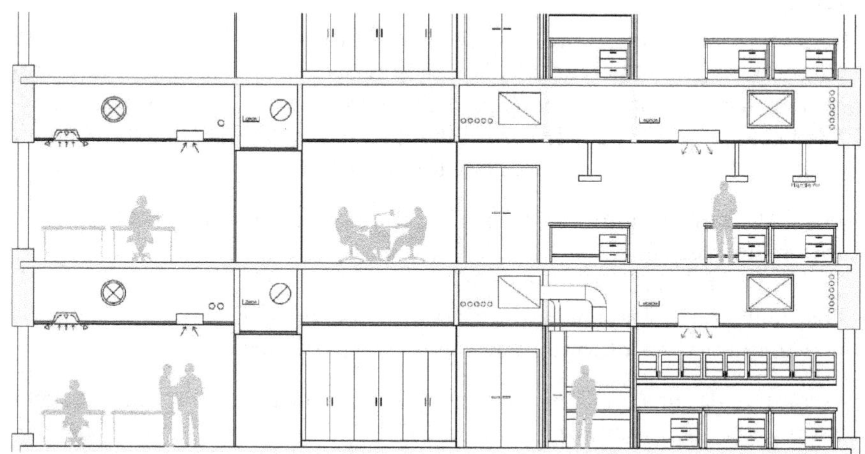

Figure 8.23 Section in laboratory building. [NNE.]

Figure 8.24 SHE areas to be considered during project initiation. [NNE.]

be decontaminated, for example between specific activities such as change of research programs, spill, or an infection situation. Depending on the technology and fumigation media used, it may impact the mechanical principles both regarding the HVAC system and the quality of the exposed components. Furthermore, room decontamination will have major impact on the architectural surfaces and building system configuration including defined tightness criteria. Finally, the SHE considerations are key to address in connection with a room decontamination strategy (see Fig. 8.24 for key SHE topics to be assessed in a project).

8.6 SHE Considerations

As an integrated part of laboratory planning and development of concepts, SHE aspects should be addressed in the earliest phase of every project.

An initial SHE screening and assessment that identifies the major SHE impacts to consider in the project going forward and to ensure that critical aspects are highlighted from the very beginning of a project (see Fig. 8.24 for SHE key areas to be assessed).

The SHE assessment will also form the basis for developing an initial SHE strategy to set the specific targets for the SHE level for the project, should it be basic level including compliance or should it include more sustainability measures, for example sustainability certification.

The SHE targets should be agreed in close cooperation with customer SHE representatives and management. If the planned laboratory operations involve handling of biological microorganisms, a comprehensive biorisk assessment will be needed [refer Section 8.4.2.3].

8.7 Glossary

Biorisk	Combination of the probability of occurrence of harm and the severity of that harm where the source of harm is a biological agent or toxin. The source of harm may be an unintentional exposure, accidental release or loss, theft, misuse, diversion, unauthorized access, or intentional unauthorized release (adapted from ISO/IEC Guide 51:1999)
BMBL	Biosafety in Microbiological and Biomedical Laboratories (US Department of Health and Human Services)
Hazard	Source, situation, or act with a potential for causing harm (adapted from OHSAS 18001:2007)
In vivo	Studies that are tested on whole, living organisms, usually animals, including humans, and plants as opposed to a partial or dead organism
In vitro	Studies that are performed with microorganisms, cells, or biological molecules outside their normal biological context.

8.8 List of Abbreviations

API	active product ingredients
BMS	building management system
BSCs	Biosafety Cabinet
BSL	biosafety level
CWA	CEN Workshop agreement
FMS	facility management system
FTE	full-time employee
GMO	genetically modified organisms
GMP	good manufacturing practice
GxP	good (anything) practice
HVAC	heating, ventilation, and air conditioning
NIH	National Institute of Health
OCM	organizational change management
OFD	operational flow diagram
PPE	personal protection equipment
QC	quality control
R&D	research and development
ROI	real operator intake
SHE	safety, Health, and Environment
TPC	total project cost
WHO	World Health Organization

References

1 ISPE. Good Practice Guide: Quality Laboratory Facilities. ISBN 9781936379422.
2 RAMM tool described in ISPE magazine Pharmaceutical Engineering, January/February 2012; 32(1).
3 International Council for Harmonisation of Technical Requirements for Pharmaceuticals for Human Use. http://www.ich.org/products/guidelines/quality/quality-single/article/quality-risk-management.html
4 BMBL section ii, 5th ed.; December 2009.
5 WHO. Laboratory Biosafety Manual, 3rd ed.; 2004, Chapter 2.
6 CEN Workshop Agreement. CWA 15793 Laboratory Biorisk Management—and CWA 16393 Guidelines for the Implementation of CWA 15793.
7 ISPE Baseline Guide. Volume 7: Risk-Based Manufacture of Pharmaceutical Products.
8 ISPE Baseline Guide. Volume 1: Active Pharmaceutical Ingredients.
9 NIH Design Requirement Manual. Appendix I: Biosafety Cabinet (BSC, Placement Requirements for new Buildings and Renovations). https://www.orf.od.nih.gov/PoliciesAndGuidelines/BiomedicalandAnimalResearchFacilitiesDesignPoliciesandGuidelines/Pages/DesignRequirementsManualPDF.aspx
10 NIH. Design Requirements Manual Section 2–3.
11 National Research Council (US) Committee for the Update of the Guide for the Care and Use of Laboratory Animals. *Guide for the care and user of laboratory animals*. 8th ed. Washington: National Academies Press; 2011.
12 DS/CEN/TS 14175 Fume Cupboards—Part 5: Recommendations for installation and maintenance.

Chapter 9

Case Study: Pharmaceutical Pilot Plant Design and Operation

Beth H. Junker

Bioprocess R&D, Merck Research Laboratories, Rahway, New Jersey, USA

9.1 Introduction

Several diverse roles, objectives, and purposes are associated with a bioprocessing, biochemical, or biopharmaceutical pilot plant. These include, but are not limited to, the manufacture of early- and late-phase clinical supplies, process development (including process characterization, validation, and comparability [1]) for new products in the company pipeline, process improvement including technical support for existing products in the market, and production of biologically produced materials (which are not intended for the clinic) to support basic research efforts such as screening, assays, and disease mechanism elucidation. Despite desires to design pilot facilities for diverse differences in scale, products, processes, clinical phases, and capacity, limited capital and operating budgets force value engineering decisions about "must have" and "nice to have" capabilities [2]. Unlike chemical pilot plants, pharmaceutical pilot plants are used not only to scale-up laboratory processes, but also to produce developmental quantities of chemicals (and biological substances) for safety, toxicological, and clinical studies [3]. Thus, balancing of the demands for scale-up and development research with those for clinical production is a key dilemma for pilot plant operation [4]. This tension has resulted in an increased recent attention to the development of scale-down models to mimic larger scale performance and thus reduce demands for pilot scale process development runs [5]. Another challenge is to determine what proportion of the material requirements for each phase of a product's development has to be outsourced to contract manufacturing organizations (CMOs). Some CMOs are even targeting clinical-scale production clients, essentially building pilot plants for hire [6]. The timing and reliability of clinical material production is critical

Process Architecture in Biomanufacturing Facility Design, edited by Jeffery Odum and Michael C. Flickinger.

as expensive and time-consuming clinical programs can be delayed, thus postponing product application filings, approvals, and launches [7]. Although traditionally pilot plants have not manufactured material for sale, accelerated launch dates, deferred manufacturing facility capital commitments, and other areas of project uncertainty have prompted consideration and use of pilot plants for manufacturing in recent years. In addition, rising fermentation titers, isolation yields, and overall process streamlining has expanded the capabilities of smaller sized facilities to satisfy larger kilogram material demands [5].

The types of products manufactured using biologically based methodologies include primary metabolites, secondary metabolites (ranging from pharmaceuticals to biopolymers), constitutive proteins, recombinant proteins, intracellular/extracellular enzymes and biocatalysts, antigens, viruses, polysaccharides, and plasmid DNA itself. These products are made using a wide range of cell cultivation and isolation techniques suitable for bacteria, yeast, fungi, actinomycetes, suspension animal cells, anchorage dependent animal cells, insect cells, and plant cells. Processing equipment considerably varies among these diverse systems.

Designing for extensive processing flexibility can be expensive, potentially in excess of $100 million, depending on the specific process requirements, size, and intended use of the pilot plant. In addition, processing requirements can change drastically as products are de-emphasized and re-emphasized in development. These decisions on priorities often are not usually caused by process robustness but rather by the state of the clinical data or potential market share. The ability to process multiple products in an acceptable manner is typically desirable to maximize facility use. Although designers may need to predict future types of processes and products, they should be cautious about over designing the facility in terms of containment and flexibility [8]. Designing for future expansion with minimal disruption to operations can help in reducing initial capital expenditures [9].

The general areas required in a pilot plant might include support labs for bench scale development and quality control, production suites for scale-up, utility areas, personnel offices, as well as ample storage areas for raw materials, supplies, product, samples, and documentation [4]. There should be a clear division between controlled and noncontrolled areas utilizing signs at a minimum and perhaps by physically restricting access. General design criteria might focus on modularity, simplicity, future growth provisions, redundancy, construction costs, operating costs, and air/water waste containment [10]. Future operating costs are important considerations for pilot plants as ample funds are necessary for raw materials, staffing, maintenance, and additional capital modifications to permit improvements.

9.2 Operational Concepts and Processing Requirements

Several operational concepts influencing design requirements need to be agreed upon prior to the initiation of design activities. Agreement should be sought among end users, operations staff, compliance auditors, environmental consultants, safety officers, as well as appropriate levels of management who are directly or indirectly influencing or controlling capital expenditures. A written document summarizing consensus on these issues may then become the basis for conceptual design activities.

Major decisions influencing pilot plant capabilities include whether the facility will produce bulk drugs (thus falling under the regulation of Center for Drug Evaluation and Research (CDER) and following guidelines from 21CFR 210/211) or biologics (thus falling under the regulation of Center for Biologics Evaluation and Research (CBER) and following guidelines from 21CFR 210/211 and 600) [11]. The level of validation and the validation philosophy are directly related to the end use of the resulting pilot plant products, specifically whether they are used for process development, clinical supplies (presafety/basic research, safety, phase I, phase IIa, IIB, phase III), or material for sale. Validation strategies, as well as other operational and design issues, are also influenced by whether the pilot plant is administratively located within the manufacturing or research division. Individual company preferences and guidelines for similarity between manufacturing and pilot plant areas need to be identified and evaluated for their impact on the proposed pilot plant facility. Typical areas targeted for standardization within a company might include vessel designs, computer systems and interfaces, equipment vendors, and construction specifications.

Design requirements center on the desired initial and end product stages of processing, specifically the forms of the pilot plant inputs and outputs, as well as in-process hold steps. Example stages include master/working seed preparation, fermentation broth, initial captured product (e.g. microfiltration retentate, ultrafiltration concentrate, cell paste), crude product (e.g. rich ion-exchange fraction), finished isolated/sterile bulk, and filled vials. Ambient, refrigerated, and frozen storage requirements based on these inputs/outputs then need to be addressed in the areas of seed, raw materials, intermediates, clinical supplies, and process samples.

The scale of the pilot plant relative to current or proposed manufacturing scale and relative to laboratory scale needs to be established. Generally, a 1:5 or 1:10 scale-up is common [1], but smaller ratios can be advantageous in terms of minimizing risk during factory start-up. The appropriate level of biosafety also must be considered with GLSP being sufficient for most recombinant work. Some pilot plant areas may require BL1-LS and/or BL2-LS, and careful consideration of multiple use/decon issues for BL1 and BL2 containing facilities is necessary. Solvent-use areas need to be identified and minimized if possible as cost implications for explosion proof design can be substantial.

Primary motivations to design a multiuse pilot plant are the conservation of capital and improved response time for clinical manufacturing requirements for new product development. The strategy for multiuse should be identified and considered throughout the design process. Common strategies [12] include (i) using totally dedicated equipment and limiting production to one product at one particular production stage, (ii) using campaigned equipment in which multiple products are sequentially processed in the same equipment "campaign style," with documented product changeover procedures completed to minimize cross-contamination potential, and (iii) conducting concurrent manufacturing in dedicated equipment in which several products are simultaneously produced in segregated areas with segregation achieved via physical separation or closed systems. These strategies result in the minimization of cross-contamination potential through engineering controls, SOPs, and/or temporal segregation [13]. Typical pilot plants combine the second and third strategies by providing segregated and dedicated areas for certain types of processing (e.g. virus and nonvirus areas, GLSP versus BL1/BL2 areas, beta-lactam antibiotic processing, heat-resistant spore former cultivation) [14], and maintaining flexible areas that can be reconfigured efficiently to suit their multiproduct portfolio needs. Example layouts for product-specific and common areas depending upon the strategy selected have been described [15].

To increase the speed and number of products entering clinical trials, platform processes are becoming increasingly more common to achieve high throughput for early-phase products, avoiding large amounts of individualized process development effort. Platform processes are based on classes of products with similar processing requirements and/or properties. Less optimized processes (e.g., lower yields, longer cycle times, higher cost) are accepted for early material deliveries (i.e. prior to Phase IIB) in exchange for reduced resource requirements. A platform for cell cultivation for monoclonal antibody production has been developed by IDEC based on their own in-house medium and feed formulations [16]. In addition, a platform for monoclonal antibody downstream processing based on Protein A capture has been described by Amgen in which a flexible, generic process is used, which predefines most operating parameters and requires only a small subset of other operating parameters to be defined via product-specific process development [17–19].

Guidelines concerning the use of open versus closed systems for various processing stages need to be devised and justified based on available and cost-effective technologies as well as quality requirements. In a closed system, the product is not exposed to the immediate environment. The quality of materials entering a closed system (such as water, steam, and air entering a bioreactor) is controlled by its quality as well as the manner in which these materials are entering the closed system (such as filter sterilization, autoclaving, and/or steam sterilization of system connections) [12]. In an open system, the product is directly exposed to the surrounding environment, thus necessitating the creation of a controlled or sometimes even a closed system around the open system. One example of this latter strategy is the transfer of seed vial contents in a biosafety cabinet. Other examples might include the supply of higher quality filtered area to areas for the cool down of autoclaved materials and assembly of clean equipment. Thus, levels of containment can be constructed until all open systems are contained within controlled or closed systems to meet quality requirements for the specific processing stage.

The specific issue of potential exposure of product to spore-forming organisms has been extensively debated. The closed system concept appears to minimize the concern for airborne spores that are typically found in the environment. Cultivation guidelines for spore-forming organisms within a multiuse facility appear to have been redefined to evaluate the robustness of the spore type formed. The careful consideration of segregation has been given when cultivating organisms that form spores that are difficult to be killed, such as *Bacillus*. A separate, self-contained area with no shared equipment, air systems, or entrances minimizes exposure concerns for other products. The relatively heat-sensitive, substantially larger, fungal spores have not been highlighted as a substantial cross-contamination concern. Thus, segregation of these types of processes is not as critical, though it still might be implemented due to company preferences for separate bulk drug and biologics processing areas.

9.3 Design

The initial generation of flow charts for a bioprocessing, fermentation, and/or isolation pilot plant is best conducted based on model processes that are developed and critiqued by several end users of the equipment and facility. These model processes are carefully selected so that they give adequate and appropriate representation to current and reasonable future needs of the pilot plant. Specifically, current and future projections for fermentation titers should be modeled to establish the potential operating range for downstream steps. Because model processes form a preliminary but sound basis for equipment and utility sizing, they should include specific unit operations, equipment scales, and expected cycle times. In some cases, simulation [17,20–23] and/or actual laboratory/pilot scale testing [24] of critical equipment under typical processing conditions or with expected process streams may be warranted to further define design specifications. Time and effort expended to identify and change the design on paper is more efficient (i.e. causing fewer delays) and less expensive than modifications during construction or after start-up [25,26].

The proper level of criticality must be assigned to each system; attention to too many critical systems results in unnecessary expenses and too few critical systems results in design changes, delivery delays, and validation challenges [25]. Example model processes and throughput evaluations have been published for vaccines, antibodies, and other intracellular/extracellular products [26–31]. Resulting flow charts can focus on raw materials for fermentation/isolation, seed development, fermentation, recovery, crude product isolation, final product isolation, product finishing, buffers, utilities, and waste (including chemical waste, biological waste and any necessary off-gas treatment) [27].

The requirements for these model processes, coupled with an estimate of the required facility output, can be used to construct a preliminary architectural layout that divides the facility into sections (commonly termed as *modules, cubes, or suites*). These sections are surrounded by controlled corridors and accessed via airlocks [8]. On the basis of this initial layout, flow patterns are developed to permit the logical, usually unidirectional, flow of personnel, raw materials, product, waste, and clean/used equipment throughout the facility. Flows are primarily designed to minimize cross contamination from multiple products and from used equipment while maintaining access and flexibility [8]. This separation is often accomplished using clean and return corridors located on either side of the processing area as well as via temporal segregation using appropriate SOPs. Evaluation of these flow patterns should be conducted carefully and incorporate end users as well as quality auditors. In some cases, companies may solicit outside GMP reviews of the project at this time from external consultants, a second design firm, or even FDA representatives. Gathering and organizing up-to-date, accurate information and comments about the design's impact on model process requirements are critical factors in forming a successful basis of design [32].

Example layouts for process flows for biotech facilities have been published [6,12,33,34]. An extensive reference (including piping and instrument diagrams (P&IDs), layouts, and flow diagrams) for the design of biopharmaceutical plant equipment, utilities, heating, ventilation, and airconditioning (HVAC), and waste treatment also has been published [35]. Although it was based on acceptable practices in 1991, the strategies and concepts for design presented are still relevant. Adequate space around the equipment needs to be reserved for operations and maintenance access as well as for process-related portable equipment. Specifically, the movement, segregation, storage, and cleaning of portable equipment such as vessels, skids, hoses and carts should be fully examined.

Various reports of successful and problematic aspects of facility design, construction, start-up, and validation have been published. Often, cost estimation for pilot plants is more difficult than that for manufacturing plants owing to incomplete understanding of processes, equipment, and technology

[36]. Pressure to meet initial schedules and budgets can result in higher initial and subsequent costs, particularly for retrofitting projects. Cost control issues and the benefits of formulating a master schedule [37] need to be reviewed in terms of their potential validation impact. Construction concerns and "lessons learned" from multiple biotech installations also have been summarized [38]. One case study of a laboratory facility highlights the need for teamwork among the client, design firm, and construction manager [39]. This trio can be extended to include validation contractor in an effort to minimize "finger pointing" and "blame storming" when unforeseen problems are uncovered. Regardless of who may actually be perceived as being "at fault," the company owning the facility ultimately is adversely affected by substantial start-up and validation delays. A comprehensive overview of biochemical pilot plant design and operational guidelines appeared in the mid-1980s [40]. A more recent review focuses on best practices for design and operation to reduce bioreactor contamination, a key type of abnormality typically resulting in batch termination [41].

9.3.1 Process Equipment

For a pilot plant, the major decision regarding process equipment is that of similarity versus diversity as it applies to multiple process steps as well as multiple products. The extent of commonality required in equipment specifications is an important operational factor to be considered, as equipment specifications are devised based on the inputs from model processes. For a pilot plant to accept a manufacturing interface role, at least a portion of the equipment (and automation) design and selection should be representative of the factory that the pilot plant is intended to support [1]. For example, if the factory is intending to use continuously sterilized medium for large scale fermentation, then there should be provisions in the pilot plant to perform continuous sterilization as products get closer to their transfer to production. Alternatively, depending on individual company preferences for the pilot plant role, later clinical batches (Phase IIB and beyond) might be manufactured in the factory to facilitate eventual process transfer upon commercialization and ensure consistency of equipment.

Skid-mounted or packaged equipment is a common selection in recent pilot plant installations because it puts the design, delivery, and start-up burden entirely on one vendor [25], who specifies the appropriate quality, supply rate, and/or pressure of the required utilities (e.g. steam, cooling water, instrument air). Skid-mounted equipment can be fixed or portable, the latter permitting different types of processing to occur in the same space as necessary. Several skid-mounted unit operations can be mounted together to create "super skids," further streamlining construction, start-up and validation, and placing even more responsibility on the vendor [9]. Customization of skid designs with respect to components and spare parts may be a drawback as vendors might not be willing to build skids to individual company specifications without added costs. This situation can be particularly problematic if the skid manufacturer has arranged a low cost deal with a vendor (e.g. diaphragm valves) or developed a programmed software on a PLC that may not be the vendor of choice for the rest of the facility. There also may be incompatibilities in piping, utility requirements, or software when the skid is connected [42] if miscommunications occurred between the skid vendor and the facility designer. Typically, a team, consisting of representatives from the vendor, facility owner, facility user, validator, procurement, engineering/construction management firms, and facility maintenance organization, is assembled.

The traditional alternative to skid-mounted designs is to request the facility designer to purchase the equipment components separately and assemble the piping and equipment on site [25]. One advantage to this "stick build" approach is that total flexibility in equipment selection now rests with the pilot plant designers who may be quite confident in their ability to design the equipment based on their extensive prior experience. Drawbacks of this approach include placement of the design burden on the facility as well as added design costs for one-of-a-kind systems. Specific references for fermentation equipment flow charts, detailed design, and qualification are available [43–46] as well as for filtration units [47], chromatographic equipment [33,48–50] and centrifuges [51,52]. A list of typical process, utility and support equipment that might be required in a bioprocessing pilot plant has been compiled (Table 9.1).

A third alternative for pilot plants is the use of disposable, single-use equipment now available for solution preparation and storage, bioreactors, filtration, and chromatography [53,54]. Several authors describe how disposables reduce fixed capital costs (often at the expense of variable operating costs), operational effort (recleaning and often sterilization are not needed which decreases turnaround time and effort) as well as space and utility requirements [2,53,55–57]. Specific analysis tools for stainless versus disposable equipment selection for clinical material preparation based on manufacturing strategy have been developed [58,59]. In fact, one company has designed a biopharmaceutical pilot plant based on disposable systems to maximize flexibility and speed of cell line process development and early-phase clinical material supply [60].

For any type of bioprocessing facility, certain aspects of equipment and piping specifications should be uniform throughout the installation where possible; this goal is particularly challenging for a pilot plant containing a large variety of equipment from multiple vendors. These specifications might include requirements for orbital welding of lines, documented weld inspections, passivation, pressure testing, self draining (sloped) piping, minimal lengths for dead legs and pockets, restricted use of flanged connections, level

Table 9.1 Types of Processing Equipment and Utilities Possibly Required for a Bioprocessing Pilot Plant

Fermentation and Harvest	Isolation	Utilities	Support
Culture storage freezer	Low pressure chromatography	Clean steam generator	Sterilizing autoclave
Incubator/shaker	High pressure chromatography	USP-purified water system	Decontamination autoclave
Biosafety cabinet	Membrane chromatograhy	WFI system	Depyrogenation oven
Laminar flow hood	Solvent blending and delivery	Deionized water	Laminar flow hoods for cool down and equip. assembly
Seed fermenter	Viral filter	Process (city) water	Glasswasher (glassware, vial, stopper)
Production fermenter	Sterile filter	Compressed clean air	Buffer preparation tank
Nutrient feed tank	Laminar flow hood	Compressed instrument air	Media preparation tank
Harvest tank	Liquid transfer pumps	Compressed plant air	In-line mixer
Microfilter	Dispensing machines	Compressed gases (O_2, N_2, CO_2, NH_3)	In-process sample refrigerator/freezer
Ultrafilter	Chemical fume hood	Electricity	Raw material cold/frozen storage
Centrifuge	Portable tanks	Chilled water	Floor and bench scale
Depth filter	Laboratory scale centrifuge	Cooling tower water	Process and equipment monitoring/alarm system
Expanded bed absorption	Ultrafilter/ultracentrifuge	Plant steam	Uninterrupted power source (UPS)
Homogenizer	In-process refrigerator/cold room/freezer	Glycol	Off-gas analyzer (mass spectrometer)
Continuous sterilizer	Product refrigerator/freezer	HVAC and building automation system	Bar code system
Filter integrity tester	Syringe/vial filling equipment	CIP skid	
Sterile tubing welder	Lyophilizer	Biowaste inactivation system	
		Environmentally potent chemical destruction system	
		Dust collection system	

of polish on product-contact surfaces, stainless steel quality (typically 316 L for product-contact surfaces) and use of FDA-approved polymers in specific applications (gasket, o-rings, valve seats, and distribution piping) [61]. Proper storage of fabricated piping awaiting installation as well as prominent identification of valves and lines also must be considered. Valve specifications can be important to determine early in the design cost estimation phase [62]. Valve capital cost contribution is high due to the large numbers required and any installed valves (either diaphragm or ball) require subsequent maintenance. Consideration should be given as to whether live steam, set up as a steam block, should be used to continually purge the back side of product-contact valves and/or whether live steam should be used for tracing between double o-rings on a port or manway [43]. Although such designs traditionally have increased sterility assurance for lengthy secondary metabolite fermentations, they may create detrimental localized hot spots during animal cell cultivations conducted in the same pilot plant fermenters. Neither the equipment itself nor its disposable/replaceable parts should release any extractable substances (after the initial postinstallation rinsing and cleaning steps) into the product [43]. An overview of sanitary piping design, installation, and validation concerns is available [63].

Transfer piping strategy is another major design decision. One common strategy for several transfer lines is to meet at a transfer panel or process manifold, typically located at a high point between two or more areas, and be connected using spool or "jumper" pieces to move material in the desired fashion [8]. Steam entry is at the high point and condensate drainage at the low point for steam-in-place (SIP) during sterile transfers. Clean-in place (CIP) solutions are passed along the direction of the previous transfer. Advantages of this approach include reduced piping distribution costs and space requirements (which can multiply fast when multiple tanks and multiple types of process transfers are required). Disadvantages center on operational restrictions concerning the elapsed time between successive transfers through a common manifold system, as well as cross contamination associated with transfers of nonsterile media/buffer, sterile media/buffer, inoculum, harvested broth, CIP solutions, and/or water through common lines. Ring headers with automated or multiport valves are one recent alternative design being introduced to avoid "jumper" connections [53]. A listing of product and nonproduct transfer line/utility piping that might be required in a bioprocessing pilot plant has been compiled (Table 9.2).

Sterilization, sanitization, cleaning, and draining sequences need to be carefully defined and documented during the equipment and transfer piping specifications because retrofits postinstallation can be time consuming and expensive. P&IDs should be traced through these process sequences prior to approval by both operational and validation personnel. Requirements for validation testing and in-use monitoring of equipment operation also need to be

Table 9.2 Transfer Lines and Utility Piping Possibly Required for a Bioprocessing Pilot Plant

Product Contact	Nonproduct Contact
Utilities	**Utilities**
Clean water	Chilled/cooling water and/or glycol
Clean air	Instrument/plant air
Clean steam	Plant steam
CIP solutions	Contained sewer
Other compressed gasses (CO_2, O_2, N_2, NH_3)	Chemical (noncontained) sewer
Process	Sanitary sewer
Sterilized medium or buffer	Vacuum system
Sterilized nutrient/acid/base/antifoam feeding	Natural gas
Nonsterilized medium or buffer	
Inoculum	
Harvest	

identified during the equipment specification phase. Examples include sizing condensate lines to permit adequate drainage of condensate when spore strips are present, specification of additional validation ports for thermocouples on vessels and condensate lines, installation of product-contact utility sampling stations (especially at worst-case locations), and installation of temperature sensors on critical lines to monitor sterilization and/or proper trap operation.

9.3.2 Utilities

Similar to any bioprocessing facility, product-contact utilities in a pilot plant might include water, clean steam, compressed air and other gases, and CIP systems. Nonproduct-contact utilities generally include chilled water, instrument air, plant steam and plant air, although for some products plant steam may be appropriate for product contact based on its quality (Table 9.3). A listing

Table 9.3 Comparisons of Example Plant Steam and Clean Steam Achievable Quality

Quality Attribute	Clean Steam	Plant Steam
Bioburden	<0.1 cfu/mL	<0.1 cfu/mL
TOC	<0.5 ppm	<2.0 ppm
Endotoxin	<0.25 EU/mL	<0.5 EU/mL
pH	5–7	5–8
Conductivity	Meets USP 23 supplement 5 standards	N/A

of product and nonproduct-contact utilities is compiled in Table 9.2. Multiple utility use points are located within and among suites for both product and non-product utilities. These are organized into utility stations for each equipment skid. To minimize piping, it might seem convenient to use product-contact utilities in nonproduct-contact applications such as compressed clean air on a vessel jacket, relying on a check valve for back flow prevention. This approach can risk disaster because the operation and quality of product-contact utilities can affect processing throughout the entire facility.

Peak utility loads should be based on carefully investigated assumptions for anticipated pilot plant processing portfolios, which are transferred from design directly to validation testing. Although it may be probable to operate several large fermenters simultaneously each at their maximum air flow rate, it may be less likely that several large fermenters will be cooled simultaneously after completion of sterilization. Thus, air compressors might be sized assuming maximum load, while process chillers might be sized based on an operationally realistic fractional load. Assumptions about processing cycle times and fermenter working volumes are important particularly when sizing CIP and liquid waste decontamination systems. Shorter (50 h) *Escherichia coli* fermentations might require a smaller tank volume (5000 L) but are run more frequently, compared with longer (400 h) animal cell or fungal fermentations, which might require a larger tank volume (20,000 L) but are run less often.

9.3.2.1 Product Contact

The most critical product-contact utility for a bioprocessing facility often is water due to the sheer volume required during typical processing steps. It is well known that the quality of water needs to be appropriate for the intended application, but higher than the necessary water quality often is designed and system capacities underestimated, particularly when actual pilot plant processing portfolios exceed the design basis. In many installations, purified water is utilized for upstream processes and early CIP steps, and WFI is reserved for cultures that are sensitive to endotoxin such as animal cells and for downstream isolation steps [64]. A brief overview of various methods to produce pure water is available [65].

Water for injection (WFI) is often favored for the entire pilot plant facility for uniformity and to avoid mix-up. Specifications are set for parameters such as bioburden (typically 0.1 CFU/mL), endotoxin (typically 0.25 EU/mL), conductivity (1.3 μScm at 25°C), and total organic carbon (TOC) (<500 ppb). WFI systems containing stills are the most predominant and are generally designed as a hot 80°C loop [66]. Recently, 65°C loops have been sanctioned to reduce cooling costs and to permit the use of thermostable plastics in construction [67,68]. Point-of-use coolers provide water at a user-selected temperature. These coolers can form deadlegs and require substantial flushing prior to use, thus increasing waste water. It is also possible to design a cold or ambient WFI loop with

Table 9.4 Comparison of Example USP-Purified Water, Deionized Water, and Process Water Achievable Quality Attributes

Quality Attribute	USP-Purified Water	Deionized Water	Process (City) Water
Bioburden	<100 cfu/mL	<500 cfu/mL	<500 cfu/mL
Coliforms	Absent	Absent	Absent
TOC	<0.5 ppm	<2.5 ppm	<5 ppm
Endotoxin	<0.25 EU/mL (for info. only)	N/A	N/A
pH	5–7	5–8	5–8
Conductivity/resistivity	Meets USP 23 suppl. 5 standards	>1 Mohm	N/A

periodic (typically daily) heating to 80°C or ozonization for sanitization purposes.

USP-purified water is an attractive alternative to WFI for various applications that do not require WFI. Specifications include bioburden (typically 100 CFU/mL), conductivity (1.3 μScm at 25°C), and TOC (<500 ppb) [69]. Ion-exchange, distillation, and/or reverse osmosis (RO) systems are used [66]. Although such systems can be operated hot or cold, success with ambient systems that are sanitized with ozone or chemically [70] has been reported. Examples of the design and validation of two ambient USP-purified water systems for pilot plant use are available [70,71].

Other noncompendial grades of water also may be utilized. In these cases, quality attributes are typically set by the facility to address the requirements of specific processing steps (Table 9.4). Examples might be deionized water with bioburden (<500 cfu/mL) and resistivity (>1 Mohm) specifications or process water (municipal water passing through a break tank) with TOC (<10 ppm) or particulate (≤ disk #1 standard) specifications. Using the lowest quality of water appropriate for the quality of the process step reduces operating costs, assuming running multiple water systems in the facility is acceptable [68].

General design elements for water systems include the desired flowrate at each use point and the expected number of simultaneous use points in operation. These values are combined to obtain the total "take-off" flow rate; the system and piping are then sized such that velocity of remaining flow rate is sufficient to minimize bioburden and biofilm growth. Suggested methods for obtaining this residual velocity have been outlined [72]. Use point positioning should consider the height of the take-off valve, automation of take-off valves, sink or sewer location for flushing, as well as cleaning, sanitization, and acceptable noncoiled storage racks for associated use-point hoses. In addition, water systems should be designed with specific sanitization procedures identified at the outset. Steam, ozone, or chemical sanitization methods are most common, although steam sanitization is generally not recommended for plastic loop systems due to concerns about distribution pipe sagging. Plastic systems utilizing PVDF instead of stainless steel are becoming more common and effective for achieving even WFI conditions. A direct comparison of plastic piping with stainless steel with respect to ion leaching, smoothness, and CFU counts has been made [73].

By using plant steam for supplying heat of vaporization, clean steam can be generated from a purified water feed. Increased heat exchanger fouling may occur if a controlled feed water supply is not utilized. Clean steam distribution piping is designed with regulators to reduce the pressure from 5 atm down to about 2–2.5 atm needed for typical SIP operations. Sanitary pressure gauges, placed near or on equipment skids, can assist in monitoring dynamic pressures during SIP cycles. Steam sampling stations generally encompass a removable sanitary trap to which a sample cooling device (typically a heat exchanger) is attached. Consequently, a source of cooling water is necessary, either piped from local lines or transported by gravity via a portable holding tank about 20 L in volume. Steam traps are placed in header branches and equipment supply lines such that when skid equipment is isolated for repair, condensate buildup does not occur. Most importantly, steam traps and pressure regulating/control valves should be selected and sized properly to remove excess condensate during SIP operation [74]. Periodic cleaning of the clean steam system assists in cleaning fouled heat exchange surfaces and minimizing rouging.

Compressed air in product contact is generally used for fermentation sparger air supply, pressurized transfers, and other positive pressure requirements. Oil-free compressors are favored with intakes positioned away from other building effluents. Heating up to and holding at 200°F (93°C), typically accomplished by design of the compression ratio and residence time in a receiver, is desirable to minimize airborne biological contaminants such as bacteriophage, although there is some evidence that submicron (0.1 μ) filtration also can be effective [41]. Humidity then can be reduced from the compressed air using refrigeration and/or desiccant dryers. Moisture removal is desirable down to a dew point of −40°C. Air then is microfiltered (0.2 μ) at the point of use to remove microbes/particulates. Compressed air may be sampled for any or all of the following depending on the application: identity, hydrocarbon level, microbial content, and particulates. Other gases may be required for the facility such as oxygen enrichment for microbial cells, carbon dioxide for pH control of bicarbonate-buffered animal cell media, nitrogen, and oxygen supplementation for animal/insect cells, along with appropriate gas blending and control systems. Planning should consider the storage of bottled gas cylinders or tanks outside the facility or in segregated maintenance areas to minimize the transport of unclean cylinders into controlled locations.

Clean-in-place (CIP) skids can be portable and moved into place in the suite at a designated utility location and/or one or more fixed supply skids can be located permanently in the utility area. CIP systems are composed of supply and return pumps, one or more cleaning agent dilution tanks, provisions for heating cleaning agents, and a PLC for sequence control and associated data acquisition. Specific aspects of CIP design guidelines, cleaning agents, and step sequences are documented [75]. Owing to the tankage and pumps required [33], these skids often are located in utility areas where acid/caustic-based cleaning agents arriving in drums can be suitably diluted. More than one CIP skid may be necessary to provide product segregation and minimize cross contamination. Multiple skid requirements depend on design decisions concerning the level of containment and product segregation [8], as well as capacity.

9.3.2.2 Nonproduct Contact

The type of vessel jacket cooling/heating system selected, although not in product contact, has widespread design implications, and ultimately may limit pilot plant fermenter versatility. Recirculating temperature control loops, containing high surface area/unit volume heating and cooling heat exchangers, an expansion tank, and a circulation pump, are common. These systems rely on indirect heating/cooling of jacket fluid so that their response can be slow, particularly if heat exchangers were not sized according to user expectations. The involvement of heat exchangers suggests that practical application may be limited to vessels of few thousand liters in size. Jacket fluids can be water, glycol, or novel fluids with wide temperature ranges. Reliable operation of the circulation pump and internal (as well as external) integrity of the heat exchangers are also critical. Cooling heat exchangers can use either chilled water and/or glycol, while heating heat exchangers generally use plant steam. Dual parallel filter housings, containing 10–50 μm elements, are advantageous to reduce cooling heat exchanger fouling. Sediment buildup on filters can be monitored using measured pressure differentials across the filters, and the assembly can be designed so that one filter set may be replaced while maintaining flow to the second filter set.

Direct application of heating or cooling fluids to process jackets is the traditional alternative. Consecutive application of chilled water and steam on the jacket is acceptable if the design evacuates the jacket using higher pressure compressed air such that banging is minimized. Regardless of the indirect/direct mode of heating/cooling fluid application, jacket design can greatly affect the effectiveness of heat transfer. Dimple or straight jackets are preferred for smaller vessels, and external half-pipe cooling coils are used in larger vessels. The largest manufacturing scale vessels utilize internal cooling coils, but these coils present additional internals for cleaning and increased sterility risks, particularly at welds, because cooling water can be at a higher pressure than the fermenter contents.

Filtered dried instrument air, typically at a pressure near 100 psi, is required for solenoids valves on bioprocessing skids. Pressure specifications for control valve operation considerably vary among vendors and inversely affect valve actuator size. Plastic hosing, although convenient and cost effective, should be carefully evaluated for instrument air lines, especially on hot equipment where heat-induced leaks can develop and in solenoid cabinets where crimping can slice the tubing. A back-up instrument air compressor and expansion tank can minimize operational consequences from supply pressure dips. A separate compressed air source, at a higher pressure than chilled/cooling water and jacket steam, is desirable for the evacuation of vessel jackets to minimize risk of backflow into the instrument air system.

Natural gas usage, if required, might be best addressed using bottled gas for safety reasons, which discourage distribution piping throughout the facility. Favored for flaming of openings in many microbiological procedures, the use of Bunsen burners or propane torches in a biosafety cabinet or laminar flow hood should be examined with company safety department representatives. Care should be taken to minimize the potential for reaching the lower explosive limit near the motor of the cabinet or hood. Compatibility of plastic laboratoryware with flames also should be insured.

9.3.2.3 HVAC

HVAC is the key and costly component of a pilot plant facility owing to its multiproduct nature. Its design and operation needs to be understood by operating, maintenance, and quality personnel. Requirements for air cleanliness, as well as the ability to achieve them, are determined by HEPA filtration, airflow volume, pressurization room design as well as the type of operation being conducted (Table 9.5) [8]. HEPA filters are often used for inlet air but only used for exit air when required by biosafety containment.

The specific classification of processing areas depends on processing step quality requirements as they are related to equipment design. Where possible, support equipment (or portions of thereof) is located in cheaper gray utility space, reducing disruptions for alterations needed as pilot plant processes change. Gray space does not require classification and thus has reduced gowning, cleaning, and monitoring requirements, as well as lowered HVAC construction and operating costs [53,56]. As closed system operation in processing spaces increases, trends toward smaller classified core areas enveloped in nonclassified space are emerging [9,53,76]. The use of contained sampling devices that permit sampling by maintaining a closed system also helps in supporting this trend.

Classification level, representing the maximum number of particles permitted per cubic foot of air sampled, of an area increases from fermentation

Table 9.5 Comparisons of Example Controlled and Noncontrolled Area Achievable quality Based On Room Design in a Pilot Plant Facility

Type of Area	Pressurization	Airflow Changes (#/h)	Inlet Filtration	Nonviable (#/ft^3)	Viable (cfu/ft^3)
Bulk fermentation processing	None	5–7	Coarse	<250,000	<25
Laboratory	Positive	60	Coarse	<100,000	<2.5
Laboratory	Positive	10	HEPA	<100,000	<2.5
Laboratory-clean construction	Positive	25	HEPA	<100,000	<2.5
Laboratory-clean construction	Positive	20	HEPA	<10,000	<0.5
Laboratory-contains autoclave	Positive	100	HEPA	<100,000	<2.5
Laboratory-contains glasswasher	Negative	50	HEPA	<100,000	<2.5

(class 10,000–100,000; 20–40 room changes per hour) to purification (class 1000–10,000; up to 100 room changes per hour) [8]. Within a room, there are cleaner areas surrounding open transfers (e.g. biosafety cabinets or laminar flow hoods) which are class 100 (540 changes per hour) [77]. For viable particles, the range is from class 100,000 which permits 2.5 cfu/ft^3 down to class 100 which permits 0.1 cfu/ft^3 with intervening levels that are generally determined by the individual processing application being conducted [77]. Sufficient air velocities also are necessary to dilute particulate contaminants. In some cases, separate air handlers are considered for critical sterile areas, inoculum preparation areas, high particle or dust areas, and segregated product areas. In most other instances, air flow is designed in a once-through pathway which may substantially increase heating and cooling utility costs.

Inlet air enters from the top of the room, flows downward, and exits from ducts near the floor. Any recirculation, which may be desirable to minimize the capital costs for heating and cooling equipment as well as operational costs, should be only from the room where the air originated, limited to areas that do not generate substantial amounts of particles, and pass through inlet HEPA filters again. This strategy is beneficial to the life span of the HEPA filters and the ease of maintaining constant temperature and humidity [8]. Exit ducts should be placed near equipment within a processing area, which generates the most particulates or aerosols so as not to spread them throughout the area. Specific examples include autoclaves, glass washers, and cell disrupting homogenizers.

Room pressurizations should be higher for cleaner areas and lower for areas that generate particles [8]. Higher pressurization levels should be assigned to airlocks between adjacent areas for which it is desired to minimize cross contamination. Relative pressurizations between rooms sharing a common wall should be considered to supplement and/or minimize reliance on completely sealing rooms during their construction. The use of door sweeps and gasketted ceiling tiles to achieve and maintain pressurization should be evaluated relative to floor cleanability and ceiling maintenance access, respectively. Pressure differentials might be as high as 0.03 inches water between classified areas and as high as 0.05 in. water between classified and nonclassified areas. Lower pressure differentials might be acceptable depending on the specific product as long as the direction of airflow remains consistent. Interlocks, control, monitoring, and alarm building automation systems may be designed to reduce pressure losses when doors are opened and to alert users to unsatisfactory conditions. When desired pressurization has been lost, affected areas are recleaned and in-process materials are evaluated for quality impact, delaying schedules, and increasing manpower.

Facility temperature is generally maintained on the cool side because gowning requirements tend to make workers warm. Inaccurate assumptions concerning the impact of radiant heating from windows may cause higher than desired temperatures on certain days of operation. Humidity should be comfortable to avoid drying of workers' respiratory membranes during extended processing hours. Care should be taken to specify tolerance limits that have quality implications appropriate for the processing steps expected to be conducted in the area to avoid unnecessary additional facility costs. Although specific example guidelines are 23 ± 3°C and 40–60% relative humidity [78], substantially larger ranges may be acceptable for many processing applications. An extensive P&ID of HVAC design for clean spaces is available as a published Ref. 77. In addition, consideration should be given to Federal Standard 209E, Airborne Particulate Cleanliness Classes in Clean Rooms and Clean Zones, as well as EC regulations which establish parallel categories of Grade D (100,000), Grade C (10,000), and Grades A+B (different designs of class 100) [77,79].

9.3.3 Containment

Similar issues exist for product, environmental, and personnel protection, and these issues are heightened due to constantly changing pilot plant processes. Often the resulting design specifications overlap with those items necessary for sterile or sanitary operation. This concept is addressed in the literature [80], which compares solutions to prevent microbial transport to attain containment with those to maintain sterility for fermenters. A unified approach is recommended to minimize unnecessarily stringent specifications, which can substantially augment capital and operating costs. Environmental and personnel protection issues are evaluated for each new pilot plant process undertaken by the facility, along with product protection quality concerns. Samples of broth, and other product, intermediate, and waste streams, are submitted for aquatic (and sometimes human) toxicity testing in advance of processing in the pilot plant when appropriate. Preliminary test data might be then available to evaluate potential hazards in conjunction with published documentation concerning the similarity of the questionable compounds, raw materials, or organism to known hazards.

9.3.3.1 Product Protection

The major concept surrounding product protection is the closed system in which all material entering or leaving the process must be controlled in a specified and documented manner. This requirement extends to product-contact utilities such as purified water, clean steam, compressed air, and controlled/monitored room air so that when equipment is opened for cleaning the quality of surrounding environment is consistently at a known and documented level. The equipment is then cleaned to an acceptable quality level. Several elements are typically included as part of closed system operation. After sterilization, material enters the closed system via a sterilizing filter, steamed connection to an autoclaved bottle assembly, and/or sterilized transfer line. Off-gas passes through a 0.2 μ steam-sterilized vent filter, often contained in a low pressure, steam-jacketed housing preceded by a coalescing filter, condenser, and/or vent heater to avoid pluggage from moisture. Double mechanical seals are utilized with a barrier fluid of a controlled quality (typically condensed clean steam) at a higher pressure than the vessel. A closed system also can be used to surround an open operation such as the use of isolator technology (i.e. glove box) as an alternative to clean rooms or biosafety cabinets [81,82].

9.3.3.2 Environmental Protection

Environmental protection concerns focus on emissions from air, liquid, and solid waste. Typically, state air permits are required for new vessels and need to be altered if additional air treatment devices such as vent filters, incinerators, or scrubbers are added to existing vessels. Hydrophobic vent filters can provide emergency foam control and prevent external product or organism release. These filters can be monitored for pluggage over cultivation time using pressure differentials with plugged filters replaceable during a run only if a bypass line exists. Liquid emissions include accidental spills/sewerings, which can be minimized by locking vessel bottom valves, blocking open sewer connections, collecting sample line flush waste, and providing clear written instructions to personnel regarding batch disposition. Dikes, troughs, and vaults can be used to contain large spills or divert inadvertently sewered material. Sample waste can be minimized using low volume sampling devices. Thermal or chemical (less common) batch or continuous kill tanks need to be considered for biological hazards. For chemical hazards, other appropriate waste destruction and minimization systems may include hot caustic hydrolysis, reverse osmosis, and evaporative condensers. Batch leakage into cooling water utility systems can be minimized by operating utility systems at a pressure higher than the batch, but this differential might introduce contamination if jacket cooling coils develop leaks. Alternatively, the utility system return headers can be isolated from general use until their contents are treated and/or tested. This solution might be attractive for cooling water as well as condensate, especially when it is returned to river sources. Care must be taken to avoid sequestering large quantities of potential waste for long periods of time. Solid waste can be controlled procedurally using clearly labeled containers.

9.3.3.3 Personnel Protection

Personnel protection requirements are reduced now that several recombinant organisms can be grown under typical operating conditions of GLSP. For other pilot plant processes requiring personnel protection, either the culture is rated at or above BL1-LS or a compound produced by the culture adversely affects human heath (e.g. a genotoxin, mutagen, teratogen, or carcinogen). Environmental monitoring of the operating area is necessary for the compound or culture of interest using air and/or swab sampling techniques. For optimal recovery, swabs should be soaked in a solvent that is known to dissolve the compound of interest. Screening and monitoring of personnel potentially in contact with the process (operator, mechanics, janitors, engineers, visitors) must be arranged and notification must be given to those personnel at unacceptable levels of risk. A key element of personnel protection is personal protective equipment (PPE). This equipment can include gowns, uniforms, safety glasses, lab coats, and gloves at a minimum, but can be extended to include face shields, goggles, and respirators as required. Spill response equipment, procedures, and training are also necessary.

Reduction to practice techniques are presented in the literature for general biocontainment issues [83] as well as biocontainment regulations and requirements [84,85]. Specific examples are available for the design of waste inactivation systems [86,87], contained facility design and operation [88,89], design details for cleanliness and containment for floors, walls, ceilings and penetrations [90], and contained sampling systems [91].

9.3.4 Instrumentation

The main purpose of a pilot plant is to collect data as comprehensively as possible. Thus, a large fraction of the pilot plant capital allocation might be devoted to instrumentation. Use of in-process measurements and controls to improve product quality is a key foundation of the FDA's recent process analytical technology (PAT) initiative [92]. Traditional instruments for on-line monitoring, particularly for fermentation, include pH [93], dissolved oxygen (DO) [94–96], temperature, agitator speed, agitator power draw, back pressure, foam detection, batch level or weight, air flow rate, and vent gas analysis via mass spectrometry [97]. Redundant instrumentation is desirable for critical parameters such as pH and DO. Instrumentation for isolation processes includes the monitoring of conductivity, pH, ultraviolet wavelength intensity, and refractive index. Useful utility instrumentation includes on-line TOC analysis and resistivity/conductivity meters. Local readouts of transmitter values can be particularly useful for pressure and possibly temperature, although the potential for differences between field and control room values may have to be addressed.

Many novel sensors have been developed including dissolved carbon dioxide probes [93,98], redox sensors [93], glucose sensors [99], near infrared detectors [100,101], fiber optic biosensors [102], fluorometry of intracellular NADPH [103] and cell density probes [104–106]. Another novel measurement device example incorporates fermenter side streams to measure viscosity on line [106]. A description of useful sensors may be found in the literature [92,107]. Characteristics of on-line and *in-situ* devices include cleanability, minimal drift, minimal interferences, versatility, sterilizability, and ease of calibration. Implementation of any sensor should be evaluated for its effect on bioreactor sterility as well as product quality.

9.3.5 Automation and Control

Automation and control can reduce manual work and operator errors, as well as raise operational consistency [9]. The need for flexibility and change within a pilot plant (as compared with production) needs to be apparent in control schemes event sequencing, data collection and presentation, and alarming [108]. Increased amounts of automation restrict the abilities of the end user to implement changes, but do reduce the number of manual operations and thus possibly headcount. Specifically, the control system should readily permit set-point and alarm changes. Control systems can be distributed or centralized with distributed control systems (DCS) favored because there are individual controllers/computers for each unit connected via a high speed network. Thus, the loss of one centralized computer is not catastrophic [108].

Although there can be "islands of automation" within a pilot plant owing to automated skids purchased from different manufacturers, connectivity in a centralized location for data monitoring, set point/alarm changes, and data archiving typically is desirable. Substantial advantages exist for localized controls on skids when there are large numbers of inputs required based on an operator's evaluation of field conditions. A local readout of data versus a centralized "control room style" monitoring area (or even both) is dependent on the expected degree of operator supervision versus intervention required during processing. This decision is directly related to the flexibility of operators to move in and out of the field in terms of gowning procedures and access restrictions. The ability to remotely input set point and/or alarm changes can be useful when containment and/or product protection levels are high. Multiple areas for setting parameter values should be carefully considered from an operational control viewpoint to avoid duplication of desired changes and to batch record log omissions. Verbal communications between the field and control room using hand-held radios, intercoms, or phones also can minimize unnecessary movement within the facility.

Distributed control systems are favored for process management in some pilot plants. System architecture requirements vary depending on the processing requirements, but examples for a non-PC-based system are present in the literature [109]. Distributed control systems also can be PC-based using commercial PC control/data acquisition software [110]. In either case, some type of process-level controller is required such as a PLC (programmable logic controller), LC (loop controller), or PCBC (personal computer-based controller) [47]. The ability for the noncomputer expert to easily modify and understand control sequences using newly developed, commercial control software is attractive but requires effective change control and documentation procedures. The ability to readily configure the data acquisition system to download and archive data is an important consideration for process performance evaluation.

There are several types of computer inputs/outputs [108] which include: (i) digital inputs in the form of an on/off signal used for foam probe, pressure, flow, and/or temperature switches, (ii) digital outputs in the form of an on/off signal used for the flow of antifoam, open/close valves, and/or pump/motor start/stops, (iii) analog inputs in the form of an amplified output in the range of 4–20 ma used for temperature, pressure, pH, dissolved oxygen, flow, and power draw, and (iv) analog outputs in the form of a proportional signal which controls the valve opening via an I/P conversion (typically used for back pressure, cooling, and/or airflow valves) or else signals such as motor speed.

The most frequently utilized type of control is the PID (proportional integral differential) control loop. It is tuned using three parameters to minimize oscillations, which can change depending on the specific process (e.g. for dissolved oxygen cascade control of a fermenter cultivating faster growing *E. coli* versus slower growing animal cells) or over the course of the same process (e.g., for glucose and ethanol consumption phases of a Bakers' yeast fermentation [111]). These three parameters are the (i) size of the error relative to proportional band with smaller bands resulting in more sensitive systems, (ii) integral of the error over time with longer error durations resulting in greater corrections, and (iii) rate of error change with time with larger increases in correction with time calculated the faster the error is growing with time [107].

Typically, the third parameter is at a minimal value near zero for the optimally tuned loop.

A conscious decision needs to be made concerning the level of automation for CIP, SIP, and vessel-to-vessel transfers desired for a pilot plant. Higher levels of automation may insure reproducibility of operation, reduce reliance on operational staff, and simplify SOPs. However, raising the level of automation increases capital, validation, and maintenance costs, while reducing flexibility to quickly adapt to changing pilot plant processes. Familiarity with the details of automated versus manual equipment operational sequences may be lower among operating personnel routinely observing automated processing steps. A domino effect may exist in that automation of one valve may lead to the automation of several related valves, severely increasing design complexity and cost. Limit switches are often installed which confirm to the control system that the desired automatic valve has opened or shut as requested. Automatic pulsed diaphragm valves have been successful for nutrient feeding as well as for sterilization of transfer lines in automatic plants [112]. A prominent example of a highly automated but apparently reliable fermentation facility is operated for the manufacture of cephamycins as well as for pilot use [113]. Typical diagrams for the monitoring and control of a fermenter [33,108] and for the monitoring and control of an ultrafiltration unit [47] have been published.

9.3.6 Data Acquisition and Archiving

The collection and management of data obtained during pilot plant operations (as well as laboratory experiments) is a critical facet of knowledge management for the speeding implementation of cost-effective manufacturing processes [114]. Capture, storage, and retrieval of critical product/process knowledge, including pilot scale process development and scale-up data, also is a key challenge to implementing quality by design strategies [115]. The data collection interval is dependent upon the nature of the processing step. For fermentations, intervals might be in every 15–20 min during secondary metabolite or animal cell cultivations but might shrink to 5–10 min for faster growing bacterial or yeast cultures, especially when a key process addition or change is planned based on off-gas or dissolved oxygen measurements. Typically, 5–10 min intervals are appropriate for microfiltration or centrifugation operations and continuously for chromatography systems especially during loading and elution. Smaller data collection intervals may create substantial data archiving demands, which may be alleviated by only saving those data substantially different from previous values based on some user-defined value.

In-process trending of data is attractive to permit the timely evaluation of process status and trouble shooting of processing problems, which are the two factors critical to continuous improvement efforts. Specifically, multivariate analyses can be performed for critical variables and results can be used to adjust the process [116]. Automatic recording of alarms and changes to the control system, preferably sorted according to the equipment and/or batch for which they occurred, also can help to trace back processing problems and actions taken. On-line data might be archived on a common server by some unique batch delimiter, according to the established facility SOP for archiving data. Two formidable challenges are to capture off-line data electronically (without manual re-entry) and then incorporate it within the same or a closely related reference database, which is preferably located in a central repository. Otherwise, this off-line data can get fragmented among the notebooks and files of individual researchers. Novel customized systems for bar-coding samples taken from the field and creating a database for results can be helpful, permitting possible use of robotic systems for sample analysis. Remote access is an attractive option for monitoring process performance from off-site locations. Remote alarming can be triggered to a satellite pager thus removing the necessity for round-the-clock on-site coverage. Remote control of operations can be administratively challenging to regulate and document as well as a potential safety issue. Security issues must always be considered for any established remote capability.

9.3.7 Warehousing

Warehousing requirements for the facility need to be determined based on expected pilot plant production capacity and desired batch lead times. Common items that might be stored include ambient, cold and/or frozen solid/liquid hazardous and nonhazardous raw materials (controlled temperature and/or relative humidity storage), disposables, equipment, spare parts, as well as sample and product storage. Computerized inventory systems, organized by process or equipment, can be useful for most of these items with reminders to reorder when supplies decline to a critical level or discard when items expire. Marker equipment also might be purchased for long lead replacement parts for which redundancy was not desired during the original facility design. Sizing of these storage areas should match typical lot volumes of prepared culture media and solids to minimize repetitive release testing for multiple lots. For a pilot plant, in which several diverse processes may be run, additional process-specific storage often is required. The relative amount of storage necessary within or near the building should be compared with other available on-site and off-site options. Vendor manufacturing schedules often require the purchasing of larger quantities of rare raw materials to avoid future shortages. In this instance, if on-site storage is not available, vendor storage, or a third party warehouse alternative may be required. Warehouse access may need to be restricted depending upon the nature of its contents. Within the warehouse, there should be segregation for released, unreleased, and quarantined materials either by physical separation or by designated labels.

Although the entire facility should have a comprehensive pest monitoring program (as well as an SOP), there may be a need to directly address pest control, particularly for flours and meals used in certain secondary metabolite fermentations. These food ingredients when purchased for pilot scale fermentation purposes may not pass through fumigation unless specifically requested on the purchase order. Typically, larger customers in the food industry have invested capital for on-site fumigators, but commercial nutrient vendors may not have this type of equipment due to their high rate of inventory turnover.

Staging areas should be available for collecting raw materials and supplies in advance of a specific batch. Some method of inventory assessment and control should be available; one convenient method is to use computerized bar coding to check inventory as well as to confirm the use of specified materials when charging occurs. In addition, controlled sampling areas (to provide raw material, environmental, and/or personnel protection) for the sampling of raw materials may be necessary.

9.3.8 Back-Up Systems/Redundancy

A critical design decision centers on the level of backup and/or redundancy required in the pilot plant. Equipment sharing to reduce services for less critical tasks can be cost effective to increase the amount of backup available for critical tasks [117]. Although typically none of the material created in the pilot plant is "for sale," mechanical problems can delay processing and thus material production, which may interrupt clinical supplies and compromise the success of ongoing studies. Even if no clinical material is at risk, process development efforts are hampered when experiments are frequently interrupted for mechanical support-type equipment failures [10]. Scenarios that might be protected against include loss of clean steam lubrication to agitator seals, positive air pressure on a sterile tank, and electrical power to a controller. Corresponding equipment needs might include redundant clean air compressors, clean steam generators, and/or purified/WFI water generation systems with the peak plant capacity requirements split between two systems rather than relying on a single system sized for peak usage. Such redundancy usually is achievable because utility costs are low compared to equipment costs [10]. A UPS system sized to include the most critical equipment (typically controllers and data acquisition devices) might also be installed. Back-up power for large electrical loads such as motors might not be economical; however, consideration might be given to installing connections for a gas-powered generator if future evaluations altered the economic assessment.

9.3.9 Future Expansion/Modification

As future processing needs may change, the pilot plant must adapt to maintain its usefulness to the company. Large initially underutilized spaces might be outfitted with utility headers to minimize processing disruption during future installation. Oversizing of utilities for anticipated future loads [10], as well as unforeseen procedural modifications warranted based on future validation testing results (such as additional rinsing cycles), assures subsequent flexibility. Selection of portable equipment permits exchanges for new pilot plant processes, provided there are procedures developed for storage, cleaning, and start-up of this equipment.

Modular design and central spine approaches facilitate expansion and adaptability with minimal impact on existing operations when new components are added [9,117]. One aggressive approach to increase versatility is a modular, mobile, validated pilot plant [118], comprised of a system of interconnected and integrated modules. Modules are selected according to processing needs and can be rapidly assembled to form a processing plant capable of undergoing validation. Example processing modules include fermentation, recovery, and purification; the utility module includes clean water and clean steam; the personnel support module includes lockers and changing areas; and the HVAC module houses the air handling and filtration equipment. Additional modules can be then added or deleted as required.

9.4 Operation

Each pilot plant facility must decide for itself whether it is required to operate in a continuous mode of GMP compliance or whether it can selectively apply GMP practices to certain batches (but not to other concurrently run batches) or during certain specific processing periods. A conservative solution might be to operate continually in GMP compliance, thus insuring facility readiness and preventing the omission of a required procedure for an actual batch destined for the clinic. Dual GMP and nonGMP operation serves to create two standards and may cause confusion among operating and maintenance staff. If dual operating modes are implemented, then a clear distinction between GMP and nonGMP operation must be established, documented, and maintained.

9.4.1 Maintenance

9.4.1.1 Preventative

Preventive maintenance, performed during an annual facility shutdown, is probably the most important factor influencing smooth pilot plant operation throughout the upcoming year. The scheduling of an annual facility shutdown is desirable for several reasons including the ease of lock out of hazardous energy sources, the time economies of concentrating several maintenance workers in a single area, and the ability to supervise and document completion of a large fraction of maintenance work at one time. This yearly PM documentation can then serve as the model for maintenance personnel for the following year's shutdown. Individual PM work orders, spread over the course of a year, result in substantial time and effort expended in scheduling, preparing, and documenting the work done. Interspersing of annual PM work among batch processing inevitably results in PM delays due to equipment and/or manpower unavailability. Guidelines for planning and executing an annual plant shutdown are available [119].

All PM work orders can be categorized and then listed according to trade, equipment, material, or time requirements. They can be reviewed prior to the facility shutdown and augmented with any modifications or "one-time" replacement work orders that are requested. As PM work is completed, documentation should be signed by the tradesperson as applicable and reviewed by mechanical and operational staff for completeness. Problems that are identified with equipment or instrumentation should be highlighted as they occur and then entered into change control work order systems so that they can be tracked for their effect on validation.

Before the annual shutdown, transfer line manifolds and/or utility supply lines, that typically cannot be shutdown entirely for maintenance throughout the year, might be leak tested to identify faulty valves in need of replacement. After the annual shutdown, equipment start-up procedures should be documented, preferably through the use of equipment start-up batch sheets. These procedures might include equipment cleaning, testing of instrument calibrations such as level, temperature, and pressure [120] and vessel integrity testing for leaky valves. Instrument calibration accuracy should be evaluated for random or systematic errors with random error within accepted tolerances being the expected outcome.

Some general PM procedures can be applied to several areas in a pilot plant, both product and nonproduct contact. Sanitary fittings, which tend to loosen during repeated heating and cooling cycles, should be checked, gaskets replaced and clamps tightened as required. Any flanges in product contact should be tightened to a specified torque. PM inspections of vessel internals should be conducted and lock washers installed on external nuts that may loosen due to repeated heating and cooling. A PM should be developed for the inspection of vessel cooling coils for leaks, which can result in inadvertent cooling water contact with product [121]. For agitated systems, there should be a regular examination of foot bearings, motor/drive vibration, oil analysis, and shaft runout. Double mechanical seals, the failure of which can lead to condensate buildup and/or batch contamination, should be tested regularly and replaced preventatively based on a minimum life expectancy [122]. An outline of specific PM procedures for laboratory fermenters can be applied to larger scale fermenters [123]. An example of the documentation for the maintenance of horizontal laminar flow hoods is also available [124] which can be readily adapted to other HEPA filter units. PMs should be conducted on all utility systems and may include the rebuild of air compressor heads, descaling of heat exchange surfaces, and replacing of air dryer desiccant, among other items recommended by the equipment manufacturer. Finally, novel predictive maintenance evaluation techniques should be explored and implemented as appropriate to identify imminent equipment failures [125].

It is most appropriate to perform any annual validation work (e.g. fermenter SIP, autoclave load pattern, controlled temperature unit) at the conclusion of the annual PM shutdown. In this manner, proper equipment operation can be documented and the effectiveness of PM procedures can be evaluated. Often inadvertent mistakes or omissions during maintenance can be identified prior to having an adverse, and possibly widespread, quality impact on pilot plant batches destined for clinical trials.

9.4.1.2 Ongoing

A reliable system of tracking outstanding maintenance by equipment or by process should be designed and implemented, particularly for a pilot plant in which equipment is used less regularly. Each piece of equipment has a clearly distinguished tag name. Work orders are initiated, evaluated for their effect on validation, marked as complete when mechanical work done, then finalized when process personnel have checked the job. A computerized database can readily track maintenance and identify repetitive maintenance problems. Such a system permits easy review of outstanding work orders for each piece of equipment prior to getting it ready for service. All maintenance work orders need to be tracked within one system; care should be taken not to bypass the operations area when maintenance needs are identified by those directly controlling the mechanical staff.

Interim maintenance work, such as calibrations and subannual preventative maintenance (PM) activities, should be documented and filed in a manner equivalent to the yearly PM results. PM schedules should be reviewed based on the frequency and results of interim maintenance work. Unusually common equipment failures can be identified through maintenance database searches and appropriate longer term repair efforts or replacement justified and undertaken.

9.4.1.3 Calibrations

Calibrations of pilot plant equipment have additional concerns associated with them because important processing decisions for early scale-up batches may be based on the accuracy of the instrument readings. In some cases, specific instruments are designated as "critical" for a processing step; in other cases, desired specification ranges for the processing step are targeted based on certain noncritical instruments. These two methods provide assurance that product quality is not compromised if any of the other instruments that are present in the equipment utilized fail.

Although some facilities distinguish between calibration procedures for "critical" and "noncritical" instrumentation, it may be prudent for a pilot plant to calibrate all instruments in a reliable and documented manner because their accuracy ensures useful process development data. Calibrations should be performed using NIST traceable instruments and include full loop calibration to the process monitor as appropriate. A specific calibration and maintenance SOP for the instrument is most desirable but a general SOP referencing vendor manuals might also be acceptable, particularly in the interim period while instrument calibration SOPs are being developed. The "as found" (or "before") and "as left" conditions should be documented along with the serial number of the standard used [126]. Establishment of an acceptable tolerance for each instrument insures consistency. When an instrument is found outside the tolerance specified, the impact of this calibration "as found" error on previously executed batches may need to be evaluated. Thus, a confirmation of the error might be warranted prior to correcting the calibration. Records should be evaluated for the amount of instrument drift, which may indicate the need to either recalibrate on a more frequent basis or else replace/upgrade aging equipment. An example calibration sheet is available [127].

9.4.1.4 Modifications/Change Control

Any validated facility needs to devise procedures for tracking changes on equipment as well as on processes and computers. This requirement is heightened for a pilot plant because a large number of changes are expected owing to its rapidly changing portfolio and processes. Incomplete documentation of changes can obviate extensive time and money spent on validation testing. Although the exact contents may change, such a procedure starts during the initial design phase as equipment specifications are altered, continues through the facility start-up and qualification phases, and then moves directly into the facility operational phase. Adherence to change control procedures may be difficult during facility start-up as many unexpected items are found to be incorrect and management pressure may mount to have the facility operational quickly. Depending upon the philosophy of the individual facility, the scope of this change control procedure may include product and nonproduct equipment to varying degrees. The procedure identifies the individuals who need to approve changes and determine prior to or in concert with the change implementation what additional validation testing is required. Often, processing personnel, being most familiar with the equipment being modified and the reason for its modification, can provide valuable leadership in this area and initiate contact with quality and validation personnel as required for input.

Generally, replacements of this kind are accepted without additional testing unless the item being replaced was specifically tested or calibrated during the initial validation effort. Example items and associated testing might include a replacement or rebuilt motor or associated drive (measure motor amperage draw at various agitator speed settings) or replacement instrumentation (calibrate sensor and transmitter). Replacement items in this category need to be tagged appropriately in the field with the same tag number as the original item and serial numbers updated in any centralized maintenance data bank as well as recorded on installation documentation.

Modifications or obvious changes in equipment or instrumentation need to be evaluated for their potential impact on the validated state of the equipment. This impact is particularly important for SIP and CIP cycles. Such modifications may be minor, such as a change in gasket or o-ring material of construction, or major, such as a change in piping. An example procedure is available which can insure proposed changes are reviewed by operations, maintenance, and quality for their potential impact [128]. Specific attention should be given to alterations needed in P&IDs, SOPs, PMs, batch sheets, and/or training programs. Equipment modifications, necessary to accommodate a new process, need to be identified early on and their impact also should be evaluated [4]. It is critical to balance the request for change with an evaluation of its validation impact. Changes cannot be discouraged in a multipurpose pilot plant, but they need to be controlled and managed in a logical yet expedient manner.

9.4.2 Staffing

Personnel requirements are best defined based on the production process [129] but in a pilot plant setting, production processes change rapidly and substantially [130]. Both short- and long-term planning tools and models are necessary to plan manpower for the preparation, execution, and finalization of various campaign efforts [130]. A decision needs to be made early in the design process concerning the level of off-shift activity that is reasonably expected in the facility as it can severely impact items such as utility capacity and operation. Round-the-clock coverage, although demanding for personnel involved, may be necessary almost throughout the process to insure success and timeliness of production. The use of unionized personnel, technicians, or process development staff should be incorporated into the educational requirements and manual workload expectations for the operating staff to ensure appropriate job fits and reduce potential drivers for turnover. A core operations group, responsible for the facility, can insure continuity in areas such as SOP updates, maintenance work order tracking, batch sheet content, and equipment operation. Such a group might be supervised by an area manager, responsible for coordination and scheduling, who might interface with those directly involved in process development. Then, the processing groups might rotate through the pilot plant as their projects demanded are accompanied by a member of the operations group. Alternatively, a dedicated pilot production process development staff might be instituted, responsible for the translation of laboratory processes into GMP batches.

9.4.3 Laboratory Support

To provide adequate and accurate sample analyses, laboratory support needs to be identified and addressed through either on-site or contract laboratories. Targets for testing might include final product, in-process samples, raw materials, sterility, culture purity, environmental quality of controlled environments, and utility quality (Table 9.6). Although the goal of these laboratories may be to work toward using validated methods and approved SOPs for all assays, this goal might not be attainable for project-specific assays until the later stages of a successful development project. Initially, in the early project stages, only notebook procedures might be available to guide laboratory testing. Information directly impacting quality should be communicated directly to both quality and operating personnel and appropriate investigations conducted both in the field and laboratory for failing or out-of-specification (OOS) results.

A microbiological laboratory is critical for fermentation-based operations to prepare seed cultures and test new culture seed vials for culture purity and productivity in lab-scale processes. An examination of up-to-date sterility/culture purity results prior to vessel-to-vessel transfers should be undertaken as close to possible to the time of transfer to minimize the adverse impacts of contaminated cultures. In a pilot plant setting, tension may exist between the time required to adequately test new cultures and development pressure to use the latest mutant or transformant in a pilot scale cultivation to evaluate its capabilities. Reasonable procedures need to be established in advance for the introduction of new processes and/or new cultures into the pilot plant. Laboratory prepared nutrients, trace ingredients, and/or sterilized additions should be evaluated for their quality impact and the procedures used for their preparation should be documented where possible.

9.4.4 Standard Operating Procedures (SOPs)

SOPs are essential for the planning and standardization of various procedures conducted by operational staff and maintenance personnel that need to be completed before, during, and after pilot plant batches [4]. Typical categories include SOPs for equipment operation, facility and equipment cleaning, sampling, sterilization/sanitization, maintenance, and documentation procedures for tasks such as product changeover, change control, and batch sheets (Table 9.7). When referencing SOPs as part of the batch record, working versions of SOPs can be helpful to track progress and ensure accuracy of step execution in the field. An SOP on the requirements for writing, issuing, and inactivating an SOP also is useful. SOPs focus operation, guide operations staff, document the order and scope of activities, standardize approaches, define regulatory needs and constraints, and increase the speed of new personnel training [4]. They should be written by a person familiar with the facility or equipment operation that is the subject of the SOP, reviewed by another

Table 9.6 Types of Testing (Sampling and Analysis) Possibly Required by On-Site or Off-Site Support Laboratories for a Bioprocessing Pilot Plant

Facility Monitoring	Sterilization Cycle Validation	Cleaning Cycle Validation	Process Monitoring
Total organic carbon (TOC)	Indicator spore strip incubation	Total organic carbon (TOC) (swab and rinsewater)	Spectrophotometer
pH	Indicator water spore suspension incubation	pH	pH meter
Conductivity	Indicator media spore suspension incubation	Conductivity	HPLC for organic acids, amino acids, sugars
Endotoxin (LAL)	Sterility	Endotoxin (LAL)	HPLC for product/impurity analysis
Bioburden and coliforms	Bacterial and fungal culture purity-	Bioburden (swab and rinsewater)	LC-mass spec
Airborne viable particles	Filter integrity testing	Ultraviolet light absorbance	Protein gel analysis
Airborne total particles	Microbial identification	Visible light absorbance	Osmometer
Surface viable particles	Gram stains	Filtered solids	Microscope
Compressed gas viable particles		Dissolved solids	Cell counter/viability monitor
Compressed gas nonviable particles			Conductivity
Compressed gas identity			Metabolite analyzer
Noncondensable gases in steam condensate			Plate readers
			Blood gas analyzer
			Laboratory scale-down control/trouble-shooting experiments

Table 9.7 List of Example Types of Procedures that are Possibly Required for a Bioprocessing Pilot Plant

Administrative	Process Equipment	Lab Equipment	Maintenance	Operations
Processing abnormality	Sterilization-in-place	Autoclave operation-sterilization and decon	Instrument calibration	Gowning
Investigation of contamination	Cleaning-in-place	Autoclave load patterns	Equipment preventative maintenance	Water system flushing/ sampling
Product changeover	Sanitization	Glasswasher operation	Utility preventative maintenance	Steam system sampling
Writing of an SOP	Vessel set-up and operation	Glasswasher load patterns	HVAC preventative maintenance	Spill containment
Personnel training on equipment/operations	Transfer line/panel operations	Working in biosafety cabinet	Equipment change control	Facility housekeeping and cleaning
Personnel training on assays	Batch harvest (noncontained)	Working in laminar flow hood	Equipment start-up after processing hiatus	Facility start-up after processing hiatus
Personnel training on computers	Batch harvest (contained)	Incubator/shaker operation	Equipment start-up after major mechanical work	Daily utility checkout
Issuing and review of batch sheets	Equipment integrity test	Freezer	Equip. storage during processing hiatus	In-process sampling and analysis

(continued overleaf)

Table 9.7 (*Continued*)

Administrative	Process Equipment	Lab Equipment	Maintenance	Operations
Raw material release and storage	Utility systems operation	Refrigerator/cold room	Computer system change control	Data archiving
Investigation of monitoring excursions	HVAC systems operation	Tubing welder use and cleaning	Computer system failure recovery	Computer system operation
Forklift certification	Media/buffer preparation	Manual washing	Shutdown/interim maintenance document	Monitoring- controlled environments
Pest control	Specific equipment SOPs	Usage of scales and checkweights	Equipment isolation for maintenance	Tracking of process/equipment status
Ongoing validation testing		Calibration of pH meter	Issuing and tracking of work orders	
Ongoing facility monitoring		Facility monitoring assays		
Introduction of new products/processes		In-process monitoring assays		
		Fume hood use		
		Seed preparation		

person who might be asked to follow the SOP, and approved by a quality group. Each SOP should have an effective date of issue, be available to those who need it, and be archived when a future revision is created.

The need for structure and consistency within an SOP must be balanced by the need for flexibility in a pilot plant when unexpected process requirements arise. Procedures must be defined for what to do when an SOP is altered intentionally to accommodate a new process or an unexpected development in an existing process. Consideration of SOPs should be given during the equipment design phase so that tasks can be performed in a logical clockwise or counterclockwise fashion [129]. SOPs should be clear so that the expected outcome is insured. Specific examples of often omitted details include (i) statements regarding settings for pressure switches and acceptable calibration error tolerances, (ii) diagrams, part numbers, and manual references to aid in following the procedure, and (iii) relying solely on valve numbers, not valve purposes, to describe the procedure. Although SOPs should be reviewed periodically in every few years to determine if they are in need of clarification, care should be taken to minimize the number and frequency of similar versions of the same SOP issued to the field. Operational clarification memos, issued to all operational staff and signed off when read, might be used to provide additional noncritical SOP clarification without reissuing the SOP during the interim period between SOP reviews. All major SOPs changes should be documented, evaluated for their effect on validation, and communicated to staff in a documented manner.

9.4.5 Safety

Because concerns regarding product exposure to personnel are similar to concerns for personnel exposure to product, GMP guidelines overlap with safety regulations in many respects such as training, SOPs, preventative maintenance including calibrations, and validation/change control [4]. In fact, specific procedures for managing change to conform to process safety requirements [131] match the rigor of those required to confirm with GMP requirements. By the definition of a pilot plant's mission, change is an integral part of its existence. Consequently, comprehensive yet reasonable safety procedures must be established.

Safety captions can be required for certain raw materials based on the review of MSDS documentation. Broths, as well as isolated compounds, might be submitted for toxicity testing as soon as they are cultivated in laboratory bioreactors so that results are available prior to pilot plant scale-up. Safety is a component of installation qualification(IQ) because pressure vessel documentation and utility connections are reviewed. HAZOP (hazard operability) reviews for pressure vessels (e.g. fermenters, autoclaves) and utilities (e.g. clean steam generators) often are required by company policy and summaries included within equipment qualification reports. HAZOP action items need to be addressed promptly and any necessary modifications to equipment and/or procedures made to minimize risk. Equipment design needs to be reviewed for the ease of lockout and removal of hazardous energy sources (e.g. bleed valves for the evacuation of process and utility lines, local field-mounted pressure gauges to confirm depressurization), as well as the individual isolation of skids/units for maintenance. In addition, equipment-specific SOPs for preparing the equipment for maintenance might be required.

Regular safety inspections with documented action items that are communicated to all affected personnel (i.e. operating staff, technicians, engineers, management) can identify and correct nonoptimal working conditions and cluttered processing areas. Tracking systems can be effective to ensure that corrective actions are taken and are effective. Specific attention should be given to the purchase of ergonomic lifting devices such as material elevators, drum lifters, and vacuum assist arms. Exposure of personnel to excessive sound, heat, or dust also should be considered during facility design as well as monitored during facility operation. Although personal protective equipment is available such as ear plugs, heat-resistant gloves, and respirators, when possible engineering controls such as sound dampeners, insulation, and dust collectors may be better solutions. Safety during validation testing should always be considered, particularly when groups outside of operations are involved in setting up/operating equipment and when sources of hazardous energy are involved. Whether during validation testing or operational use, fittings of any sort should be tightly secured and should not be tightened when equipment is under pressure and/or hot, specifically during SIP and CIP procedures. Summaries of generally accepted laboratory and pilot plant safety procedures to consider are available [132].

According to ASME code, overpressurization protection using relief devices is required for vessels over six inches in diameter. The background for this regulation as well as details of pressure relief valves and rupture disks is available [133]. Although sanitary rupture disks are most desirable for product contact, once blown, processing must stop until they are replaced. Sometimes, a rupture disk may be installed followed by a pop-safety valve rated at a slightly higher pressure with a pressure gauge or other type of sensor in between to indicate rupture of the disk [84]. If this arrangement is used, then the cleanliness of the area between the rupture disk and the pop-safety valve must be demonstrated for the batch quality not to be adversely impacted if the disk ruptures but the pop-safety valve does not release. Removal of entire rupture disk assembly (not the removal of the disk out from the assembly) for cleaning should be checked with company policy but may be acceptable in most cases.

Another key concentration is that the outlets of safety relief devices (which might be unexpectedly spewing steam, hot water, or broth after relief) should be oriented such that personnel safety is considered for those who may be standing near them. One method is to pipe outlets to outside containment troughs, but it may not be desirable to have a potential connection to the outside. A second alternative is to create a designated area within the pilot plant processing area. Although piping outlets to the floor beside the equipment should be minimized, if they are found to be necessary, they should be carefully situated outside the normal path of personnel.

9.4.6 Training

Initial and ongoing training programs need to be developed and documented for each individual associated with the facility based on the facility needs. The number of different individuals associated with pilot plant processing can be high unless a dedicated group is established regardless of product. A training SOP should be established, which might include a training manual containing useful items such as copies of key SOPs, facility layouts, equipment diagrams, and policies. Initial training might include items ranging from vendor supplied training sessions to validated load pattern arrangements. Often this training falls at a peak workload period during the facility start-up and validation (or during subsequent batch preparation) when attentions can be diverted into seemingly more urgent areas. In other cases, more individuals require training than can be reasonably accommodated in the area surrounding the equipment. When permitted, such training might be recorded for later review by the staff once they have fully assumed responsibility for operation. Another alternative is for those just trained to provide training to others, which may be suitable for certain operations. For best results, training should occur with written reference documentation available to those being trained.

Ongoing training should be done on a regular basis with sessions once in every one or two months, which seems to be most appropriate for an active pilot plant facility. Training should include supervisors, operators, engineers, and maintenance personnel, as well as validation and GMP staff. Training might include a process review that highlights critical steps and equipments, SOP reviews, identification of personnel safety and environmental containment issues, review of PM and ongoing maintenance needs, as well as what operational personnel should anticipate with respect to proper equipment function [127]. Training sessions are one method to communicate recent SOP changes and to explain why certain procedures are necessary. They also might be used for periodic SOP reviews to obtain feedback from those using the SOP frequently. Attendance should be required, sessions recorded when possible, and those new to the facility should review prior training sessions for the past few years as applicable.

9.4.7 Validation

When manufacturing clinical materials, pilot plants are evaluated under the guidelines for investigational new drug products, among other regulations, which clearly state that GMPs applying for drug products are approved for clinical trials in humans and animals [134]. The recent guideline issued by the FDA [135] makes pilot plants eligible for product licensure should company needs require a product launch from a pilot plant facility. Individual company philosophies vary on the suitability or desirability of their pilot plants for this task because using a pilot facility for manufacturing reduces its availability for the production of clinical material. The level of validation effort required is best considered during the initial facility design. Prospective validation is the preferred option, owing to the association of document collection and equipment testing with facility start-up. Validation can be done retrospectively although the collection of the required documentation may be more challenging and modifications that are necessary for testing may be costly.

The major steps for validation ensure that installed equipment and utilities operate in a manner acceptable for the process to be run in that equipment. Because multiple processes, many of which are not necessarily identified at the time of validation, may be conducted in a pilot plant, operational ranges need to be set to accommodate a wide range of reasonably expected processes. For example, an incubator might be validated for temperatures between 25 and 37°C to accommodate common bacterial, yeast, and fungal cultures, but not necessarily at 45°C to accommodate unusual thermophiles. The selection of "worst-case" scenarios can be difficult for a pilot plant if the associated "worst-case" conditions are unduly burdensome for normal operation. Specifically, the cleaning procedures required after a three week fungal production cultivation are significantly more involved than those required after a two day fungal seed cultivation of the same culture or for a two week animal cell cultivation. Similarly, the sterilization cycle hold time for a concentrated sugar solution is substantially longer than that required for a heat-sensitive waterlike growth medium [136]. Approaches for efficiently resolving these multiproduct validation issues can be complicated by desires to follow manufacturing validation guidelines, which most likely were established based on a single product.

Often validation, qualification, verification, commissioning, and start-up are used with specific definitions in mind for individual facilities and organizations. Generally, validation implies the highest rigor of direct reproducible testing of a specific procedure. Qualification encompasses a wider range of documentation collection and testing of capabilities. Verification might imply a single test against an identified standard. Commissioning can refer to documented trouble shooting of newly installed equipment, and start-up might refer to initial equipment checkouts directly after installation. The major validation phases are divided to focus on qualification followed by validation. IQ focuses of the verification of proper installation, appropriate utility connections, and adherence to manufacturers' and purchase order specifications [137]. Operational qualification (OQ) includes testing of equipment function and an acceptance criteria for equipment performance [135]. Performance qualification (PQ) centers on process-specific testing for SIP, CIP, and autoclave sterilization and decontamination load patterns. For computers, a system life cycle methodology incorporating specifications, program design, coding, testing, start-up, operation, and maintenance has been adopted [137,138].

Process validation centers on the reproducibility of the process in the manufacturing area [139]. Parameters influencing this reproducibility might be initially identified at the laboratory scale then studied in the pilot plant to obtain acceptable operating ranges for those processing parameters, such as cultivation temperature, pH, and dissolved oxygen, are found to influence product quality. The set of criteria for proceeding to the next processing step, specifically volumetric/specific productivity, yield and purity, must be validated in a documented fashion. The critical processing steps important to process quality should be identified and validated [137]. Other noncritical process steps may be studied for their impact on manufacturing productivity. Process validation may require multiple identical pieces of equipment (such as fermenters) at the laboratory and process scales to define acceptable processing ranges in a parallel rather than sequential manner.

9.4.8 Facility Records and Manufacturing Execution Systems (MES)

Documentation associated with a new pilot plant facility and its equipment can be extensive initially and increases substantially during the time that the facility operates. Although centralizing all documentation in a single archive can be attractive, duplicate copies of some items are often needed in other locations during every day operation. For example, equipment manuals and facility drawings may be needed near the actual location of the equipment, in the maintenance shop, and archived in the centralized documentation repository. Validation documentation, consisting of IQ/OQ, PQ, and load pattern/SIP/CIP reports, is generated both during the initial validation and during annual retesting and ongoing environmental monitoring programs. Storage of change control documentation might be accomplished through notebooks/file folders designated for individual pieces of equipment or individual suites. SOPs might be archived in a centralized location but also posted near equipment for ready reference. Access and sign-out procedures for the control of documentation need to be developed. Electronic on-line access is one way to minimize space requirements and facilitate locating documentation.

A robust system for batch records needs to be created, based on individual batch sheets for processing steps or for individual pieces of equipment or for some appropriate combination of the two approaches. Issued batch sheets should be tracked by a unique sequential number with the product, processing step, and/or equipment number clearly identified. There is a need to track the use of equipment by product and batch, particularly for multipurpose pilot plant facilities. Although this might be done through individual equipment use logs, it is also helpful to include such tracking within the batch record itself.

Newly authored or completed batch sheets should be reviewed by the pilot plant operations area and may be required to be approved by a quality group prior to execution. Batch sheets should have clear written instructions as well as a signature, date and time, as required, for each step. Handwritten instructions, likely to be added during the processing of pilot scale batches, may need to be reviewed by representatives from operation and/or quality groups. Brief justifications for alterations in batch sheet instructions also should be considered. A written explanation documenting that the material from the current step is acceptable for continued processing in a subsequent step is recommended. Care should be taken to use common templates wherever possible to simplify successive batch sheet differences and minimize the review effort. Executed batch sheets should be reviewed for completeness by the operations area, by those in charge of the batch, and by a quality group as required. A tracking system should be instituted to monitor the review of completed batch sheets as well as to ensure that all issued batch numbers are accounted for in the batch record filing cabinets. Photocopying, microfilming, or scanning of completed batch sheets serve to provide a backup in case of loss or damage. Paperless electronic batch records, as well as electronic signatures, can minimize documentation volume and guidelines for implementation and validation are published [140].

Manufacturing execution systems (MES) control manufacturing processes. These include areas such as inventory, batch sheets, scheduling, and calibration management [141], as well as mechanical and operational status of equipment. Bioprocess facilities in general have been slow to implement these systems owing to their large initial costs, need for knowledgeable IT support, challenges with melding with paper-based systems, and validation requirements [55]. Bioprocessing pilot plants have been even slower because their inherent need for flexibility can make scoping of MES systems difficult. However, there are several benefits deriving from MES tools and they should attain successful implementation and be readily maintainable [142]. In the case of a pilot plant, frequent changes in batch-associated procedures make it challenging to gain the benefits of electronic batch records that utilize computerized cross-checking to ensure that steps are completed in the proper order [55].

References

1 O'Leary RM, Etcheverry T, Bezy P, Anicetti V, Burton LE. *PDA J Pharm Sci Technol* 2001;**55**:230–234.
2 Levin J. *Bioprocess Int* 2004; **2**:26, 28, 30, 32, 74.
3 Basu P, Quaadgras J, Holleman J, Mack R, Noren A. *Chem Eng Prog* 1997;**96**(6):66–75.
4 Basu P, Quaadgras J, Mack R, Noren A. *Chem Eng Prog* 1998;**94**(2):67–74.
5 DePalma A. *Genet News* 2007;**27**(9):48–52.
6 Lias RJ, Perry SD. *Bioprocess Int* 2006;**4**:12–18.
7 Burnett M, Santamarina V, Omstead D. *Ann NY Acad Sci* 1991;**646**:357–366.
8 Tulsi B. *Pharm Process* 2007;**23**(6):10–12.
9 Barrer P. *Biotechnology (NY)* 1983;**1**(8):661–666.
10 Bader F. In: Ladisch M, Bose A, editors. *Harnessing biotechnology for the 21st century*. Washington (DC): American Chemical Society; 1992; p 228–231.
11 Odum J. *Pharm Eng* 1995;**15**:8, 10–18, 20.
12 Bader F, Blum A, Garfinkle B, MacFarlane D, Massa T, Copmann T. *BioPharm* 1992;**5**(7):32–40.
13 Hamers M. *Biotechnology (NY)* 1993;**11**:561–570.
14 Fitzpatrick S, Ma'ayan A, Wagget J. *Chem Eng Prog* 1990;**86**(12):26–31.
15 Narodoslawsky M. *Chem Biochem Eng Q* 1991;**5**(4):183–187.
16 Samsatli N, Shah N. *Trans IChemE* 1996;**74**(4)C:221–231.
17 Sofer G, Chirica LC. *Biopharm Int* 2006;**19**(11):48–54.
18 Shukla A, Hubbard B, Tressel T, Guhan S, Low D. *J Chromatogr B* 2007;**848**:28–39.
19 Chang DYH, Garza P, Huang Y-M, Talabardon M, Rahmati S, Fallon E, Noe W. In: Godia F, Fussenegger M, editors. *Animal cell technology meets genomics*. Netherlands: Springer; 2005; p 459–464.
20 Samsatli N, Shah N. *Trans IChemE* 1996;**74**(4)C:232–242.
21 Shanklin T, Roper K, Yegneswaran PK, Marten MR. *Biotechnol Bioeng* 2001;**72**(4):484–489.
22 Bader RA, Weiss J, Rumsey T. *Bioprocess Int* 2004;**2**(5):66–73.
23 Schell D. *Appl Biochem and Biotechnol* 1995;**51**/**52**:549–557.
24 Marks DM. *Bioprocess Int* 2003;**1**:50–57.
25 Cannales M, Enriquez A, Ramos E, Cabrera D, Dandie H, Soto A, Falcon V, Rodriguez M, de la Fuente J. *Vaccine* 1997;**15**(4):414–422.
26 Eliezer E. *Biopharm* 1993;**6**(4):24–29.
27 Sato R, da Costa M. *Biotechnol Lett* 1996;**18**(3):275–280.
28 Petrides D, Sapidou E, Calandranis J. *Biotechnol Bioeng* 1995;**48**:529–541.
29 Titchener-Hooker N, Gritsis D, Mannweiler K, Olbrich R, Gardiner S, Fish N, Hoare M. *Biopharm* 1991;**4**(7):34–38.

30 Kemp G, O'Neil P. In: Subramanian G, editor. *Antibodies*. New York: Kluwer Academic/Plenum Publishers; 2004; p 75–100.
31 Zawistowski J, Rago J. *Pharm Eng* 1994;**14**:24+.
32 Shahidi A, Torregrossa R, Zelmanovich Y. *Pharm Eng* 1995;**15**:72–83.
33 Doblhoff-Dier O, Huss S, Litos R, Plail R, Unterluggauer F, Reiter M, Katinger H. *Process Biochem* 1991;**26**(4):201–207.
34 Nelson K. In: Prokop A, Bajpai R, Ho C, editors. *Recombinant DNA technology and applications*. New York: McGraw-Hill Inc.; 1991; Chap. 17.
35 Palluzi R. *Chem Eng* 2005;**112**(11):40–45.
36 Leach K. *Pharm Eng* 1993;**13**:54–62.
37 Odum J. *Biopharm* 1992;**5**(6):36–38.
38 Scutti L, Stark S. *Pharm Eng* 1993;**13**(1):47–52.
39 Hamilton BK, Schruben JJ, Montgomery JP. In: Demain A, Solomon N, editors. *Manual of industrial microbiology and biotechnology*. Washington (DC): American Society for Microbiology; 1986; p 321–344. Chap. 24.
40 Junker B, Lester M, Leporati J, Schmitt J, Kovatch M, Borysewicz S, Maciejak W, Seeley A, Hesse M, Connors N, Brix T, Creveling E, Salmon P. *J Biosci Bioeng* 2006;**102**(4):251–268.
41 James P. *Pharm Eng* 1998;**18**:72–82.
42 Wisniewski R, Burman C. In: Lubiniecki A, Vargo S, editors. *Regulatory practice in biopharmaceutical production*. New York: Wiley-Liss; 1994; p 407–445.
43 Charles M, Wilson J. In: Lydersen B, D'Elia N, Nelson K, editors. *Bioprocess engineering: systems, equipment and facilities*. New York: Wiley; 1994; p 3–67.
44 Chisti Y. *Chem Eng Prog* 1992;**88**(1):55–58.
45 Alford J, Allen B, Bisch S, Brill L, Clapp D, Danaher L. *Ann NY Acad Sci* 1994;**721**:326–336.
46 Ransohoff T, Murphy M, Levine H. *Biopharm* 1990;**3**(3):20–26.
47 Barry A, Chojnacki R. *Biopharm* 1994;**7**(9):43–47.
48 Fulton S, Shahidi A, Gordon N, Afeyan N. *Biotechnology (NY)* 1992;**10**:635–639.
49 Rathore AS, Kennedy RM, O'Donnell JK, Bomberis I, Kaltenbrunner O. *Biopharm Int* 2003;**16**(3):30–40.
50 Aronsson G, Zadorecki P. *Aust J Biotechnol* 1987;**1**(2):17–20.
51 Mahar J. *Biopharm* 1993;**6**(7):42–51.
52 Huang W. *Genet Eng News* 2005;**25**(19):42–44, 46.
53 Houtzager E, van der Linden R, de Roo G, Huurman S, Priem P, Sijmons PC. *BioProcess Int* 2005;**3**:60–66.
54 Hodge G. *Bioprocess Int* 2004;**2**(5):74–80.
55 Carson KL. *Nat Biotechnol* 2005;**23**(9):1054–1058.
56 Pora H. *Biopharm Int* 2006;**19**(6):72–75.

57 Farid SS, Washbrook J, Titchener-Hooker NJ. *Biotechnology* 2005;**21**:486–497.
58 Sinclair A, Monge M. *Pharm Eng* 2002;**22**(3):20–34.
59 Brecht R, Sandig V, Koch S, Riedel M. *Biopharm Int* 2005;**18**(7):22–30.
60 Stadler EL, Henon BK, Koiro D. *Bioprocess Int* 2007;**5**(6):78–82.
61 Sahoo T. *Chem Eng* 2004;**5**:34–39.
62 Odum J. *Pharmaceutical Engineering* 1992;**12**(1):8–12.
63 Nelson K. *Biopharm* 1988;**1**(3):34+.
64 Williams J. *Biotechnology (NY)* 1989;**7**:75–76.
65 Collentro A. *Pharm Process* 2004;**20**:14–20.
66 Dabbah R. *Bioprocess Int* 2006;**4**(5):18–23.
67 Glaser V. *Genet Eng News* 2006;**26**(1):38,40,41,43.
68 Baird A, Williams R. *Chem Eng* 2005;**112**(5):36–43.
69 Junker B, Stanik M, Adamca J, LaRiviere K, Abbatiello M, Salmon P. *Bioprocess Eng* 1997;**17**:277–286.
70 Tunner J, Katsoulis G, Denoncourt J, Murphy S. *Pharm Eng* 2006;**26**(4):58–70.
71 Gray G. *Pharm Eng* 1997;**17**(6):28–33.
72 Burkhart M, Wermelinger J, Setz W, Muller D. *PDA J Pharm Sci Technol* 1996;**50**(4):246–251.
73 Grunenberg D. *Valve Mag* 2006;**18**(2):12–19.
74 Chisti Y. In: Robinson R, Batt C, Patel P, editors. *Encyclopedia of food microbiology*. London: Academic Press; 1999; p 1806–1815.
75 DePalma A. *Pharm Manuf* 2003;**2**:32–38.
76 Del Valle M. *Pharm Eng* 1995;**15**:14–22.
77 Petrossian A, Smart N, Projetto R. *Biopharm* 1993;**6**(6):40–45.
78 Sharp J. *Good pharmaceutical manufacturing practice rationale and compliance*. New York: CRC Press; 2005; p 358.
79 Brooks CH, Russell PD. *Process Biochem* 1986;**21**(3):77–80.
80 Akers J. *Biopharm* 1994;**7**(6):43–47.
81 Akers J. *J Pharm Technol* 1995;**19**(3):28+.
82 Miller S, Bergmann D. *J Ind Microbiol* 1993;**11**:223–234.
83 Van Houten J, Fleming DO. *J Ind Microbiol* 1993;**11**:209–215.
84 Chisti Y. In: Subramanian G, editor. Volume **2**, *Bioseparation and bioprocessing: a handbook*. 2nd ed. New York: Wiley-VCH; 2007; p 533–574.
85 Janssen D, Lovejoy P, Simpson M, Kennedy L. In: Yu P, editor. *Fermentation technologies: industrial applications*. New York: Elsevier; 1990; p 388–393.
86 Kossik J. *Genet Eng News* 1998;**18**(6):21.
87 Kennedy L, Boland M, Janssen D, Frude M. In: Yu P, editor. *Fermentation technologies: industrial applications*. New York: Elsevier; 1990; p 383–387.
88 Maigetter R, Bailey F, Miller B. *Biopharm* 1990;**3**(2):22–29.
89 Odum J. *Biopharm* 1993;**6**(1):42+.

90 Hodgson J. *Biotechnology (NY)* 1994;**12**(10):983–987.
91 Junker BH, Wang HY. *Biotechnol Bioeng* 2006;**95**(2):226–261.
92 Gary K. *Am Biotechnol Lab* 1989;**7**(2):26–33.
93 Lee Y, Tsao G. In: Ghose T, Fiechter A, Blakebrough N, editors. Volume **13**, *Advances in biochemical engineering*. New York: Springer-Verlag; 1979; p 35–86.
94 Johnson M, Borkowski J, Engblom C. *Biotechnol Bioeng* 1964;**6**:457–468.
95 Elsworth R. *Chem Eng* 1972;**258**:63–71.
96 Salmon P, Buckland B. In: Rhodes M, Stanbury P, editors. *Applied microbial physiology. Oxford: Oxford University Press*; 1997; p 131–163.
97 Uttamlal M, Walt D. *Biotechnology (NY)* 1995;**13**:597–601.
98 Bradley J, Kidd A, Anderson P, Dear A, Ashby R, Turner A. *Analyst* 1989;**114**:375–379.
99 Macaloney G, Hall J, Rollins M, Draper I, Thompson B, McNeil B. *Biotechnol Tech* 1994;**8**(4):281–286.
100 Macaloney G, Hall J, Rollins M, Draper I, Anderson K, Preston J, Thompson B, McNeil B. *Bioprocess Eng* 1997;**17**(3):157–167.
101 Agayn V, Walt D. *Biotechnology (NY)* 1993;**11**:726–729.
102 Humphrey A, Brown K, Horvath J, Semerjian H. In: Fiechter A, Okada H, Tanner R, editors. *Bioproducts and bioprocesses.* Chapter 5.3. Berlin: Springer-Verlag; 1989; p 309–320.
103 Kell D, Markx G, Davey C, Todd R. *Trends Anal Chem* 1990;**9**(6):190–194.
104 Olsson L, Nielsen J. *Trends Biotechnol* 1997;**15**(12):517–522.
105 Groot W, Mijnhart R, van der Lans R, Luyben K. *Biotechnol Tech* 1991;**5**(5):371–376.
106 Collins S. In: Rhodes M, Stanbury P, editors. *Applied microbial physiology—a practical approach.* Oxford: Oxford University Press; 1997; p 75–101.
107 Petersen J. *Pharm Technol* 1988;**12**(7):42+.
108 Doblhoff-Dier O, Huss S, Litos R, Plail R, Unterluggauer F, Reiter M, Katinger H. *Process Biochem* 1991;**26**:201–207.
109 Blachere H. *Genet Eng News* 1998;**18**(7):23.
110 Jones K, Williams D, Phipps D, Montgomery P. On-line automatic tuning PID control of dissolved oxygen IFAC Symp. Ser. 8 (Adv. Control Chem. Processes). 1992, p 35–40.
111 Wilde F. *Genet Eng News* 1998;**18**(3):24.
112 Eiki H, Kishi I, Gomi T, Ogawa M. In: Ladisch M, Bose A, editors. *Harnessing biotechnology for the 21st century.* Washington (DC): American Chemical Society; 1992; p 223–227.
113 Hill C, Sinclair A. *Biopharm Int* 2007;**20**(7):38–42.
114 Kataria AR. *Pharm Process* 2007;**23**(7):14–16.
115 Dittmer M. *Pharm Technol Eur* 2006;**18**:48–40.
116 Yakren M. *Biopharm Int* 2006;**19**(5):34–38.

117 Bardone E, Lani B, Cassani G. *Pharm Eng* 1994;**14**:34–38.
118 Williams G. *Plant Serv* 1998;**19**(3):115–118.
119 Junker B, Lynch J, Leporati J, Schmitt J, Gieger J, Garah T, Stober M, Salmon P. *Bioprocess Eng* 1995;**17**:279–287.
120 Perkowski C. *Biotechnol Bioeng* 1984;**26**:857–859.
121 Todhunter D. *Bioprocess Eng Symp* 1989;**20**:97–103.
122 Kleppinger F. *Int Biotechnol Lab* 1987;**5**(3):28.
123 Kleppinger F. *Int J Pharm Compound* 1997;**1**(5):344–345.
124 Mobley K. *Plant Serv* 1998;**19**(4):147+.
125 Sorensen C. *Pharm Process* 2005; 21.
126 Wade D. *Maintenance Technol* 1989:44–46.
127 Kennedy C. *Biopharm* 1995;**8**(3):34–37.
128 Perkowski C. *Biopharm* 1987;**1**:62–65.
129 Mockus L, Vinson JM, Luo K. *Comput Chem Eng* 2002;**26**:697–702.
130 Eastman MMR, Sawers JR. *Chem Process* 1998;**61**(11):15–22.
131 Singh V. *Chem Age India* 1988;**39**(6):387–388.
132 Kossik J. *Genet Eng News* 1998;**18**(4):25.
133 *Guideline on the preparation of investigational new drug products (human and animal).* Washington (DC): CDER, FDA; 1991.
134 *Fed Regist* 1995;**60**(132):35750–35753.
135 Junker B. *Biotechnol Bioeng* 2001;**74**(1):49–61.
136 Werner RG, Langlouis-Gau H, Walz F, Allgaier H, Hoffman H. *Arzneim Forsch* 1988;**38**(6):855–862.
137 Junker B, Kardos P, Smizaski W, Brix T. In: Langer ES, editor. *Advances in large-scale biopharmaceutical manufacturing and scale-up production.* Washington (DC): ASM Press; 2004; p 510–554.
138 Junker B. In: Gad SC, editor. *Handbook of pharmaceuticals biotechnology.* New York: John Wiley and Sons; 2007; p 319–370.
139 Junker B. *Fed Regist* 1997;**62**(54):13429–13466.
140 Fry S. *Quality* 2005;**44**(1):56–61.
141 Vinson JM, Keck DD, Basu PK, Houston RB, Mockus L, Noren AR. *AIChE Symp Ser* 1998;**320**:171–177.
142 Junker B, Reddy J, Gbewonyo K, Greasham R. *Bioprocess Eng* 1994;**10**:195–207.

Chapter 10

Addressing Sustainability in Biomanufacturing Facility Design

Josh Capparella[1], Samuel Colucci[1], Daniel Conner[1], Robert Dick[1], and Amanda Weko[2]

[1] *Precis Engineering, Inc., Ambler, Pennsylvania, USA*
[2] *AGW Communications, Haddonfield, New Jersey, USA*

10.1 Introduction

Sustainability is defined as meeting the needs of the present without compromising the ability of future generations to meet their own needs. The engineering approach to sustainable biopharmaceutical and vaccine manufacturing facility design examines holistic optimization and integrates two primary concerns—using the least amount of energy and resources while generating as much biopharmaceutical or vaccine product as efficiently as possible. Sustainable engineering design is quite simple: evaluate consumption and maximize efficiency of output. The integration of sustainability into both process technology and business productivity yields the most positive outcome for the triple bottom line of people, planet, and profits (Fig. 10.1).

The triple bottom line comes from an understanding and appreciation of several justifications for sustainability in the biopharmaceutical industry [1,2].

The Ethical Basis Sustainability contributes to a healthy life science and healthcare community, which in turn supports and advances global health. Ethical sustainability decisions are grounded in the moral value of a healthy global population [3].

The Environmental Basis Sustainability supports energy and renewable resource conservation, health of a building site, and global environmental health. Environmental sustainability decisions meet the needs of the present without compromising the conditions of future generations [4].

Process Architecture in Biomanufacturing Facility Design, edited by Jeffery Odum and Michael C. Flickinger.

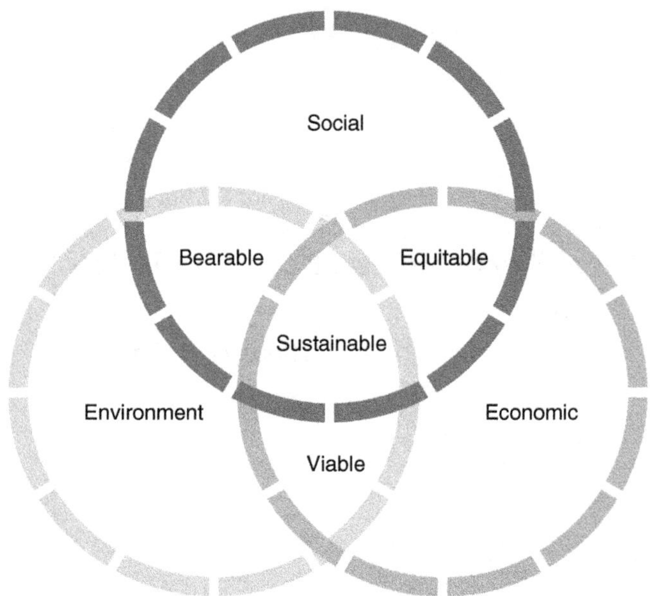

Figure 10.1 Triple bottom line Venn diagram.

The Operational Excellence Basis Sustainability contributes to a more efficient, flexible, and profitable facility. Operational excellence sustainability measures seek business success through innovation.

The Measurable Basis Sustainability can be measured by evolving industry standards, for example, US Green Building Council's Leadership in Energy and Environmental Design (LEED), UK-based Building Research Establishment Environmental Assessment Methodology (BREEAM®), Green Globes, Life Cycle Assessment, and others. The ease of measuring sustainability efforts justifies implementation.

The Regulatory Basis Sustainability has become a mandate. FDA, EMA, and other regulatory bodies and ISO compliance make sustainability a critical component of facility design.

Regardless of justification for sustainability, a biopharmaceutical facility can achieve significant process improvement by evaluating energy consumption. The complex engineering systems integrated into biomanufacturing facilities presents the depth of opportunity [5–8].

10.1.1 Economics of Sustainability

The economic impetus for sustainable design includes lower construction costs—reduced site preparation and landscaping and lower waste disposal costs of 50–98%; reduced operating costs—lower utility costs by 20–50% and reduced maintenance expenses; and a more productive environment—better tenant and/or staff attraction and retention, less absenteeism by up to 45%, and higher productivity up to 16%.

10.1.2 Energy Benchmarking in the Biopharmaceutical Industry

You cannot manage what you cannot measure. This axiom holds true in all aspects of life; it is hard to manage one's weight without a scale and it is hard to manage one's speed without a speedometer. Similarly, it is hard to manage the energy performance of a building without a similar indicating gauge.

There is no better way to assess a building's overall performance than by comparing it to other buildings similar in form, function, and location. Unfortunately, this can be an acutely difficult task to perform in the biopharmaceutical sector. While evaluation tools exist, the approach for comparison is often overgeneralized. Facility design and layout play a dramatic part in the overall utility consumption. Unlike in the commercial sector, the majority of the energy consumption in biopharmaceutical manufacturing buildings lies within the manufacturing process and supporting systems.

This chapter describes the roles of sustainability and reduced carbon footprint on the design and operation of biopharmaceutical facilities. It examines how to drive meaningful results into a process-driven environment by evaluating benchmarking, building envelope and materials, water, and environmental air quality [9].

10.1.3 Integrating Sustainability into the Design Process

In a conventional, linear design process, a biopharmaceutical client engages an architect. Once the building design is conceptualized, an engineer gets involved. When architecture and engineering design are completed and documented, a construction manager is brought on board. Progressive involvement of teammates lessens the impact each subsequent participant can have on the overall design. Changes become increasingly time-intensive and costly to incorporate.

10.1.3.1 Building Sustainability into the Process Early

An integrated team provides a much more efficient alternative for building sustainability into the process early. Early in the design process, all critical teammates—owner, architect, engineer, construction manager, and other key stakeholders—participate in a collaborative meeting or *charrette*. The charrette gets everyone into the room at the same time, for meaningful dialogue about the purpose and goals of the facility. The modern design principle, *form follows function*, is illustrated in this method. Design is optimized at the front end, minimizing effort and maximizing results.

10.1.3.2 Building Information Modeling

The use of Building Information Modeling (BIM) software enables an advanced level of coordination and project organization. BIM is an intelligent, 3-D modeling tool with which design and construction teammates can collaborate. The tool enables detailed visualization of the proposed facility to reduce uncertainty. From project conception, all members of the design team can use BIM to develop and visualize a biopharmaceutical facility in three dimensions. BIM is a means for communicating design intent to stakeholders and the construction team and is the tool by which construction can be verified. Clash detection using BIM can prevent problems during construction when the myriad building systems are installed. BIM not only promotes efficiency within a project but also supports sustainability by reducing printed drawings and lessening coordination-related travel fuels and expenses.

10.1.3.3 Integrated Utilities Approach

In a traditional engineering approach, systems are considered individually. A boiler plant is designed to achieve a target heating load. A chiller is designed to achieve a desired cooling load, and so on. In the complex and energy-intensive biopharmaceutical industry, this method creates redundancies.

An integrated utilities approach examines all utilities together to improve efficiency. With integrated utilities, the whole is greater than the sum of its parts. For example, a chiller might be designed to function as a heat pump instead of rejecting heat to the atmosphere via a cooling tower. The integrated approach evaluates how to generate all utilities with the smallest possible energy footprint.

On-site renewable energy sources are a component of integrated utilities. Most common are photovoltaic arrays to capture solar energy. The solar energy parallels and supplements incoming city electric service, offsetting some of the energy expense and supplying usable power. On-site solar generation offers low risk, as any failure in the photovoltaic system will be compensated by the city system, and vice versa. Wind turbines offer similar low risk energy generation but are less commonly used in the biopharmaceutical industry.

Cogeneration and trigeneration are the primary examples of integrated utility approach. Both processes utilize waste energy for additional energy usage and consequent energy efficiency. Cogeneration is a process to simultaneously produce electrical and thermal energy. Also known as combined heat and power (CHP), cogeneration uses the waste heat from generating electricity for other uses, such as HVAC and water heating. Utilizing the waste product makes cogenerating an extremely energy-efficient process. Trigeneration is a similar waste-capturing process that generates electricity, heat, and cooling energy with a single process. Trigeneration is also called combined cooling, heat, and power (CCHP) [10,11].

10.1.4 Sustainable Building Benchmarking

In the US, two primary rating systems exist to benchmark and certify sustainably designed and operated buildings. It is important to note that even when certification is not pursued, the criteria for either or both systems can still be used as a design rubric for driving sustainability in facility design.

The US Green Building Council's LEED program has certified over 44,000 buildings. Upfront costs for certification can be expensive, adding an average of 4–10% to overall design fees. Particularly relevant to the high-energy biopharmaceutical industry, LEED scoring considers a performance-based approach to credentialing, placing emphasis on total building energy modeling, water efficiency, and smart grid thinking [8].

The Green Globes program is a faster and less expensive alternative to LEED. The US General Services Administration (GSA) has approved Green Globes as an equivalent certification system to LEED. The program awards one to four globes based on a 1000-point system that addresses site, energy, water, emissions, indoor environment, materials and resources, and project management [6].

UK-based Building Research Establishment Environmental Assessment Methodology (BREEAM®) is a global benchmarking program that has certified over 540,000 buildings in over 75 countries since its launch in 1990. It takes a comprehensive, scientific approach to reviewing projects across 10 categories (energy, health and well being, innovation, land-use, materials, management, pollution, transport, waste, and water) [5].

10.1.4.1 Commercial Building Benchmarking

In the commercial building sector, there are extensive resources that make energy benchmarking data readily available. The Commercial Buildings Energy Consumption Survey (CBECS), managed by the US Energy Information Administration under the Department of Energy (DOE), has collected and analyzed data for many different types of commercial buildings. The CBECS database is the most robust resource currently available, in both quantity of buildings surveyed and variety of ways the data has been broken down for comparative purposes.

The CBECS is a national survey that collects information on US commercial buildings including energy-related building characteristics, energy consumption, and energy expenditures. The broad categorization of commercial buildings includes any in which at least half of the floor space is used for a purpose that is not residential, industrial, or agricultural. By that definition, commercial buildings include office buildings, retail, restaurants, warehouses, schools, hospitals, correctional institutions, and religious buildings [12].

As of the time of this writing, CBECS data for 2012, which was released in 2016, was the most current information available. The combination of building surveys and energy supplier surveys makes the CBECS database one of the best sources for reliable energy benchmarking data [12].

CBECS data provides the basis for the Environmental Protection Agency (EPA) ENERGY STAR building performance rating system. ENERGY STAR's energy performance indicator (EPI) tools allow users to input simple statistics about building size, type, location, occupancy, and overall utility consumption in exchange for a 1–100th percentile rating that benchmarks the performance of the user's building against similar buildings. Buildings scoring in the 75th percentile or higher are eligible for ENERGY STAR certification. The assessment provides a normalized model that adjusts for inconsistencies in form, function, and location based on the CBECS database [9].

Commercial buildings are similar in that their primary engineering design function is to maintain occupant comfort. Lighting and space conditioning (heating, ventilation, and air conditioning, or HVAC) energy consumption details are relatively similar and vary predictably based on a few conditions (Fig. 10.2). In the Energy Information Administration Annual Energy Outlook 2012, lighting and space conditioning account for 13.6% and 47.3%, respectively (60.9% combined) of the total building energy consumption footprint.

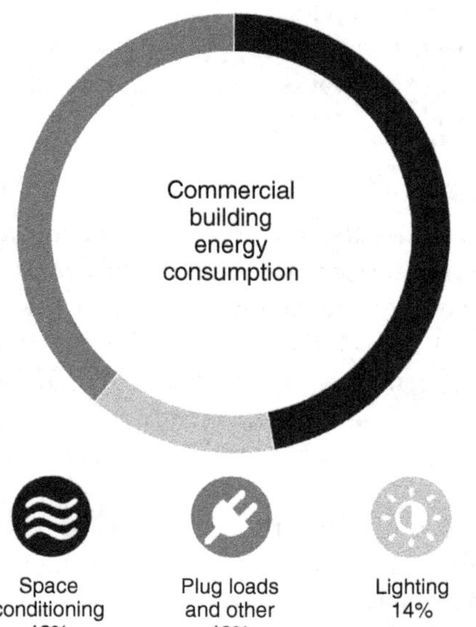

Figure 10.2 Commercial building energy consumption by end use.

The remaining consumption is allocated between various plug loads and water heating requirements [12,13].

As most commercial building energy consumption is dedicated to lighting and environmental comfort systems, the ENERGY STAR tool needs only to prompt the user for a small number of inputs to be able to quickly normalize the subject building to the model and develop a highly reliable energy performance benchmark.

10.1.4.2 Biopharmaceutical Building Benchmarking

Moving out of the commercial building realm into industrial and manufacturing buildings, a problem arises. In general manufacturing facilities, energy is consumed for standard commercial uses (e.g. lighting and HVAC) and also to conduct the processes required to manufacture product. The consumption frequently goes beyond powering the production equipment. It encompasses increased base building utility system demands for additional or more stringent HVAC requirements; additional lighting intensity requirements; and additional specialized utilities such as compressed air and vacuum systems. The manufacturing energy consumption is highly variable and difficult to normalize within a benchmarking model.

The variability problem is exacerbated in the biopharmaceutical industry owing to the need to maintain critical environments for production with respect to temperature, humidity, room pressurization, cleanliness, containment, and other factors. Building HVAC loads that support specialized processes are many times greater than in a similar-sized commercial building. As a result, overall building energy usage intensity (EUI) is typically an order-of-magnitude larger or more. The average recent commercial building (constructed after 2000) has an average EUI of 81.4 kBtu per square foot (257 kWh/m^2) [14]. The average biopharmaceutical facility has an average EUI of 1210 kBtu per square foot (3819 kWh/m^2) [15].

Increased airflow quantity, ventilation, and filtration requirements coupled with tighter environmental controls for temperature and humidity combine for higher building HVAC loads. The abovementioned factors lead to higher energy consumption by increasing the cooling, preheat, and reheat loads; fan energy consumption; and dehumidification and humidification loads.

For simple comparative purposes, a standard air-handling unit serving an office area will condition and supply between two and six air changes per hour (ACPH) using a variable air volume control strategy. An analogous air-handling unit serving an ISO 14644 Class 8 (EU Grade C in operation) clean room in which pharmaceutical product is being manufactured would typically condition and supply between 20 and 35 ACPH in a constant volume control strategy. This represents a four-to-six-fold increase in HVAC energy expenditure before tighter and lower temperature and relative humidity control requirements, increased outside air requirements, and additional filtration are considered. The additional energy consumption of the manufacturing systems and processes easily illustrates how the overall energy consumption of biopharmaceutical manufacturing facilities is far more intensive than their commercial facility counterparts.

On the basis of the unique nature of the biopharmaceutical industry and the inability of existing benchmarking tools to effectively gauge energy performance, the EPA ENERGY STAR program developed a meaningful comparative tool uniquely dedicated to the biopharmaceutical industry. Led primarily by energy managers from several major drug manufacturers, the ENERGY STAR program began a biopharmaceutical industry focus early in 2005 [16]. From those initial efforts, the first version of the pharmaceutical industry EPI tool was developed and published at the end of 2008. At the time of this writing, the last update was in the summer of 2012.

The tool seeks to normalize and benchmark energy performance of biopharmaceutical facilities located in the US across four major categories:

- Bulk Chemical (active pharmaceutical ingredients and excipients): Areas where both active and inactive ingredients are prepared in bulk form, including mixing, milling, and drying of powders, and the mixing of liquids, gels, and creams [15].
- Fill/Finish: Areas used for fill or finish processes or other manufacturing, production, or warehousing with climate-controlled environments due to product requirements, including tableting; encapsulation of powders or liquids; the filling of liquids, gels, or creams in their consumer packages; and final product packaging [15].
- Research and Development: Laboratory buildings including animal laboratories, storage space, in-process labs, quality assurance labs, and pilot plants [15].
- Other: Any area that does not fall into one of the main categories [15].

As data regarding space allocation in the abovementioned categories is not collected in the Census of Manufacturers, data had to be provided directly from participating entities in the program. Similar to the commercial buildings program, the pharmaceutical EPI takes several inputs regarding percentage of facility floor space allocated to the above functions, hours of operation, location, and utility costs and, in return, provides a 1–100th percentile rating for the facility [15].

This tool also provides the EPA with benchmarking data to grant the ENERGY STAR rating to facilities scoring in the 75th percentile or above, which was not previously possible for biopharmaceutical facilities [9].

10.1.4.3 Variations in Benchmarking Data

When the EPA initiated discussions about developing a plant-level benchmarking tool with biopharmaceutical manufacturers, most initial reactions from experts within these companies were skeptical about whether a useful benchmark could be developed.

The typical approach for developing EPIs for other industries is to relate plant input to plant output as expressed as a unit of production. The pharmaceutical EPI does not. It was decided that the value of product shipments would not provide a uniform measure of activity since, as discussed above, while the level of production is not insignificant, much of the energy use in this industry is devoted to environmental controls [15].

Those involved in the tool's development admitted that normalization of all facilities against the four allocated categories might not be entirely appropriate. It is recommended that additional categories be developed in accordance with the International Society of Pharmaceutical Engineering (ISPE) Baseline® Guides at a minimum. For example, separate categories could differentiate between the large energy consumptions that exist in oral solid dose, aseptic, and biopharmaceutical processing. The engineering best practices that govern design for these types of manufacturing processes are vastly different. Yet, under the current model, the processes are grouped.

Even among the subsets of biopharmaceutical processing as described in the ISPE Baseline Guides, certain design factors drastically affect energy consumption from one facility to the next. For example, cross contamination concerns at a biopharmaceutical facility handling multiple products may encourage the HVAC system designer to utilize 100% outdoor air, whereas a single product facility may utilize a recirculation HVAC system with supplemental filtration to minimize the risk of airborne cross contamination. The impact of a 100% once-through system versus a recirculation system can be as much as three times the energy consumption [9].

10.1.4.4 Making a Meaningful Impact to Facility Energy Reductions

The main purpose of the examples presented is to show that once a facility is designed, its energy performance is largely locked in, but significant variations can exist. Therefore, it comes into question what the benchmarking data actually shows; is a 75th percentile score depicting the performance of a building that display significant energy management a result of best practices or is it simply a result of the inherent facility design driven by less stringent process requirements? And does it matter?

While the type of facility analyzed may be precluded from ultimately achieving an ENERGY STAR certification, the better use of the EPI tool is as a gauge to measure a facility's energy management strategies over time. A reduction in annual score in an otherwise unchanged building may hint that a tune-up is needed via retro commissioning (RCx). The average biopharmaceutical facility would need to reduce overall energy consumption by a third to reach the ENERGY STAR threshold—the equivalent of the total EUI of five commercial office buildings of equal size.

While such a task may seem daunting at the onset, most biopharmaceutical entities have aggressive internal mandates to reduce energy expenditure year over year. With the current downward pressure on internal costs, a 5% target reduction annually is not uncommon. It may place the ENERGY STAR certification within reach in the not-distant future. However, achieving this lofty annual energy reduction goal goes far beyond conventional energy retrofit projects.

On the basis of the breakdown of biopharmaceutical energy consumption by end use, it appears that common energy retrofits such as lighting upgrades or motor replacements do not make a significant impact. Lighting accounts for only 2–3% of overall energy usage in biopharmaceutical manufacturing; even a 50% reduction in lighting usage will barely yield significant savings as it pertains to overall facility EUI. Lighting retrofits certainly should not be discounted, especially as utility rebate programs frequently subsidize them, which enhances the overall project payback.

In the biopharmaceutical industry, energy is tied into the process and supporting systems. For example, reducing the overall ventilation rate by only 5–10% in a pharmaceutical facility is the equivalent of eliminating the total lighting energy use in the facility.

Significant reductions cannot be achieved unless the processes are analyzed, challenged, and optimized. Doing so is a task that requires significant engineering expertise and knowledge regarding the regulatory and cGMP (current good manufacturing practice) requirements of the manufacturing processes. Maintaining an environment to ensure product integrity, operating personnel protection, and comfort are crucial factors that must be accommodated when making any changes to the process to save energy [9,17].

10.1.4.5 Energy Efficiency: Current Trends

In the past, energy expenditures have been a small portion of the total cost of goods in the biopharmaceutical industry and thus were not a major concern. Over the past decade, with increased economic headwinds and rising energy costs, energy consumption has been scrutinized. The impetus to lower internal and operating costs combine with an industry push toward environmental stewardship and carbon reduction. As a result, energy efficiency has been driven into technologies used in the industry. For example, isolator technologies currently in production utilize catalytic converters to break down vaporous hydrogen peroxide (VHP) used during decontamination cycles into nonharmful constituents. The use of catalytic converters speeds up the aeration process and negates the need for the HVAC system serving the isolator to go into purge model (100% once-through) during aeration.

As the industry progresses, energy consumption and cost will continue to face downward pressure. Until or unless a low-cost, clean energy source is developed and industrialized, there is no near-term end in sight. Many large biopharmaceutical companies have implemented energy management programs to varying degrees of success. For some, the challenge remains in how to continue to make improvements once all of the quick wins have been garnered, while still remaining fiscally responsible. For others, mainly contract and generic manufacturers, efforts are ramping up to begin to manage and control energy costs. Where normal business practice has always been first-cost sensitive, operating costs are quickly rising to the point where they can no longer be ignored.

Normalizing energy consumption across the biopharmaceutical industry as a whole has proven difficult. Placing all of the operations that occur within the manufacturing processes into four separate categories for analysis may not always be specific enough for a true apples-to-apples comparison between facilities. Unlike in the commercial building sector, a pharmaceutical facility meeting the ENERGY STAR performance threshold may have achieved the certification simply as a result of the process requirements driving the design of the facility, without any significant introduction of energy management best practices. Similarly, a facility with many energy optimization strategies in place may ultimately be precluded from achieving certification based on the original factors considered during its design.

Benchmarking data using the EPI proves valuable if the results are interpreted in the proper way and, even more so, if tracked over time. Using the benchmarking tool to establish baseline energy performance and updating it on a regular basis allows the user to monitor how well a facility performs against its own baseline, which is key to validating the efficacy of the overall energy management program. While the energy performance range of a biopharmaceutical manufacturing facility is often determined even before the facility is built, through consistent energy management best practices, the facility can optimize the hand it was dealt. This not only is the environmentally responsible thing to do, it is now recognized as good business practice as well [18].

10.1.5 Cost of Utilities

Costs of generated or secondary utilities should be considered during the design of any facility. When analyzing the operating costs of equipment, it is important to capture all associated costs so that a realistic picture of the overall operating expense is generated. It is also important to consider not just the peak operating condition, but the anticipated operational profile over the course of an entire year. This information is especially important when navigating the countless options during the design process. An understanding of the costs of utilities and secondary/generated utilities informs the design process. The knowledge brings context to the decision-making process by providing perspective and scale to compare the incremental costs of various options or design philosophies and their impact on the life-cycle cost of the system.

Primary utilities include natural gas, electricity, and water—those utilities that are purchased directly by a company from public or private utility providers. Water usage typically incurs costs for the water and wastewater that is discharged to the drain. Wastewater is typically equal or more expensive than the cost of water, making wastewater reduction an important consideration. In some cases, facilities purchase chilled water, hot water, and/or steam from a utility company, in which case they are handled like primary utilities because the costs associated with generation are neatly packaged into the cost per unit charged. However, chilled water, steam, and hot water are typically considered secondary utilities.

Secondary utilities are generated by the systems and technologies that consume primary utilities. These include chilled water, condenser water, hot water (heating or domestic), steam (for heating, humidification, or process), clean utilities [WFI (water for injection), WPU, clean steam, etc.], and conditioned air. Conditioned air may have different costs for each room classification based on different temperature and humidity requirements. It follows that each room classification will have its own operating cost per square foot based on the quality and quantity of air required.

Often, when making design decision to minimize energy, water consumption, or utility costs, it comes down to either pay now (first cost) or pay later (operating costs) because more efficient technology is often more expensive upfront but justifies the higher initial investment with lower operating costs. Life-cycle cost analysis should be performed to make design decisions that will optimize the balance between minimizing first-cost and minimizing life-cycle operational costs. A complete picture of the total costs of each primary and secondary/generated utility is critical to achieving the most energy-efficient, cost-effective design.

Total utility costs should be developed that take into account the total operational costs associated with the generation and delivery of each utility. The costs should include primary utilities required to operate major equipment (e.g. boilers, chillers, and steam generators), primary utility costs to operate supporting equipment (e.g. pumps, fans, and cooling towers), and costs associated with maintaining the operational integrity of the systems (e.g. water treatment, makeup water, and maintenance).

For example:

- Chilled water should take into consideration the evaporation rate from cooling towers, pump electricity consumption, cooling tower fan energy consumption, water treatment and chemical costs, and chiller energy consumption over the annual load profile. To measure cost per ton, measure chilled water flow and temperature at the inlet and outlet of each chiller.

- Condenser water should take into consideration the evaporation rate from cooling towers, pump electricity consumption, cooling tower fan energy consumption, and water treatment and chemical costs. To measure cost per BTU (British Thermal Unit) rejected, measure condenser water inlet and outlet flow and temperature. To measure evaporation rate, evaluate makeup water flow.
- Hot water from a natural gas-fired boiler should take into consideration natural gas, boiler combustion fan electricity consumption, boiler exhaust fan electricity consumption, water costs (including return percentage, blow down rates, leaks, and losses), water treatment and chemical costs, feed water pump electricity consumption, and distribution pump energy consumption.
- Steam should take into consideration natural gas, boiler combustion fan electricity consumption, boiler exhaust fan electricity consumption, water costs (including return percentage, blow down rates, leaks, and losses), water treatment and chemical costs, feed water pump electricity consumption, and condensate pump energy consumption.
- Water treatment parameters include chemicals and pumping. Wastewater costs are typically greater than costs for city water itself.
- The cubic feet per minute (CFM) of conditioned air to each room classification should factor in chilled water, humidification steam, heat (steam, hot water, or direct-fire natural gas), and fan energy consumption. To calculate per unit of consumption, measure airflow from the air-handling unit and CFM at the air-handling unit outlet. Measure enthalpy of return/mixed air before and after cooling and heating coils [19].

10.1.6 Is Net Zero a Possibility?

Interchangeably known as zero-energy, zero net energy, net zero energy, or the more common and simplified net zero buildings use an annual total amount of energy less than or equal to the amount of renewable energy generated on site. A zero-energy campus is an energy-efficient group of buildings owned by a single institution where actual annual delivered energy is less than or equal to on-site renewable exported energy. A zero-energy portfolio is an energy-efficient group of buildings or sites owned or leased by a single entity in which actual delivered energy is less than or equal to on-site renewable exported energy [20].

Achieving net zero in a biopharmaceutical facility poses a greater challenge than a typical commercial building owing to relatively large energy consumption and corresponding space requirements of renewable energy sources. The challenge is how, in a limited footprint, to generate enough energy to balance what is consumed. A partial net zero approach adopted by some biopharmaceutical companies involves buying energy from a renewable resource (e.g. an unaffiliated solar farm) in exchange for energy credits (e.g. Solar Renewable Energy Credit or SREC) and federal, state, or utility rebates or incentives. For example, in the state of New Jersey, which offers one of the higher value SRECs, it is not uncommon for biopharmaceutical companies to purchase energy from independent solar energy providers. On the individual company electric meters, the facilities are not net zero; however, the purchased solar energy helps counterbalances the grid usage on a portfolio level.

In 2011, biopharmaceutical company GlaxoSmithKline committed to carbon net neutrality for all of its global facilities by 2050, with a goal of 10% reduction in energy usage by 2015 [21–23].

10.1.7 Process Drives the Design

Several decisions and details of the biopharmaceutical manufacturing process have major impacts on a facility's carbon footprint or net neutrality. Two of the most critical decisions—open versus closed process and single-use versus fixed process equipment—drive much of the design development and corresponding facility design and room environment conditions.

A closed process protects the product and requires a less energy- and resource-intensive facility. Once a process is open, then the facility environment, production flows, and personnel gowning are required to protect the product. The ISPE Biopharmaceutical Manufacturing Facilities Baseline Guide offers strategies to close the process, thus reducing risk, HVAC energy, and disposable gowning and cleaning resources.

Fixed process equipment typically has a high associated energy use. Single-use or disposable technology has less demand for energy but creates more material waste. From a sustainability perspective, a balance of resources and energy and risk assessment should be reviewed on a product and facility basis [24].

10.1.8 Risk-Based Approach to Sustainability

The risk-based approach to sustainability is grounded in the FDA's Quality by Design (QbD) concept. This concept states that quality should be built into a product based on a comprehensive understanding of the product, the process by which it is developed and manufactured, the risks involved in the process, and how best to mitigate those risks.

In simpler terms, the FDA offers biopharmaceutical companies flexibility to drive toward innovation, so long as the company can prove its products and processes remain safe. It puts the burden of responsibility on the company to prove benefits and the same or less risk than in traditional methods (e.g. cleanliness and efficacy). The guidelines are left intentionally vague, leaving the industry to self-regulate [25].

10.1.9 Risk in a Closed Process

In biopharmaceutical facility design, there has been a push to close the process so product is rarely exposed to the room environment. A closed process isolates both process and product from people and the environment. When the process is closed, the amount of clean room space required for manufacturing can be reduced. Consequently, associated airflow, airlocks, energy, and waste from disposable gowning are also reduced.

In a typical facility, operations take place in Grades B, C, and D clean room classifications. A closed process can be operated in Grade D or a controlled environment because everything is piped or tubed from one piece of equipment to another.

This "ballroom" concept applies an idea popular in research laboratories to the cGMP manufacturing facility—multiple activities occurring in a shared space. Instead of the traditional clean rooms divided by airlocks, the ballroom concept used closed process equipment grouped into a single room. This larger, clean process room contains multiple types and sizes of isolators that enclose each process while offering flexibility to be moved around the room. The ballroom is engineered to support proper function of the closed process equipment but is a less restrictive environment overall. Several processes might be running at once. The approach affords flexibility to change over time, accommodates the manufacture of different products, and improves communication and interaction among those who work in the environment. Finally, the entire ballroom can be viewed as a showcase of a company's manufacturing activities [26].

Despite its appearance as a valid approach, with sustainable energy and waste reducing effects, closing the process is not yet a common practice. First, it can be difficult from an engineering perspective to fully enclose the entire process. Equipment connections may be open for even a short period of time. The ballroom concept puts multiple products in the same space, meaning that maintenance on a single machine may affect multiple pieces of equipment or products.

The biopharmaceutical industry has been averse to any potential risk, whether real or perceived. Human health remains the priority with sustainable concerns valued second. As a few early adopters pave the way, engineers anticipate a change in the next 5–10 years [27].

10.2 Process Architecture

10.2.1 Process Technology Impact on Footprint

Process technology creates complexity with respect to the building footprint. As companies move to hybrid single-use (disposable) and fixed technology or as companies move to 100% single-use technology, facilities are becoming more flexible and scalable.

Single-use (disposable) and hybrid technology create more opportunities for flexibility and scalability. Single-use technology offers ease of operations, no risk of cross contamination, and no need for cleaning (or the associated energy and water). The process uses much less WFI and chilled water (with reduced energy for less cooling need) and from a utility standpoint, is much more efficient. However, there is the associated medical waste product of disposable technology. The design team should balance waste of materials versus waste of energy, water, detergents, and so on, while giving consideration to how and where the wastewater drains and how and where the medical waste is disposed (e.g. incineration or landfill).

The use of disposable technology in combination with a closed process results in smaller overall building footprints with respect to processing area (Fig. 10.3). The suite gets smaller but the air lock gets bigger to enable moving material in and out of the suite. A fixed technology environment requires tanks, piping, and all support components in the room to accommodate clean in place (CIP) and steam in place (SIP). In a disposable scenario, the tank remains but everything else enters and exits the room via the airlock. Fewer fixed components required smaller area [29].

Another impact to process technology comes via the cleaning strategy. Most commonly, a CIP strategy is utilized. With CIP, a hard-piped network of stainless steel pipes extends to a central CIP mechanical rom. All chemicals and WFI are transferred via the network direct to vessels. It is an expensive strategy but is not uncommon in large commercial facilities.

CIP carts provide a sustainable alternative. Instead of the infrastructure associated with traditional CIP, a facility can use mobile carts. The carts function like mobile pressure washers, hooking to the utilities in each room to clean vessels. Once hooked up manually, the carts are fully automated. CIP carts are not appropriate for high-volume facilities with high turnover, where cleaning occurs more frequently but may make smart economic sense for smaller or multiuse facilities. In general, bioreaction growing times average 30–45 days, with another 4–5 days downstream. For those 45 days, the CIP system is idle. CIP carts save on the water and chemicals that would have gone through the system and down the drain. However, they do present the associated trade-off of labor to manually operate each cart versus a single operator of CIP mechanical room technology.

10.2.2 Tech Transfer and Scale Up

When a process begins at the bench scale in a laboratory, it is typically performed in a glass line reactor. When the process is scaled up from 1 to 10 to 50 L or more, process results will vary due to subtle differences in quantity, temperature, and vessel material (glass to stainless steel to disposable plastic bag). As

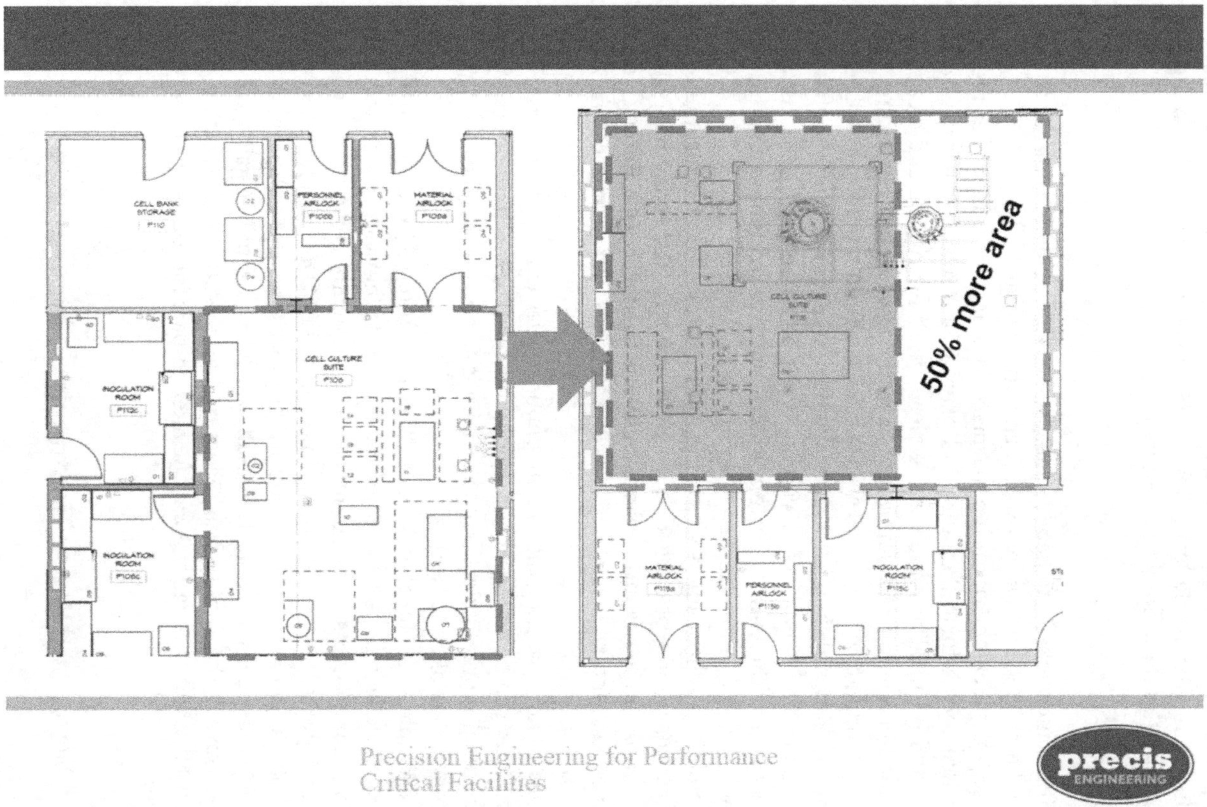

CELL BANK STORAGE
PERSONNEL AIRLOCK
MATERIAL AIRLOCK
CELL CULTURE SUITE
INOCULATION ROOM
INOCULATION ROOM
MATERIAL AIRLOCK
PERSONNEL AIRLOCK
INOCULATION ROOM
50% more area

Precision Engineering for Performance
Critical Facilities

Figure 10.3 Floor plan diagram illustrating a fixed 1000-L bioreactor space with the outline of two disposable suites, showing 50% more area available [28].

technology scales up or varies, it is important to analyze the application and ensure consistent results. An understanding of both performance and product are required. For example, cell production is higher in stainless steel tanks. If a facility using disposable technology wants a certain cell density in the final product, more volume might be required.

10.2.3 Water

The ISPE Handbook: Sustainability suggests that the industry should "challenge the need for significant use of process and utility water" [30].

Water is the lifeblood of the biopharmaceutical facility. It is used as the basis of nutrient media for cell culture and plays a vital role in downstream separation and purification processes. Water is used for rinsing and cleaning in place, for sterilization and sanitization, and for black utilities including boiler makeup, cooling tower condenser water makeup, and humidification. Incoming water quality and chemistry varies significantly depending on variables such as geography and time of year. Varying levels of water quality require different treatments. Therefore, a custom water solution is recommended for every facility. There is no one-size-fits-all approach to water.

As buildings reliant upon water, biopharmaceutical facilities should make water conservation a priority for both environmental and economic responsibility. Conservation supports sustainability and business goals. The engineering design process should include analysis of water flow into and out of a facility to minimize waste.

Owing to large water usage, many opportunities exist to reduce and reuse water. Water can be collected at waste points for reuse. Where direct reuse is not permissible or feasible, thermal reclaim is also available. The economics of any scenario are often favorable.

For example, the Morphotek Pilot Plant in Exton, Pa., was designed to be 30% more water efficient as per LEED standards; however, when individual building components are examined independently, efficiency increases. Note that process water is not factored into LEED efficiency calculations. Morphotek's water-efficiency measures for process water further enhance the building's water conservation. The result is just shy of 2 million gallons of water per year saved. The facility was recognized as the 2013 ISPE Facility of the Year Sustainability Category Award winner, in part for its water conservation program [31,32].

10.3 Water and Water Treatment

Treatment methods remove impurities from water and impact sustainable objectives in terms of water usage, filtration and purification device expenses, and energy from associated treatment technology.

Water treatment removes impurities including suspended particles—silt, debris, and organic or inorganic colloids—that cause turbidity in water. Treatment also removes dissolved organic compounds including calcium and magnesium (that contribute to hardness), silicates, ferrous compounds, chlorides, aluminum, phosphates, and nitrates and also decayed live matter, fats and oils, and solvents; microorganisms and bacteria; and dissolved gases [33].

10.3.1 Incoming City Water

Incoming water analysis will determine the type of pretreatment required. Typically, a media filtration and softening process is utilized at a minimum, both for compendial (USP purified) and noncompendial (potable) applications. Compendial water meets requirements for total organic carbon (500 ppb or less), pH between 5 and 7, conductivity less than 1.3 μS/cm at 25°C, and a maximum bioburden action limit of 100 colony-forming units per milliliter. Noncompendial potable water meets EPA minimum national primary drinking water regulations and includes process or utility water meeting the requirements of potable water but with additional treatment to meet chemical and/or microbiological requirements [identified by final treatment steps, e.g. reverse osmosis (RO)].

These differ from USP WFI, which meets all requirements for USP purified water (PW) and the final treatment process of distillation or RO, at a minimum. WFI has a maximum bioburden action level of 10 colony-forming units per 100 mL and a maximum endotoxin level of 0.25 EU/mL.

10.3.2 Filtration and Softening

Filters are used to remove particulates from the water stream. Filter materials range from multimedia to carbon to sand. Water treatment filters and equipment can significantly offset the amount of chemicals used to treat water downstream and should be considered as a sustainable initiative via cost-benefit analysis.

Water softeners almost always make economic sense. Water softeners should be used if the water hardness exceeds 5 ppm. The softener prevents scale build-up on the heat transfer section of the boiler, which can decrease the boiler's efficiency. The softener also decreases the amount of chemicals that must be used. Softeners require regeneration, a process that can take several hours. As such, they are available in twin and triplet units so the softener is not off-line while regeneration occurs.

Most water softeners operate using the sodium zeolite process, which uses an ion exchange to remove hardness and replace it with highly soluble sodium ions. These ion exchange softeners remove hardness ions Mg^{++} and Ca^{++} and replace with the less harmful Na^{+} ion via ion exchange processes. Calcium and magnesium would otherwise react with carbonate ions and precipitate out of the water as scale [34].

Dealkalizers are used in applications where the water is known to have high alkalinity. The dealkalizer reduces bicarbonate alkalinity, which minimizes carbon dioxide production (the primary cause of condensate line corrosion), decreases chemical consumption, and lowers blow down rates to reduce fuel consumption. The dealkalizer operates on a similar process to a water softener, except instead of using a cation resin, it utilizes a strong anion resin to remove negatively charged ions from the water via a chloride cycle process. It replaces bicarbonate sulfate, nitrate, and silica with chloride ions.

Softeners and dealkalizers reduce the use of chemicals such as antiscalants and polymeric dispersants designed to further reduce water hardness. Softeners and dealkalizers both must go into regeneration to remove unwanted ions when saturated. Brine solutions are used to recharge resin beds. Regeneration waste streams are typically not candidates for capturing owing to harmful chemistries. However, they can potentially be combined with other waste streams, such as WFI blow down and clean steam condensate, to create blends of reusable water. A holding tank should be considered to capture waste streams. Mass balance is important to understand and predict comprehensive water quality [35].

10.3.3 Deionization and Reverse Osmosis

Deionization (DI) is a similar process to water softening in which ion exchange resins remove ionized impurities by exchanging them with H+ and OH− ions. The DI process removed ionized impurities to the lowest possible levels. DI equipment is available in mixed bed or separate bed (anion and cation resins separated) systems. Beds are grounds for microorganism growth and must be accommodated downstream in the purification process. The separate bed DI system uses twice as many regeneration cycles. A mixed bed is a more economical solution and saves on water regeneration. Usage versus capacity should be considered when designing the system.

RO is used to remove contaminants less than 1 nm nominally. The process removes approximately 90% of ionic contamination, including most organic and nearly all particulate matter. As it does not remove dissolved gases, RO equipment is often paired with a membrane degasser to remove CO_2. Oxidizing agents can damage RO membranes, so upstream chlorine removal is paramount. Chlorine can be removed via carbon bed or sodium metabisulfate injection. Carbon beds are more expensive and provide breeding grounds for bacteria, requiring weekly steam or ozone sanitization.

A cost-benefit analysis should be conducted when choosing sterilization technology. Steam sanitization requires energy to heat the media and chilled water to cool it. Ozonated systems with UV (ultra violet) destruct can perform the same process with much less energy but require more significant initial equipment costs.

10.3.4 Water for Injection (WFI)

USP WFI, the purest form of water used in biopharmaceutical applications, is used for diluting buffers, hydrating dry powder media, and making solutions. WFI can be generated via three primary methods: vapor compression (VC), multieffect still (ME), or RO. To generate RO, water flows through a series of membranes until it becomes super filtered. VC and ME stills make WFI by evaporating water under pressure to separate out impurities. RO is appropriate for smaller usage; ME works best for mid-range usage; VC is best for very large WFI requirements. The different grades of WFI and their generation methods should be evaluated based on a facility's process and demand [30].

10.3.4.1 Ambient, Intermediate, and Hot WFI Requirements

Almost all process uses besides cleaning require ambient WFI at 25°C. When preparing media and buffers, WFI temperature requirements fluctuate; temperatures range from 30 to 32°C down to 25°C. The problem at 25°C is that the WFI is no longer self-sanitizing. Most facilities use an ambient loop that can be increased to 80°C at regular intervals in order to sanitize the loop. At that hot temperature, the WFI becomes self-sanitizing. Alternatively, the facility could maintain a consistent 80°C and spot cool it to utilization temperature. In either case, the associated energy for heating and cooling becomes excessive.

Ambient loops risk contaminant growth in water that is not hot enough to self-sanitize. Regularly increasing the temperature is mandatory. Keeping the water at 80°C all the time is a safer method. In either case, when the WFI needs to be cooled or heated—flowed through a heat exchanger or flowed through a chiller—the water runs down the drain until it reaches desired use point. WFI in general and wasting it down the drain due to temperature are expensive.

Individual point-of-use heat exchangers are an option. However, this becomes extremely expensive owing to the associated automation, drain valves, piping, trim, and so on, that are required.

A better solution is to employ subloops. Multiple use points are grouped together based on the process. On these miniambient loops, temperature can be lowered to circulate and flush and then raised as needed. Subloops provide the best balance of first cost and operating cost [30].

10.3.5 Clean Steam

Clean steam is used for CIP and SIP sanitization treatments. The HVAC impacts of steam should not be overlooked. SIP heat gain from reactors changes the room from air change driven to load driven at 20 ACH. Maintenance of the space temperature during SIP raises the air change rate to

25 ACH. Moisture increases when vessels are opened. Substantial loads are associated with CIP items and parts cleaning [30].

10.3.6 Black Utilities

PW treatment design typically drives water system design within a biopharmaceutical facility. Design of water treatment for black utilities is as simple as determining the point in the water treatment train at which the feedwater will be taken.

Cooling tower water makeup requires minimal treatment. Softened water is usually acceptable. This water could be reclaimed from other waste streams. Chemical-free treatment technologies exist that use magnetic fields, but the science behind the chemistry is not well understood and appears to work on very specific water chemistries only. Those technologies should be explored with caution. Higher water quality will allow for higher cycles of concentration and less blow down water required; consequently, less makeup water is needed.

Boilers can benefit from additional makeup water treatment. Softened water at a minimum is usually acceptable. RO and DI water systems are worth considering for makeup if the aggressive nature of these water streams is considered and taken into account. Using either RO or DI water will reduce surface blow down requirements. Boiler makeup water can also be reclaimed from waste streams. Blow down heat recovery typically minimizes or completely removes the need for quench water for blow down. In general, it is best to maintain high levels of condensate return to minimize the makeup water requirement and to conserve as much heat as possible in the system.

Humidification steam can be captured from chemical-free steam systems at a minimum. Plant steam is not typically used. Chemical-free steam can be generated from RO or DI makeup. Clean or pure steam (validated for endotoxins and generated from WFI or PW) is not required.

10.3.7 Wastewater Treatment

Wastewater used in biologics plants typically needs extensive treatment before discharging into the city sewer system. The water must first be biologically inactivated, pH neutralized, temperature quenched, and contaminants removed as required per the facility's discharge permit.

Facility designers must challenge the process into order to minimize wastewater. A typical cleaning regiment for a bioreactor at the beginning and end of a process might include the following rinse cycles: ambient PW rinse, caustic rinse, ambient PW rinse, acid rinse, ambient PW rinse, and hot WFI rinse. A facility might instead utilize final hot WFI rinse at the end of one cycle as the prerinse for the next cycle. This is typically acceptable if the same product is being repeated. If wastewater cannot be reused, heat might be extracted from the cleaning process [30].

10.4 Energy Efficiency

10.4.1 Building Envelope and Materials

Although examining biopharmaceutical facility processes and utilities can achieve the most significant sustainability results, the building envelope and building materials also contribute to energy efficiency [36].

One of the first sustainable considerations comes in site selection and preparation. Here, the environmental and human impact of sustainability may provide the highest level of motivation. The Morphotek pilot manufacturing plant was constructed on a remediated brownfield site. Morphotek targeted remediation at EPA Act 2 Residential Standards, a level exceeding what was required for the commercial property, to ensure a completely clean site for the plant and its employees [32]. When siting the building on the property, consideration should also be given to orientation in relation to the sun, so as to provide passive solar efficiency.

When selecting materials for building envelopes, the operation and application of the building should be considered. High-performance products are recommended. A biopharmaceutical manufacturing facility has different requirements (e.g. air change rates, humidity, and temperature controls) than a typical commercial building. Building envelope materials should optimize the balance between skin loads and HVAC loads (e.g. overcooling in the summer months often occurs to control humidity, and therefore, more insulation does not necessarily minimize annual energy consumption in all cases). Building envelopes should be considered holistically along with requirements for the processes and technology within them [37].

Advanced technology and prefabrication make it easier than ever to provide a high level of insulation value while minimizing construction cost and schedule. Modular exterior wall panel systems with integral insulation built into their cores may be factory assembled and delivered to a project site ready to be installed. Installation is faster than traditional stick-built construction and with less construction waste. High-performance glazing, when used alone or in combination with a unitized, prefabricated system, can also afford construction efficiency, cost savings, and ultimate sustainable benefit.

Maximizing daylighting—the amount of natural light that enters a building—minimizes demand on electric lighting and provides an optimal work environment. Daylighting also has the potential to maximize sun exposure during cooler months and minimize exposure during warmer months, reducing demand on heating and cooling systems. Strategic placement of window louvers, light shelves, skylights, and shading devices can maximize the passive solar benefits of a facility design.

Building materials provide an opportunity for reducing carbon footprint. The entire life cycle and environmental impact of a material should be evaluated (i.e. cost and energy to produce and deliver to a project site). For example, site-cast concrete offers a significantly reduced impact versus precast or standard concrete. Wood products may be sourced from sustainably managed forests. The Forest Stewardship Council ensures environmental, social, and economic benefits through a rigorous certification process for wood products and the service providers and processors who construct with wood. Reused, recycled, and repurposed materials should be considered where possible [8,38].

10.4.2 Heating, Ventilation, and Air Conditioning (HVAC)

According to the ISPE Good Practice Guide: Heating, Ventilation, and Air Conditioning, pharmaceutical HVAC should control airborne contamination and ensure product purity, identity, and quality. Temperature, humidity, and cleanliness affect the quality of air. Room layout, velocity, and volume affect air quantity. Quality and quantity in combination affect a room's ability to maintain environmental conditions.

HVAC systems support environmental air quality, keep building occupants comfortable, and provide protection from airborne hazards. Biopharmaceutical facilities have the added concerns of temperature, humidity, and contamination that can affect both process and product. As with other engineering design described earlier, understanding of product and process are paramount for effective HVAC design and maintaining environmental air quality [39].

HVAC systems are major energy consumers in any biopharmaceutical facility consuming up to 60% of the energy in the facility. HVAC design involves several strategies to reduce energy consumption without adversely affecting product or process. With any of these strategies, a risk assessment should be performed to fully understand the impact based on the individual facility's needs and process implications.

One of the greatest opportunities to reduce a biopharmaceutical facility carbon footprint is to review and challenge the design criteria. Room classification and associated cleanliness requirements impact HVAC decisions. Cleanliness classifications are typically achieved by numbers of air changes for filtered supply air. The intent is to flush the manufacturing environment with clean air to maintain the cleanliness classification. The cleanliness classification has clearly defined testing specifications in ISO 14644. ISO 14644 also has recommended air changes per hour without any specific knowledge of the unit operations, gowning procedures, outside air quality, or any other source of particulate that may warrant a greater air change design criteria and therefore are conservative in nature. New and existing biopharmaceutical facilities can challenge design criteria through risk assessment and testing and iteratively reduce the air change to reduce energy consumption with little or no risk to the operation [39].

10.4.2.1 Once-Through HVAC Versus Recirculation

Some designs utilize 100% outside air. This is a risk-averse strategy focused on avoiding cross contamination between products or stages of the process. This strategy addresses concern that aerosols from Room A might contaminate air in Room B. Particularly when considering biologics with associated hazards, 100% air that travels once through the HVAC system maintains a very low risk. However, once-through HVAC involves significant energy usage. Strategic air system zoning using recirculated air, considering ventilation, pressurization, and local removal of any contaminants, uses less energy and is a more sustainable choice with no additional risk.

10.4.2.2 Filtration

Different types of filters can achieve the same level of air quality with fewer air pressure drops. HEPA filtration is usually the final filtration process for biopharmaceutical air. However, prefilters before that costly filter can offer intermediate benefits. Minimum efficiency rating value (MERV) is an industry standard rating system comparing filters from various manufacturers. MERV ratings of 1 (least efficient) to 20 (most efficient) identify filter performance. For example, some facilities will do six to seven intermediate filtration processes (with an associated energy and pressure drop for each).

A life-cycle cost analysis should be performed to determine the right balance of filter, pretreatment, and overall life-cycle cost. Better filters are more expensive. Alternatively, lesser quality filters require multiple treatments [40].

10.4.2.3 Primary–Secondary Air

A common practice in biopharmaceutical and sterile manufacturing facilities is the use of primary–secondary air systems or a series configuration that uses a primary air-handling unit to precondition outdoor air as makeup air to one or more recirculating air-handling units downstream. The approach uses less energy because only the primary air unit conditions the outside air required for makeup, pressurization, or ventilation (typically 10% of overall biopharmaceutical manufacturing systems airflow).

The primary–secondary approach lends itself to sustainable design because the outside air is the only air that gets humidified, dehumidified, or warmed in the primary treatment. The secondary systems are used to recirculate and heat or cool air from room to room. The approach means a much smaller air-handling unit and overall lower energy consumption. It is typically employed in areas with 50–60% relative humidity or lower and where people are gowned. On a 95°F outside air day, it becomes much easier and requires less energy to cool one area than the entire facility.

10.4.2.4 Setback Strategies

Most clean rooms' HVAC systems function as if the room is in operation 24 h a day, 7 days a week. However, most clean rooms are not occupied around the

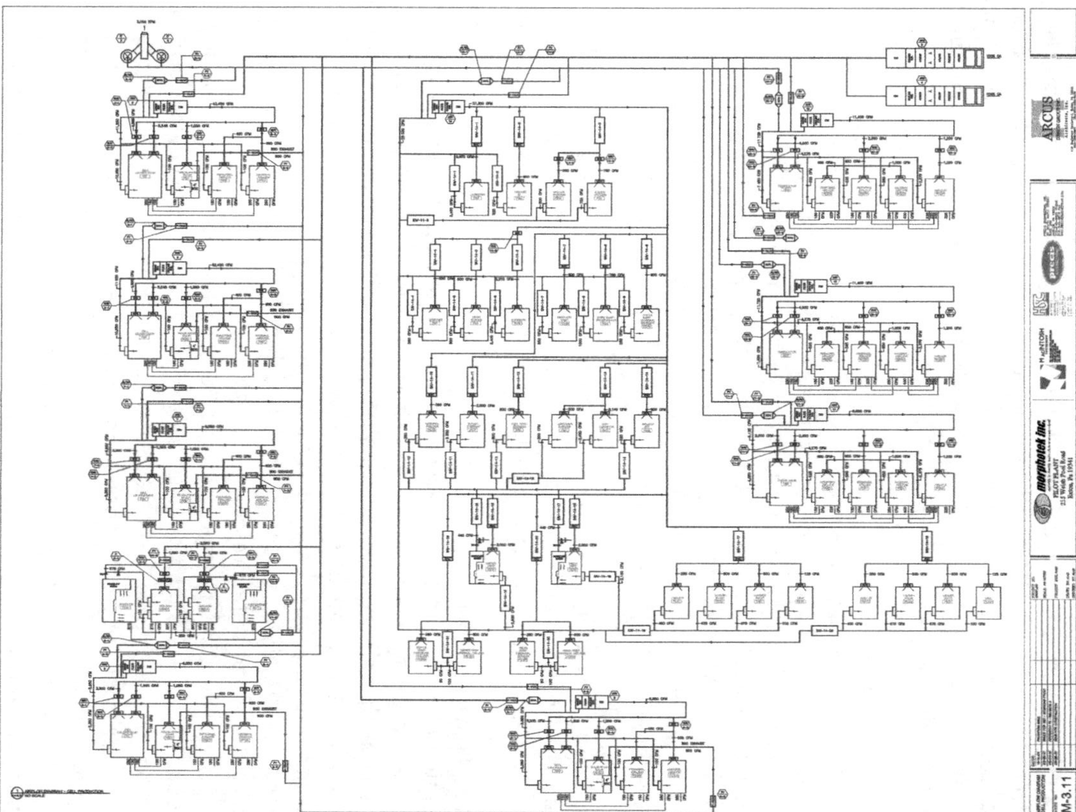

Figure 10.4 Airflow diagram in cell production suite [28].

clock. While there may be a requirement of 40 air changes during the day, that can reasonably reduce to 20 air changes at night when work is not being performed and particulates are not being generated. Setback is an automatic tactic to reduce the amount of airflow without reducing the environmental air quality (for an example of an airflow diagram in a cell production suite, see Fig. 10.4). It presents an opportunity for improvement at very little capital expenditure. Commissioning of this functionality should be performed and standard operating procedures developed to ensure no impact to manufacturing and critical environmental parameters such as cleanliness classification, temperature, relative humidity, and differential pressure.

10.4.3 Chilled Water

Sustainable design of a facility's chilled water system should follow the typical approach of generation optimization coupled with needs minimization. Basically, the goal is to utilize as little chilled water as necessary to meet the facility's needs and to generate the chilled water as efficiently as possible. This leads to the most efficient and sustainable approach to chilled water.

The most significant analysis must take place with respect to the HVAC and coincident process loads. Given the large loads associated with cooling process vessels and WFI heat exchangers, it is critical to understand the coincident process uses and overall energy consumption necessary to support the operation. Design of the process cooling system, based on peak load of all process operations, will result in a drastically oversized system that ultimately will not operate properly. It is important to review the proposed operating schedule of the facility to understand the maximum coincident loads and the total amount of energy necessary to support the process cooling needs. An operating model allows for a detailed calculation of chilled water system diversity.

A primary concern of the process chilled water system is its ability to provide cool enough water to support minimum operations, while being able to provide warmer water where necessary. Ultimately, the system will be most efficient when providing the warmest water possible at all times. Process vessel cooling typically requires the coolest chilled water temperatures. Return water temperatures are also of concern given the limitations of returning very warm water to the chillers. There are two issues associated with warm return water. The first is the possibility of exceeding the chiller capacity and losing the supply water temperature when a slug of very warm water hits the chiller. The second issue is possible rapid pressure increase within the evaporator bundle, resulting in rupture of the chiller bundle pressure safety device. Storage capability is typically provided to allow for variability of return water temperature blending to smooth the low profile on the chillers. Most chiller manufacturers provide a minimum recommended amount of water in the system per ton of cooling.

10.4.3.1 Chilled Water and the HVAC System

A similar process should be undertaken for the HVAC chilled water loads in consideration of facility operations. Integrated HVAC design is critical to effective facility operation. Utilization of a primary–secondary air-handling unit scheme allows for the most efficient chilled water utilization.

Once load profiles for the process cooling load and the HVAC cooling load are prepared, they can be overlapped to determine if a single system is appropriate. This may be acceptable in small facilities. However, for most facilities, a segregated process chilled water system and HVAC chilled water system are more common.

10.4.3.2 Chilled Water Generation

Numerous options are available for generating chilled water. The first decisions to be made are the trade-offs between air-cooled package chillers and water-cooled chillers. The difference between first cost and energy consumption is significant between the two options. Air-cooled chillers are typically limited to approximately 400 tons in a single machine. Water-cooled chillers can exceed 7500 tons in very large applications but, in general, are typically seen in 1500 tons and smaller sizes.

10.4.3.3 Chilled Water Analysis and Design

A sustainable design practice for the chilled water system starts with a holistic view of overall energy conservation by source. Creation of a low profile or chilled water load duration curve offers insight into the overall system operation. When evaluating load and analysis, it is typical to look at peak tonnage, minimum tonnage, and overall annual load in ton-hours. These provide a load duration curve that is used to right size the chiller selections. The overall chiller performance map needs to be reviewed for robustness of operating capability throughout the performance map. Typical chiller selections are based on standard ARI wet bulb suppression, which is accurate for a single chiller in a single building using standard office-type HVAC.

Part load operation analysis is absolutely critical to sustainable selection of building chillers. Variable frequency drive and variable primary chilled water flow provide the greatest capacity for energy minimization; however, significant optimization occurs with the selection of the chilled water and condenser water heat exchangers. A trade-off exists between first cost and energy consumption associated with the selection of the chiller heat exchangers. Larger, more efficient heat exchangers require less water flow pressure drop and closer refrigerant approaches, resulting in overall reduction in energy consumption but requiring higher first cost.

Given that the chilled water system is one of the largest energy consumers in any biopharmaceutical plant, significant effort should be placed in the analysis. Care must be taken to not select the most efficient chiller at peak operating conditions; rather, it is the chiller that provides the overall most efficient operating map of points that should be selected, understanding that the chiller with the lowest kilowatt per ton at peak load may have rather poor performance conditions where it will operate for the majority of the time.

Significant energy is expended in the support or auxiliary components of the system including chilled water pumping, condenser water pumping, and cooling tower fans. Each of these components needs to be examined with its individual impact to overall energy consumption as well as to the integrated effect of the systems. Variable primary flow allows for chilled water distribution and consumption at the used points to closely match. Any differential between chilled water produced and chilled water consumed negatively impacts the energy consumption of the plant in two ways. First, the pumping costs are higher than necessary owing to the excess flow and its associated pressure drop. Second, diluting is when the chilled water supply mixes with chilled water return, resulting in a lower than design chilled water ΔT. Low chilled water differential temperature is a common problem in all operating facilities. It is not uncommon to see operating temperature differentials at half of the design parameters. Consequently, the chillers operate at part load conditions, and the facility requires more chillers to meet the load. It is common to see plants that are flow limited to the point where new chillers are needed to meet the tonnage because the existing installed machines cannot be fully utilized.

Options for chilled water distribution include fixed primary, primary–secondary, and variable primary. Fixed primary flow is antiquated and not typically utilized in modern design. Primary–secondary uses pumps dedicated to the generation of chilled water and pumps dedicated to the distribution of chilled water. In a fully optimized system, the primary and secondary water flow rates are equal. Whenever there is a difference between primary and secondary flows, there is bypass flow between the two systems, which is a measure of inefficiency. Given that in a fully optimized system the flows are equal, it is common to have a single set of pumps that vary primary flow both through the chillers and out to the chilled water users [41].

The condenser water system, including its distribution pumps and cooling tower fans, also uses significant energy. Optimization of the tower fans and condenser water pumps needs to be reviewed on an individual basis and in a holistic way. Using a higher-than-normal condenser, water temperature split between supply and return may provide decreased pumping costs and a smaller tower footprint; however, it is at the penalty of higher chiller operating energy owing to the increased differential head between the bundles. Similar to the earlier chiller heat exchanger example, there are dramatic trade-offs between first cost and operating energy. Sustainable design provides the best balance between first cost and operating energy.

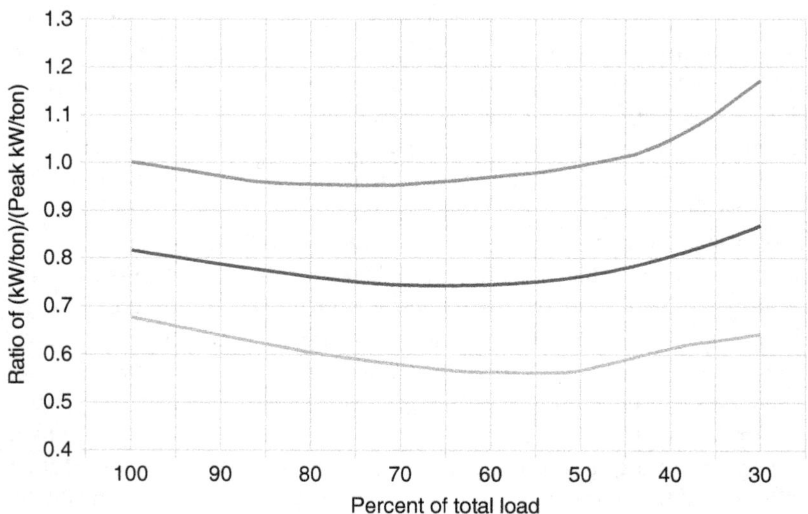

Figure 10.5 Typical effect of load/ECWT-VFD drive. Red = 85F ECWT; Blue = 75F ECWT; and Green = 65F ECWT [42].

Typically, providing the coldest condenser water available based on ambient wet bulb temperatures will result in lower chiller kilowatt per ton as shown in Fig. 10.5. The increase in energy of providing colder condenser water is offset by the reduction in chiller energy. There is an optimized balance point to be found for each of these components and for the overall system.

There are other opportunities to optimize chiller utilization. For process applications that may experience relatively high chilled water temperature differentials compared to those in a typical HVAC system, chillers are commonly installed in series to allow for the temperature to be split between two machines. This provides overall better efficiency than would be possible with a single machine operating at the overall split. This arrangement also provides safety against losing chilled water supply temperature as machines are sequenced on and off. It is, however, more typical to see parallel operation of chillers, pumps, and cooling towers to allow greater flexibility and operation.

10.4.3.4 Free Cooling Opportunities

Free cooling using a plate and frame heat exchanger, which uses condenser water to cool chilled water when outside ambient conditions are capable, provides significantly lower generation energy consumption. There are limitations to the application of free cooling in the biopharmaceutical industry without significant upfront planning. System uses must be designed for warmer entering conditions in order for free cooling to be applicable.

10.4.3.5 Cooling Tower Design

No discussion of sustainability would be complete without reviewing cooling tower makeup water. Makeup water to the cooling towers requires significant domestic water flow to compensate for the effects of evaporation, drift, and blow down. Given the amount of water that is used in a biopharmaceutical facility for laboratory grade water and WFI, there is significant opportunity to use water other than raw city water for cooling tower makeup.

10.4.4 Steam

The primary sustainable design consideration for steam involves matching steam pressure generation to the application. Higher pressure requires more resources to produce the steam—more natural gas to boil the water at high pressure, more feed water pumping energy to get the water into the boiler, and more insulation and/or heat loss because of the higher temperature.

Various technologies should be considered in selecting steam generation equipment, again, matching the technology to the application. Steam generators are better for intermittent process loads, while conventional boilers are more efficient at producing steam when the load fluctuates over a longer period of time.

Considerations include:

- Matching the steam generating technology to the pressure, cleanliness, quality, and quantity required for each application as closely as possible. Different technologies are better suited to different applications.
- Capturing waste heat wherever possible and wherever a practical use for it can be found. Hot waste water often must be cooled before being sent to the drain, so capturing waste heat also offsets the auxiliary cooling cost.
- Returning as much condensate as possible to reduce makeup water and treatment costs. This has the added benefit of increasing boiler efficiency because the less makeup water used, typically the higher the boiler feed water temperature [43].
- Optimizing natural gas consumption to avoid waste using the best available technology for the application.
- Matching the water treatment approach to the specific water quality available at the site.
- Challenging steam consumption and evaluating ways to reduce it through efficiency modifications [34,44].

10.4.4.1 Steam Optimization

Boiling water to make steam requires natural gas (heat source) and water treatment chemicals depending on the hardness of the city water coming into the process. The more treated water that can be returned to the steam system, the less makeup water that needs to be treated. Water that returns to the boiler is also closer to the high steam temperature.

It is important to ensure that water used in the boiler is prepared before it being turned into steam. This includes removal of harmful substances that could damage the boiler and steam plant by causing corrosion or depositing solids that could result in inefficiencies and possible heat exchanger and boiler tube damage. Generally, when water enters the boiler room, it travels through a water filter, softener, and dealkalizer. Next it travels through a continuous blow down heat recovery unit to the deaerator and through an economizer. It then passes through a chemical feed system and into the boiler.

The overarching sustainability objectives in steam optimization include recovering as much heat back to the boiler as was initially imparted on the incoming water by the burners, recovering exhaust and blow down heat, and reducing heat loss. Heat loss reduction can be achieved by insulating piping and identifying ways to transfer heat to other processes that need it. The latest controls serve to optimize combustion of natural gas. Boilers should be sized to appropriately match loads; steam-generating equipment that is too large for the connected load will not operate efficiently. Finally, a robust maintenance plan ensures all equipment functions in accordance with design intent and manufacturer requirements.

Harvesting waste heat requires evaluation of all processes and how they can be integrated. For example, waste heat from the steam system can be used to generate domestic hot water. Exhaust heat can be recovered via conventional or condensing economizers installed on the exhaust stack. Conventional exhaust economizers are less expensive but do not achieve the most heat. Condensing economizers remove more heat but require stainless steel pipes because the chemicals used in the water are corrosive when taken out of solution. It almost always adds value to incorporate an economizer; analysis should be performed to evaluate if the extra heat is worth the extra cost for the condensing economizer.

It is generally expensive to produce a unit of steam; it makes smart sustainable and economic sense to return or reuse the water and heat generated in steam production [44,45].

Blowdown Heat Recovery (BHR) The minerals in steam feed water will accumulate in the boiler as only steam goes out the steam header. The minerals would eventually fill the boiler if not blown down. The blow down water contains a significant amount of energy. In BHR, it can be used to preheat makeup water, thus saving energy. The blow down water also flashes a percentage of steam that is used for deaeration, thus saving energy and water in that process [34,46].

O_2 Trim and Parallel Positioning O_2 Trim dynamically matches the amount of air necessary for complete combustion. It is an advanced combustion system that moderates airflow to optimize the flame. As the technology associated with O_2 trim progresses, it is becoming a more economical choice that should be considered. With this method, the natural gas coming into the boiler should also be optimized so that gas is not being wasted through poor air-to-fuel ratios.

Similar to O_2 Trim, but in a less-effective manner, parallel positioning modulates the air intake louver with the natural gas valve. This increases the amount of air into the boiler along with the amount of natural gas, generally with a mechanical linkage [34,44,45].

10.4.5 Compressed Air

Compressing air requires an energy (usually electricity) input that should not be ignored. As shown in Fig. 10.6, the efficiency of compressed air generation is very low with only 10% of the energy input actually making its way to the points of use. Once again, matching generating equipment capabilities to the needs of the application as closely as possible will inform the most sustainable design decisions. The higher the pressure of the compressed air, the more energy (and therefore cost) required to produce it. Therefore, consideration should be taken to ensure that the pressure generated is in alignment with the pressure required by the end users [47].

Sometimes, the pressure set point of compressed air generation equipment will be considerably higher than most end users need, simply to satisfy a single end user with a high pressure requirement; these situations should be avoided.

Instead, the compressed air system design should be evaluated to determine if lowering overall compressed air generation pressure set point is an option that could reduce operating expenses. The smaller number of end users with high pressure requirements could be served using distributed high pressure compressed air boosters to keep pressure of the majority of the compressed air generated as close as possible to the majority of the compressed air demanded by the system.

Additional optimization techniques for compressed air include variable frequency compressors that modulate speed to match required loads, and dynamic pressure and staging controls.

A lower pressure system also reduces compressed air leakage if or when leaks occur. Compressed air leakage is a major source of wasted energy and can lead to inflated compressor run times as the system cannot maintain pressure, and the compressor must run. Leaks may make operators increase pressure set points to satisfy end users, further increasing energy consumption. A system should be in place to ensure leak testing occurs regularly. Testing leads to significant energy savings with very little cost and is a no-brainer when it comes to optimizing system performance.

In a recent example for a biopharmaceutical client, engineers were engaged to perform an ultrasonic leak detection study. The study team observed several

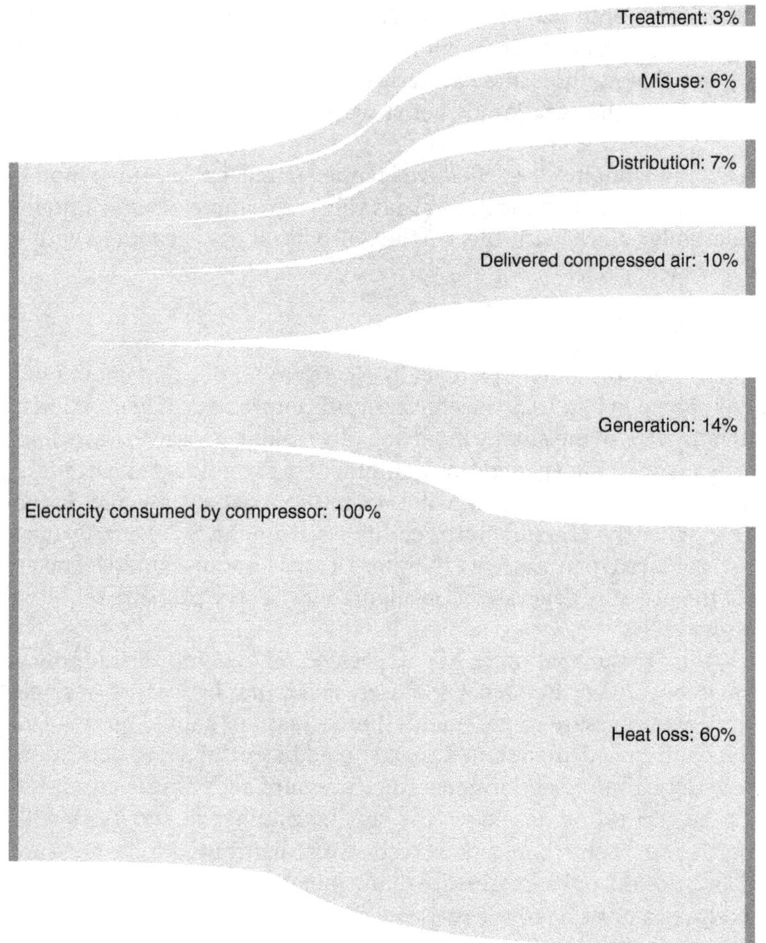

Figure 10.6 Compressed air diagram illustrating that only approximately 10% of the energy used by the compressed air system is embodied in the air delivered to the end user. The rest is lost in the form of heat, misuse, treatment, or distribution loses. Simple, cost-effective measures can be used to reduce losses resulting in energy savings of approximately 30% without major equipment replacement.

leaks at the air compressors in the mechanical room. With the use of an ultrasonic leak detector, the entire compressed air system was investigated for leaks. Leaks were identified and remediated. The estimated savings totaled $32,000 with an implementation cost of $50,000 netting a simple payback of 1.6 years. The cost for this example was based on an outside contractor hired to perform the work; however, a leak detection program can be implemented and carried out by site personnel. Costs vary with geography and specific scope;

Table 10.1 Annual Cost of Compressed Air System Leaks [50]

Annual Cost of Compressed Air System Leaks	
Leak Size (in.)	**Annual Cost (USD/year)[a)]**
1/16″	$1046
1/8″	$4190
1/4″	$16,785

a) Based on electricity cost of $0.10/kWh, assuming constant operation and perfect orifice at 100 psi, and modern air compressor.

this example is provided for reference only and not as an indicator of actual cost to replicate [48–50]; see also Table 10.1.

10.4.5.1 Air- and Water-Cooled Air Compressors

Where a source of cooling water is readily available, water-cooled air compressors are significantly more efficient at generating compressed air than air-cooled air compressors. This is related to the thermodynamic fact that water is a better heat transfer medium than air. However, there is an energy penalty on the chilled water and/or condenser water system associated with supplying cooling water to the compressors. This trade-off should be evaluated.

Temperature requirements of the air compressor cooling water should be coordinated with the available cooling water so that cooling water can be pulled from the point with the least impact on overall operating costs. If chilled water is used for a water-cooled air compressor, the cost and energy implications on the sizing and operating costs of the chilled water system should be evaluated. Typically, air compressors are cooled with water from the condenser water loop of a chilled water system. The impact this has on the operating costs and sizing of the cooling tower should be evaluated [48–50].

10.4.5.2 Drier Technology

Just as air should not be overcompressed, it should not be overdried. Drier technology should be aligned to room dew point requirements for humidity. Drying air imparts heat, a pressure drop, or both, which add to the operating costs of the system. Matching drying technology as closely as possible to requirements of the application is an important step in ensuring the most efficient delivery of compressed air to the end user.

In a recent example for a biopharmaceutical client, the compressed air dryer dew point in a new facility was set to −40°C. The URS for the dryers only requires a −20°C dew point. By resetting the dryer leaving air dew point to

−28°C, the client saved electricity while allowing ample safety margins to ensure that the −20°C dew point can be met. Implementing this opportunity required installation of a pressure dew point control on the new compressors that uses the heatless blower purge dryers. The sensor controls the purge based on actual leaving air dew point as opposed to the current method of time-based purge, which saves compressed air. As with the example provided in the Compressed Air section, costs vary with geography and specific scope; this example is provided for reference only and not as an indicator of actual cost to replicate [48–50].

10.4.6 Nitrogen

Nitrogen use is another sustainable consideration. Used as a blanket in bioreactors, nitrogen can be an expensive product in biopharmaceutical manufacturing. In most cases, companies purchase nitrogen in bulk liquid form from third-party providers. Nitrogen is provided and stored on-site in Dewars (aluminum canisters intended for low pressure transport and storage of liquid nitrogen) rented from the provider. Larger microbulk tanks can hold up to about 2000 L. Depending on the usage, the nitrogen may be delivered once or twice a month in dewars or microbulk size. In addition to the expense of the nitrogen, there is an associated carbon footprint of regular transportation.

For companies with higher nitrogen demand, on-site generation is a possibility. Pressure swing absorption methods are used to generate pure nitrogen from the atmosphere. Although capital equipment costs, maintenance expenses, and energy are associated, the unit cost per liter of nitrogen is reduced. Analysis should be performed during facility design to evaluate nitrogen usage and determine if purchase or self-generation is the better choice [19].

10.4.7 Retro Commissioning

RCx, also known as continuous commissioning, is a process to identify operational improvements in environmental comfort and energy performance. RCx may be performed alone or as part of a retrofit project and should be part of an overall energy management program. RCx is beginning to be mandated in certain areas [51].

RCx is not an energy audit or facility assessment. Instead, the process identifies operational savings and improves facility performance by:

- Enhancing documentation of the operational and maintenance (O&M) requirements for mechanical, electrical, plumbing, and process automation systems.
- Documenting baseline operating conditions through trending and performance measurements.
- Optimizing control systems through calibration of critical sensors, reviewing metered data and trend logs, and functional equipment testing.

- Identifying O&M enhancements that improve energy efficiency, occupant comfort, and indoor air quality.
- Identifying O&M staff training needs.

RCx benefits biopharmaceutical companies by reducing operating costs (e.g. energy and maintenance), prolonging equipment life, improving overall building performance, increasing asset value of the property, increasing occupant comfort, identifying indoor environmental quality issues, ensuring building operations meet the owner's requirements, and providing building facilities operator training.

During the planning phase of RCx (Fig. 10.7), the commissioning team will define objectives, interview facilities personnel, and develop a scope based on a list of systems, reviews of utility data history, energy consumption, and other documentation. The planning phase identifies potential improvements, operational strategies, and maintenance procedures that could improve facility performance and reduce operating costs.

The investigation phase determines design intent, develops and executes diagnostic monitoring plans, develops and executes functional tests, and monitors operations. Then, a list of prioritized deficiencies and energy conservation opportunities (ECOs) is assembled. The investigation is based on an understanding of the systems and the O&M energy savings potential. Cost estimates are developed for each measure, and recommendations indicate the most cost-effective improvements for implementation.

During the implementation phase, measures from the investigative report are executed. The highest priority operating deficiencies are corrected and proper equipment operation is verified, with retesting and monitoring as needed. Operator training is verified, O&M manuals are reviewed, and improvements are fine-tuned, including revisions to estimated energy savings.

Handoff takes place during the verification phase, when performance monitoring is used to confirm proper operations. Energy savings and lessons learned

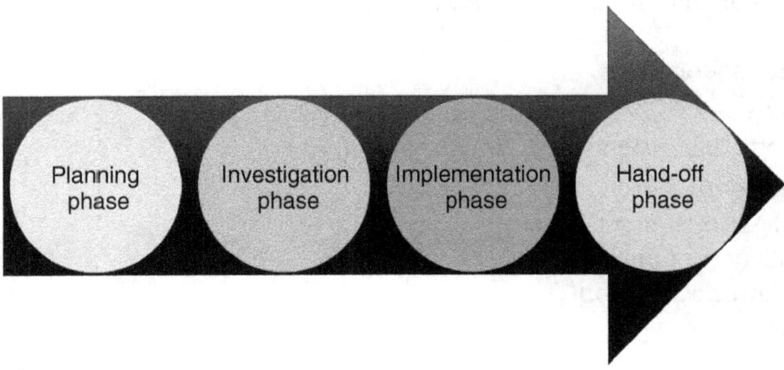

Figure 10.7 Phases of retro commissioning.

are documented as part of a final report, owner/operator training is provided, and a future RCx plan and schedule are established.

Typical opportunities for improvement identified during RCx include the following:

Chiller plant/cooling towers:

- Primary/secondary chilled water pump sizing
- Poor chiller performance
- Chilled water return temperature/system balancing
- Load profile/chiller sizing (leads to chiller cycling at low loads)
- Tower and chiller staging sequencing
- Variable water flows and variable speed drive operation
- Valve performance
- Simultaneous heating and cooling
- VFDs for cooling tower fans
- Free cooling heat exchange

Boiler plant:

- Inefficient staging and sequencing of boilers
- Inefficient combustion (excess air)
- Failed steam traps (trap monitoring program)
- Heat exchanger performance
- Valve performance
- Set points incorrect or not optimum for conditions
- Sensors out of calibration, broken, or poorly placed

Air distribution systems:

- VAV/CAV boxes accessibility/calibration
- Fans operating at higher/lower capacities than necessary
- Poor sensor locations, dirty filters, and coils
- Simultaneous heating/cooling and humidification/dehumidification
- Control valves passing or incorrect balancing
- Sequence of operations/reset schedules not functioning or set at inefficient levels
- Economizer operations
- Challenge design airflows

Building automation systems:

- Controls tune-up
- Trend and optimize system
- Verify sequences of operations
- Confirm sensor calibration
- Loop checks

RCx is a team effort. Depending on the level of scope, the team could include the commissioning agent, one or two operations personnel, or an entire team (e.g. owner or owner's representative, design engineer, controls/maintenance contractors, and utility representative). As part of an energy management program, RCx offers a typical 5–20% energy savings with payback in less than 2 years [52,53].

10.4.8 Maintenance and Operations Best Practices

Although most sustainability decisions should happen during the design phase—when they can have the greatest impact on process and energy efficiency—maintenance and operations best practices cannot be overlooked.

Common maintenance items include insulation repair, steam trap repair, compressed air leakage repair, lighting management, HVAC night setbacks for occupied and unoccupied spaces, and heating and cooling dead band implementation. Energy management is an ongoing process. Metering and submetering can help a company understand and measure its utility costs.

A best practice energy management approach comes from the Six Sigma process and is used to understand the cost of poor quality. Define, measure, analyze, improve, and control (DMAIC) seeks to increase performance and reduce cost (Fig. 10.8). In facility operations, issues are best understood when they are measured, alternatives are evaluated, and metering and monitoring are employed to determine if the solution works.

Almost all facilities monitor primary utility costs (i.e. water, electricity, and natural gas) and most monitor primary utility cost on a unitary basis (i.e. US dollar per kilowatt-hour, US dollar per hundred cubic feet of natural gas). This is fundamentally important because it forms the basis for most of the operating budget of any facility. However, a biopharmaceutical company has no real control over primary utility costs beyond what they are able to negotiate with the utility company. Monitoring primary utility costs, although important, does not provide any real leverage. The information does not empower the energy manager to make any major decisions beyond choosing fuel type for an application.

By comparison, measuring, tracking, and evaluating trends for secondary or generated utilities can glean real insights. Synergies between efforts to determine secondary utility costs and efforts to measure and record them should be implemented when pursuing metering and submetering initiatives.

A facility with a thorough understanding of all costs associated with delivering a unit of a particular secondary or generated utility is empowered with that knowledge in two fundamental ways. First, they have data to continually challenge the unit cost and make efficiency upgrades to drive the cost per unit down as available technology improves over the life of the facility. Second, they are empowered to challenge operational decisions to increase—and incentivize—consumption of a given secondary utility. When a facility is empowered and incentivized to reduce cost per unit of a secondary utility and

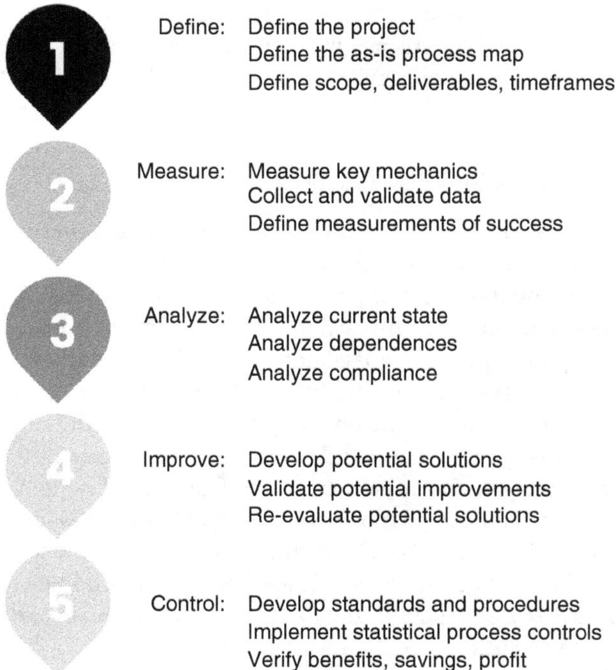

Figure 10.8 Five steps to best practice metering and submetering.

to reduce consumption of the utility, energy efficiency becomes ingrained into the operational culture in a profound and lasting way.

Designers of biopharmaceutical facilities have the opportunity to facilitate a positive culture of continuous improvement. They must do so by considering all of the parameters that need to be monitored to enable the measuring, tracking, and trending of secondary and generated utilities. The granularity of the monitoring should be sufficient to identify unit costs of major secondary and generated utilities, including chilled water, condenser water, hot water, steam, clean utilities (e.g. WFI, WPU, and steam), water treatment, waste water, and CFM of conditioned air to each room classification (for an example of facility energy consumption, see Fig. 10.9).

Monitoring data at the equipment level is valuable to target maintenance efforts. It becomes possible to accurately state that "Chiller 4 produces chilled water at $X/ton, while Chiller 3 produces at $Y/ton. The design documentation of Chiller 3 indicates that it should be running at $Y-1/ton. We should check that chiller and see what's wrong with it."

Data monitoring is also valuable in quantifying the cost impact of an upgrade or repair. "Now that we fixed Chiller 3, we saved $Z+1 in the past month alone and the repair only cost us $Z. That upgrade saved us $Z*12/year. We should

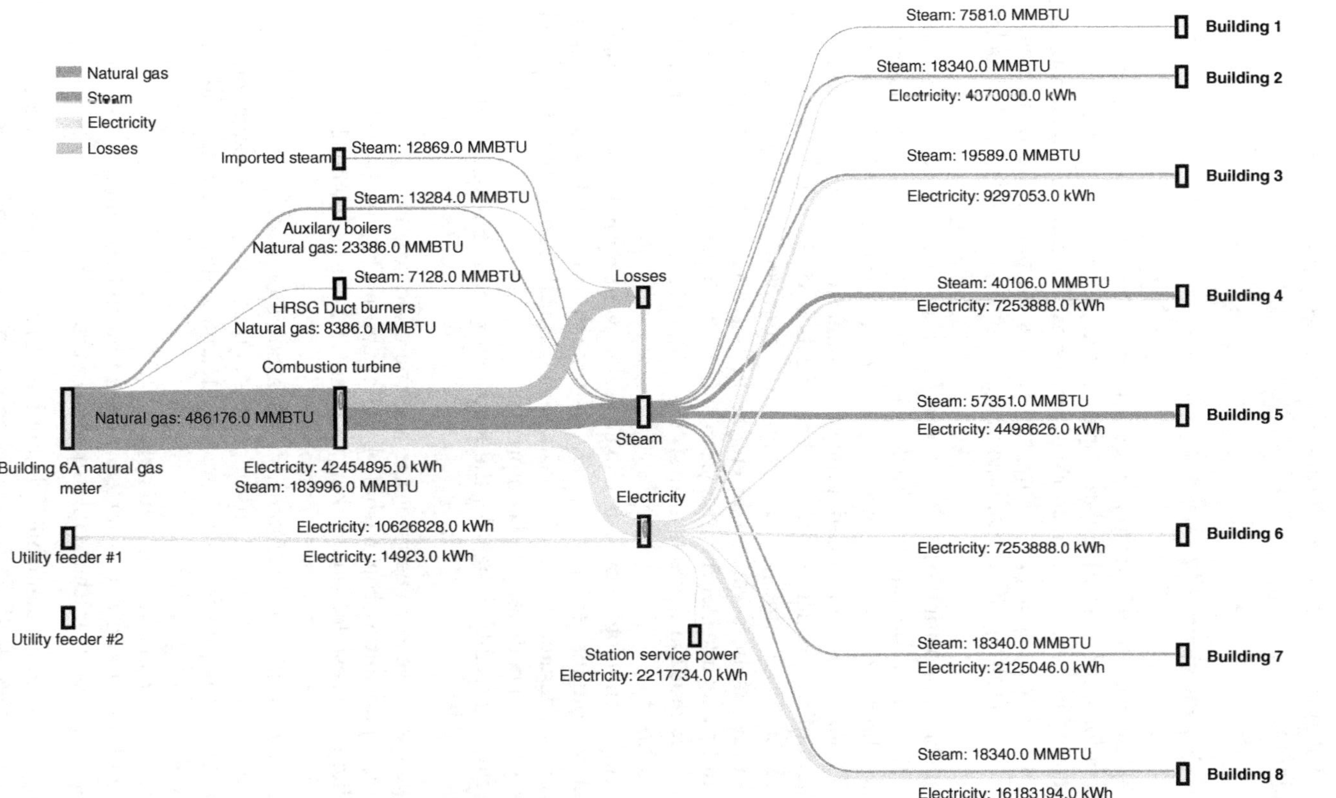

Figure 10.9 Understanding facility energy consumption.

include that in our standard specifications so that we capture these savings from the beginning."

10.5 Conclusion

The biopharmaceutical industry faces growing cost pressures for funding and operations, increased regulatory parameters, competition, and complex and energy-intensive manufacturing processes. Corporate sustainability goals now must be met in tandem with targets for personalized medicine, flexibility in multiproduct manufacturing, improved yields, and smaller volumes.

Sustainable opportunities exist in the areas of technology, design and construction, and operations. Single-use technology, microenvironments (i.e. globe box and isolators), process analytical technology (PAT), process intensification (PI), and closed processing offer technology improvement potential. Building benchmarking and certification via LEED, Green Globes, or ENERGY STAR; integrated design and construction; BIM; and lean construction practices can be leveraged in the design phase. Operations opportunities include incorporating lean manufacturing, leveraging the supply chain, failure mode and effects analysis (FMEA), RCx, and energy monitoring and assessment.

Emerging technologies include chilled water plant optimization packages, continuous commissioning engines, real-time utility purchasing systems, return of cogenerated energy/fuel cells, once-through boiler technology, and chillers with magnetic levitated compressors.

In evaluating facility design, let the following sustainable design principles serve as a guide:

- Adopt aggressive low-energy design and operation targets.
- Assess opportunities from a whole-building approach.
- Use life-cycle cost decision making to assess total cost of ownership.
- Commission equipment and controls.
- Benchmark facility energy intensity and carbon footprint.
- Measure energy and water consumption and track emissions reductions.
- Evaluate on-site power generation, CHP technologies, and solar and other renewable power purchases.
- Specify green construction materials.
- Promote energy and water-efficiency operation and training.
- Incorporate lean workshops into the design process.

Conservation and sustainability do not have to be viewed as daunting standalone challenges. With an integrated approach, sustainability and business objectives can be achieved together.

Acknowledgments

All diagrams without specific reference credits were developed for the purpose of this chapter and should be credited to Precis Engineering, Ambler, PA.

A warm *thank you* also goes to the authors who helped in the project:

Josh Capparella, PE, LEED AP BD+C, Director, Utilities & Infrastructure, Precis Engineering, Inc., Robert Dick, PE, Principal, Precis Engineering, Inc., Sam Colucci, PE, CEM, LEED AP BD+C, Principal, Precis Engineering, Inc., Daniel W. Conner, Energy/Mechanical Engineer, Precis Engineering, Inc., Amanda Gibney Weko, Principal, AGW Communications.

References

1 Portney KE. *Sustainability*. Cambridge, MA: MIT Press; 2015.
2 Robertson M. *Sustainability principles and practice*. New York, NY: Routledge; 2014.
3 What is ethical investment? definition and meaning. n.d. Available at http://www.businessdictionary.com/definition/ethical-investment.html. Retrieved 2016 Jun 13.
4 Our Common Future, Chapter 2: Towards Sustainable Development—A/42/427 Annex, Chapter 2—UN Documents: Gathering a body of global agreements; 1987. Available at http://www.un-documents.net/ocf-02.htm. Retrieved 2016 Jun 13.
5 BREEAM. Supporting health & wellbeing through buildings and communities. n.d. Available at http://www.breeam.com/. Retrieved 2016 Jun 13.
6 Green Building Initiative. Green Globes Certification Overview and Why Green Globes? n.d. Available at https://www.thegbi.org. Retrieved 2016 Jun 13.
7 McDonough W, Braungart M. *The upcycle*. New York, NY: North Point Press; 2013.
8 U.S. Green Building Council. n.d. LEED v4 Building Design + Construction Guide. Available at http://www.usgbc.org/guide/bdc#mr_overview. Retrieved 2016 Jun 13.
9 Capparella J. *Energy benchmarking in the pharmaceutical industry*. Tampa, FL: Pharmaceutical Engineering; 2013.
10 Cogeneration. n.d. Available at https://en.wikipedia.org/wiki/Cogeneration. Retrieved 2016 Jun 13.
11 Combine cycle power plant—how it works. n.d. Available at https://powergen.gepower.com/resources/knowledge-base/combined-cycle-power-plant-how-it-works.html. Retrieved 2017 June 8.
12 U.S. Energy Information Administration—EIA—Independent Statistics and Analysis. 2012. Commercial Buildings Energy Consumption Survey

(CBECS). Available at http://www.eia.gov/consumption/commercial/about.cfm. Retrieved 2016 Jun 13.

13 U.S. Department of Energy. 2012 US. DOE Buildings Energy Data Book. Available at http://web.archive.org/web/20130214191129/http://buildingsdatabook.eren.doe.gov/TableView.aspx?table=3.1.4, Table 3.1.4. Retrieved 2013 Jun 28.

14 U.S. Department of Energy. 2012 US. DOE Buildings Energy Data Book. Available at http://web.archive.org/web/20130218134645/http://buildingsdatabook.eren.doe.gov/TableView.aspx?table=3.6.1, Table 3.6.1. Retrieved 2013 Jun 28.

15 Boyd, G.A. n.d. Development of a Performance-based Industrial Energy Efficiency Indicator for Pharmaceutical Manufacturing Plants. Available at https://www.energystar.gov/ia/business/industry/in_focus/Pharmacuetical_EPI_documentation.pdf. Retrieved 2013 Jun 28.

16 Thomas P. *Will pharma wear the energy star*. Pharma Manufacturing; 2006. Available at http://www.pharmamanufacturing.com/articles/2006/046.html. Retrieved 2013 Jun 28.

17 Sartor D. *5 big hits in laboratories*. U.S. Department of Energy; 2010. Available at https://fimsweb.doe.gov/fimsinfo/Workshop/2010/Sartor_Lab.pdf. Retrieved 2013 Jun 28.

18 Capparella J. *Energy management: trends in the pharmaceutical industry*, PowerPoint presentation. Blue Bell, PA: IPS; 2013. Developed and Presented While an Employee of IPS.

19 Tredinnick S. Benefits of economic analyses (part 2): real-world examples. *District Energy* 2011; 66–69.

20 National Institute of Building Sciences. *A common definition for zero energy buildings*. Washington, DC: National Institute of Building Sciences; 2015.

21 Carmichael, C. 2012. Healthcare Giants Leading the Way to Sustainability. Available at http://blox.rmi.org/blog_Healthcare_Giants_Leading_Way_Sustainability. Retrieved 2016 May 18.

22 Li A. *Closing in on net-zero energy for labs*. Laboratory Design. Available at http://www.labdesignnews.com/articles/2015/05/closing-net-zero-energy-labs. Retrieved 2016 May 18; 2015.

23 Zero-energy building. n.d. Available at http://http:s//en.wikipedia.org/wiki/zero-energy_building. Retrieved 2016 Jun 13.

24 International Society for Pharmaceutical Engineering. *Pharmaceutical engineering guides for new and renovated facilities*. 1st ed, Volume 4, Water and Steam Systems. Tampa, FL: International Society for Pharmaceutical Engineering; 2001.

25 U.S. Food and Drug Administration. n.d. Available at http://www.fda.gov/AboutFDA/CentersOffices/OfficeofMedicalProductsandTobacco/CDER/ucm128080.htm. Retrieved 2016 Jun 13.

26 Perciali M. *Case study: pharmaceutical lab construction.* Controlled environments. 2007. http://www.cemag.us/article/2007/06/case-study-pharmaceutical-lab-construction. Retrieved 2016 May 24.
27 Wolton DA, Rayner A. *Lessons learned in the ballroom.* Tampa, FL: Pharmaceutical Engineering; 2014.
28 Precis Engineering. *Airflow diagram illustrating a fixed 1,000-L bioreactor space with the outline of two disposable suites, showing 50 percent more area available.* Ambler, PA: Illustration Graphic Developed for Firm Use; 2010.
29 International Society for Pharmaceutical Engineering. *Baseline guide volume 6: biopharmaceuticals.* Tampa, FL: International Society for Pharmaceutical Engineering; 2013.
30 International Society of Pharmaceutical Engineering. *ISPE sustainability handbook.* Tampa, FL: International Society for Pharmaceutical Engineering; 2015.
31 *Pharmaceutical pure water guide.* Vandalia, OH: Veolia Water; 2005.
32 Precis Engineering. *Morphotek Pilot Manufacturing Plant.* Entry submitted for consideration. International Society of Pharmaceutical Engineers 2013 Facility of the Year Awards; 2013.
33 Collentro WV. *Pharmaceutical water: system design, operation, and validation.* New York, NY: Informa Healthcare; 2011.
34 Capparella J. *Steam 201—boiler room equipment and function,* PowerPoint presentation. Ambler, PA: Precis Engineering; 2016.
35 Peairs D. *Water softeners.* San Jose, CA: California Water Service; 2004.
36 Patrick DR, Patrick SR. *Energy conservation guidebook.* Lilburn, GA: Fairmont Press; 2007.
37 American Society of Heating, Refrigeration, and Air-Conditioning Engineers. *2009 ASHRAE handbook: fundamentals.* I-P ed. Atlanta, GA: American Society of Heating, Refrigeration, and Air-Conditioning Engineers. 2009.
38 International Well Building Institute. *Our standard: WELL building standard.* n.d. Available at https://www.wellcertified.com. Retrieved 2016 Jun 13.
39 International Society for Pharmaceutical Engineering. *ISPE good practice guide: heating, ventilation, and air conditioning.* Tampa, FL: International Society for Pharmaceutical Engineering; 2009.
40 The MERV Rating System for Air Filters. n.d. Available at http://www.ontimeairfilters.com/air-filter-merv-rating. Retrieved 2016 Jun 13.
41 Hansen EG. *Hydronic system design and operation: a guide to heating and cooling with water.* New York, NY: McGraw-Hill; 1985.
42 Precis Engineering. *Typical effect of load/ECWT – VFD drive.* Ambler, PA: Illustration Graphic Developed for Firm Presentation to ImClone Systems; 2012.

43 U.S. Department of Energy. (2006). *Steam tip sheet #8—return condensate to the boiler*. Washington, DC: Industrial Technologies Program Energy Efficiency and Renewable Energy Information Center.
44 *Improving steam system performance: a sourcebook for industry*. Washington, DC: Industrial Technologies Program, U.S. Dept. of Energy, Energy Efficiency and Renewable Energy; 2012.
45 Woodruff EB, Lammers HB. *Steam plant operation*. 6th ed. New York, NY: McGraw-Hill Interamericana; 1992.
46 U.S. Department of Energy. *Steam tip sheet #10—recover heat from boiler blowdown*. Washington, DC: Industrial Technologies Program Energy Efficiency and Renewable Energy Information Center; 2006.
47 Frankel M. *Pharmaceutical facilities plumbing systems*. Chicago, IL: American Society of Plumbing Engineers; 2005.
48 Bureau of Energy Efficiency. n.d. 3. Compressed Air System. Available at http://www.em-ea.org/guide%20books/book-3/chapter%203.3%20compressed%20air%20system.pdf. Retrieved 2016 May 19.
49 *Energy efficiency best practice guide compressed air systems*. Victoria, Australia: Sustainability Victoria; 2009.
50 *Improving compressed air system performance: a sourcebook for industry*. Washington, DC: Industrial Technologies Program, U.S. Dept. of Energy, Energy Efficiency and Renewable Energy; 2003.
51 Summary of the Energy Independence and Security Act. n.d. Available at https://www.epa.gov/laws-regulations/summary-energy-independence-and-security-act. Retrieved 2016 Jun 13.
52 American Council for an Energy-Efficient Economy. n.d. Commissioning and Retrocommissioning. Available at http://aceee.org/topics/commissioning-and-retrocommissioning. Retrieved 2016 May 2.
53 U.S. Environmental Protection Agency. 2008. *Energy star building upgrade manual: chapter 5: retrocommissioning*. Washington, DC: U.S. Environmental Protection Agency. Available at https://www.energystar.gov/sites/default/files/buildings/tools/EPA_BUM_Full.pdf. Retrieved 2016 May 2.

Chapter 11

Technology's Impact on the Biomanufacturing Facility of the Future

Jeffery Odum and Mark F. Witcher

NNE, Durham, North Carolina, USA

11.1 Introduction

Concerns about product quality, reducing environmental impact, cost of goods (COG) pressures, increasing product demand, more sophisticated products including cellular and genetic therapies, and recent advances in manufacturing technologies have added significantly to the challenges and opportunities of designing, building, and operating biopharmaceutical manufacturing facilities in the twenty-first century. Recent technical advances have created new enabling technologies that can be exploited to improve flexibility, reduce energy consumption, decrease COG, and increase the throughput of a biopharmaceutical manufacturing facility. For this discussion, an enabling technology is equipment or methods that provide a means to significantly improve the performance and capabilities of the process and the facilities required to run them.

Facilities that are able to rapidly achieve first-time high product quality, continuous improvement, and adaptation of manufacturing operations; minimize environmental impact and energy consumption; and have flexibility in initiating and completing manufacturing campaigns, along with high throughput to reach competitive COG targets, should be developed. With medical science increasing *in vivo* insights into product requirements, developing more sophisticated adaptive clinical trial designs, and improving *in vitro* analytical product characterization methods, the critical path for the development of new products will inevitably shift more and more toward the process development and manufacturing timelines.

The common approach of locating biopharmaceutical manufacturing in classified clean rooms is also being reexamined as the industry aims to improve patients' access to products by reducing costs, while maintaining assurance

Process Architecture in Biomanufacturing Facility Design, edited by Jeffery Odum and Michael C. Flickinger.

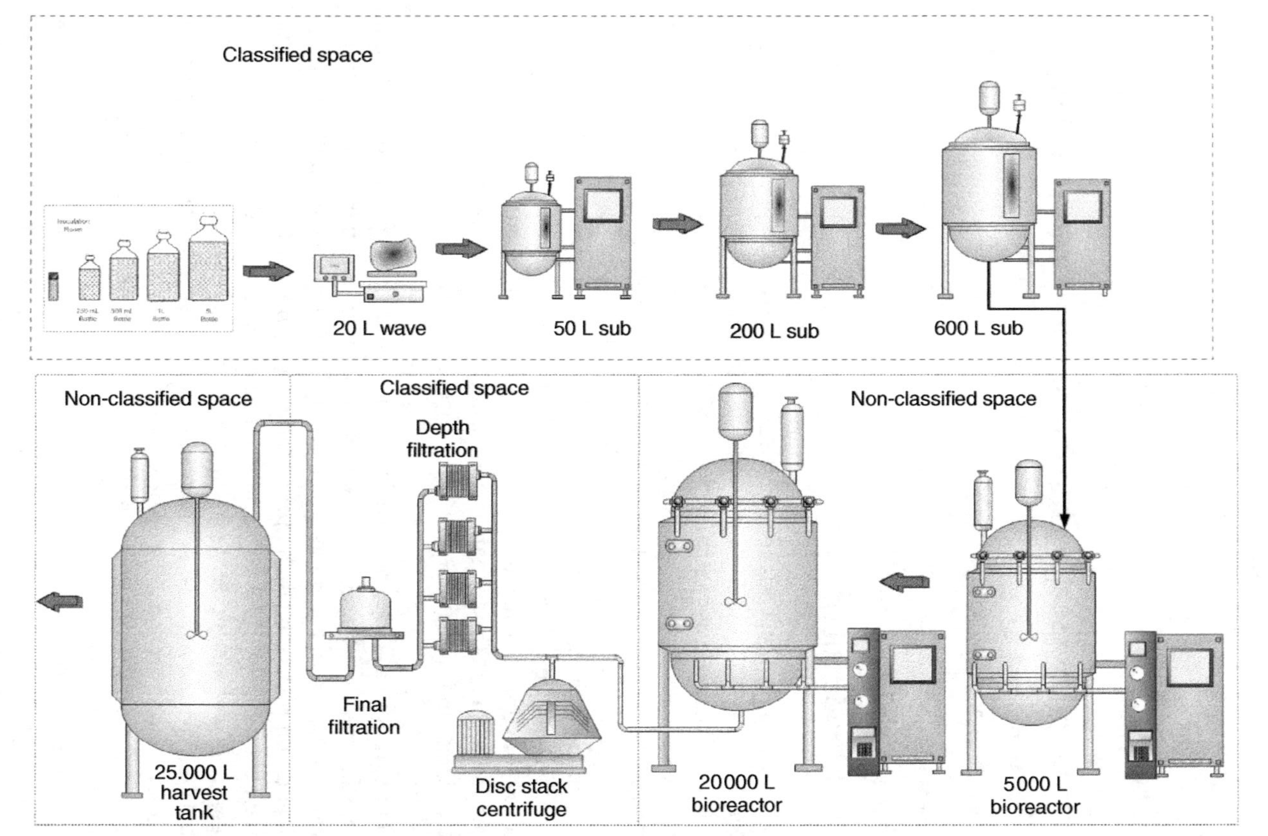

Figure 11.1 Typical fixed SS bioreactor systems integrated with the facility. [Courtesy IPS.]

of product quality and patient safety. The evolution of verifiably closed manufacturing technologies is replacing open systems, thus removing the need for the extensive use of costly to build and operate classified environments as a risk mitigation strategy. By putting technical and procedural advances together with scientific and quality risk-based approaches, many industry leaders are improving process equipment to reduced area classifications (grade D/ISO 8 at rest) and controlled nonclassified (CNC) environments that are now deemed appropriate for the functionally closed systems and closed bulk drug substance bioprocesses. The approach is one step toward helping reduce capital and operating costs, and thus product costs that are a factor in improving patient access.

We begin by looking at typical systems used for large-scale manufacturing of biopharmaceutical products. Stainless steel (SS) systems have been used from the beginning to manufacture safe and effective products requiring sterile or low bioburden environments. A typical SS system is shown in Fig. 11.1.

The systems are very expensive to design, build, validate, operate, and maintain. They are very inflexible and difficult to change or modify to accommodate changing needs, requirements, or to adapt to new products. The key to improving manufacturing facilities is to examine new enabling technologies to identify and use possible synergies from these technologies to realize more flexible facilities to meet the challenges the industry faces.

11.2 The Enabling Technologies

The biomanufacturing facility of the future is being currently shaped by a number of new technologies that are allowing companies to meet critical needs for improving flexibility, reducing COG, keeping manufacturing and development off the critical path, and improving operational efficiency. These technologies are also having an impact on the design elements and design execution, regardless of facility type. The three technologies that we will look at are manufacturing process platform improvements, single-use (SU) technologies, and process automation. We begin by looking at process technologies and how they impact facility designs.

11.2.1 Process Platform Improvements

Significant improvements in cell culture processes have increased cell culture yields at least 10-fold, with another 2–5 times improvement on the horizon from improvements in media formulation, specially designed production cell systems, and from further advances in molecular biology. These process improvements have resulted in significantly smaller bioreactors and fewer lots required for upstream processes to meet capacity requirements. In addition, the higher upstream efficiencies have significantly reduced the volumes required in downstream purification. This translates to smaller space requirements for process equipment, less air handler capacity, reduced facility costs, and potentially smaller operating personnel requirements.

Further improvements in downstream processes will come from integrating downstream unit operations, continuous or high cycle operations, higher column resin loading, and greater product-resin specificity and selectivity. Additional improvements in cell lines that reduce contaminating and impurity protein burdens will further increase the efficiency of downstream operations. These advances will continue to reduce downstream production space requirements, thus reducing facility and operating costs as described above.

11.2.2 Single-Use Technology

The second enabling technology is the SU or disposable component technology. SU technology was originally developed to reduce the need for cleaning and sanitization activities, along with their associated validation requirements. The advantages often cited for SU implementation include the following:

- Lower capital costs due to simpler, independent systems and reduced physical footprint.
- Lower operating costs due to the reduction/elimination of cleaning times and raw material expenses.
- Lower utility costs associated with elimination of CIP/SIP (clean in place/steam in place) cycles.
- Lower costs and less time to implement process changes.

The advantages of SU technology, however, go far beyond simple process economics. Because single-use systems (SUSs) are closed, or functionally closed, processes can be safely and effectively operated in less stringent environmental classifications such as controlled, not classified spaces. When combined with movable skids, SU technology can be exploited to provide considerable flexibility for replicating or moving the process element to achieve significant advantages in operating the manufacturing enterprise. It can also significantly reduce energy costs and utility requirements. Thus, portable SU process equipment, as shown in Fig. 11.2, permits the building of smaller and simpler manufacturing facilities.

However, SUSs result in several complex operational issues that must be identified, understood, and resolved. Operation of closed SUSs must be broken down into the following parts [1]:

- *Equipment Assembly.* The system must be reliably assembled from sterile components and readied for use. Examples of complex process equipment systems include bioreactors up to 2000 liters, complex harvest filtration systems, and chromatography column systems including the packed columns.

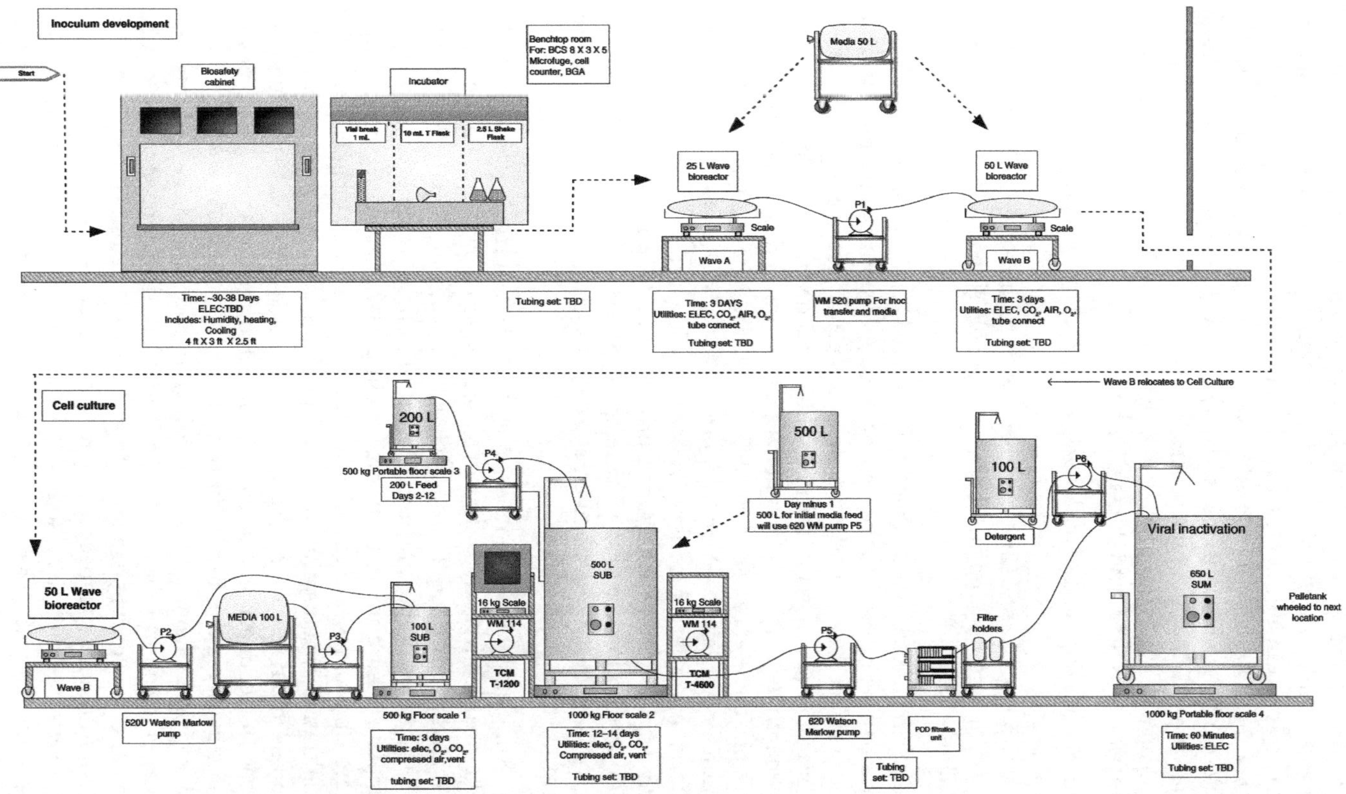

Figure 11.2 Portable single-use biomanufacturing system process flow diagram. [Courtesy of IPS.]

- *Input and Output Accessible.* Media and buffers must be attached and removed without the possibility of contamination. The systems must be validated and verifiable as closed at all times.
- *Component Integrity.* While the system is being operated, valves, double-block valves, agitator seals, and connects must remain sealed to leaks either in or out of the process system.
- *Sampleable.* Samples must be taken by mechanisms that assure the integrity of the primary system and the sample taken.

Significant development effort by SUS vendors is making progress in resolving these issues.

There are a number of definitions currently recognized by the Industry that will impact how system and facility Designs are developed and executed. They are as follows:

- *Closed System.* A system designed and operated such that the product is isolated and never exposed to the environment. Additions to and effluents from closed systems must be performed in a completely closed manner. Transfers into or from these systems must be validated and verified as closed.
- *Functionally Closed.* Closed systems that are opened between processing operations but are "rendered closed" by a validated handling, cleaning, sanitization, or sterilization process appropriate or consistent with the process requirements, whether sterile, aseptic, or low bioburden.
- *Briefly Exposed Operations.* Open processes containing process materials and/or product intermediates. These open processes are rendered closed by means of appropriate environmental controls surrounding the open process system.

For SUS implementation, another key element in the introduction of the technology has to do with risk reduction of real or perceived product contamination during manufacturing and the approach taken to validate SUS closure. Definition and validation of the "preclosure" assembly phase are critical.

A complete closure analysis thus plays an important part of the implementation approach. Examples of complex system to be evaluated include bioreactor vessels, filtration systems, chromatography column systems, buffer, and media makeup and holding systems. Examples of specific components include SU connectors for connections, disconnections, and reconnections; all types of valves; and bags, mixers, and manifold systems.

The focus is to demonstrate the risk mitigation for these systems and components to confirm that the SUS operates in a closed manner that can be validated as closed during qualification and verified as closed during and after operation. Another key aspect is to have clear agreement on the definition of system closure based on the system implementation within the facility design. Closure should be neither assumed or taken for granted, and must be verified through appropriate controls and practices for every batch.

For many products, highly productive processes and the SU technologies provide a wide variety of opportunities. The final enabling technology, process automation, can be used to increase the range of opportunities for simpler, more efficient facilities.

11.2.3 Process Automation

The final enabling technology is the automation of process and operational activities and functions. Process functions are monitored and controlled using the principles behind process analytical technology (PAT) and direct digital control (DDC). Using PAT in SU-based processes requires the development and use of cost-effective, disposable, SU sensors or sensor interfaces. Development of these sensors is currently an area of significant research and developments resulting in many new sensors becoming available. DDC systems provide PAT capability to monitor and control processes in real time. As SU sensor technologies evolve, the ability to monitor processes in the SU world will improve. This also facilitates real-time monitoring and assurance of quality by monitoring critical quality attributes (CQA) and critical process parameters (CPP) to facilitate real-time release testing (RTRT).

The tools for monitoring and controlling business functions include MRP-II (materials resource planning), LIMS (laboratory information management systems), and EBR (electronic batch records). These technologies provide powerful tools for managing information and monitoring procedural processes to assure and, in most cases, guarantee adherence to procedures and ultimately assure high product quality. Positional control systems are used to control the location and flow of material, people, and equipment throughout the facility at all times for all operational and transition functions. Collectively, the systems provide comprehensive infrastructure element support in the form of an MES (manufacturing execution system) that provides tools for assuring continuous control over all aspects of the manufacturing enterprise.

As we shall see, these three enabling technologies can be combined to provide many options for building and running manufacturing facilities. To understand how these enablers are combined, we must first understand manufacturing facilities and how they operate the process within the surrounding facility.

11.3 Elements of a Biomanufacturing Enterprise

To understand technology's impact on biomanufacturing facility design and operations, it is important to understand the relationships among the three major elements of a biomanufacturing enterprise (process, facility, and infrastructure) [2]. The three elements are defined as follows:

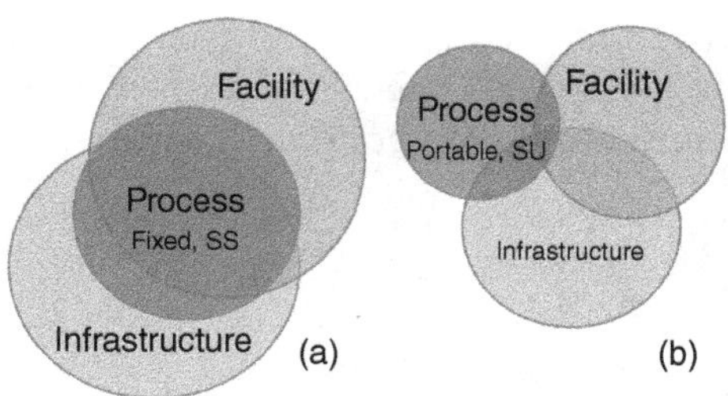

Figure 11.3 (a) The fixed SS process system is designed and connected to the facility and infrastructure as three highly integrated elements. (b) The single-use process systems are much less connected to the facility and infrastructure elements.

The *process* element includes the process unit operations, equipment required to run the unit operations, and process-specific operating procedures. In the past, the process was defined during development and then implemented during scale-up into specifically designed fixed SS systems.

The *facility* element is the manufacturing environment and layout required to support the process and process equipment and includes the building, operating rooms/suites, and support functions such as supplying raw materials and components, utilities, and HVAC (heating, ventilation, and air conditioning).

The *infrastructure* element is composed of the people, practices, procedures, and polices used to run and control the process's overall operation and support functions within the facility.

A typical conventional relationship between the three elements is shown in Fig. 11.3a. The process is operated within the facility under the control of the Infrastructure.

The baseline relationship between these elements for decades was one of highly integrated, interdependent elements where the facility and its support functions were hard piped to the process equipment and therefore completely interdependent and operated together as shown in Fig. 11.1. All of the major SS equipment shown in Fig. 11.1 are fixed in place and supported by facility elements such as utility and CIP systems in order to operate. These facilities are relatively inflexible and must often be significantly modified to run alternative processes. Owing to the nature of the process equipment, the facility also required significant utility and energy resources in order to meet production demands.

To meet the challenges of the future, a facility described in Fig. 11.3b is required. The facility is smaller and more flexible because of the reduced overlapping and interdependence of the three enterprise elements. The facility required to meet the challenges of the future requires decoupling the facility elements as shown in Fig. 11.4. Using the enabling technologies described earlier, the elements can be separated into distinct units that can be quickly and efficiently integrated, rearranged, and modified to create a highly effective flexible manufacturing enterprise.

With the use of skid-based SUSs, it is possible to have movable process equipment. If the skid-based SUSs are closed or functionally closed, the process is

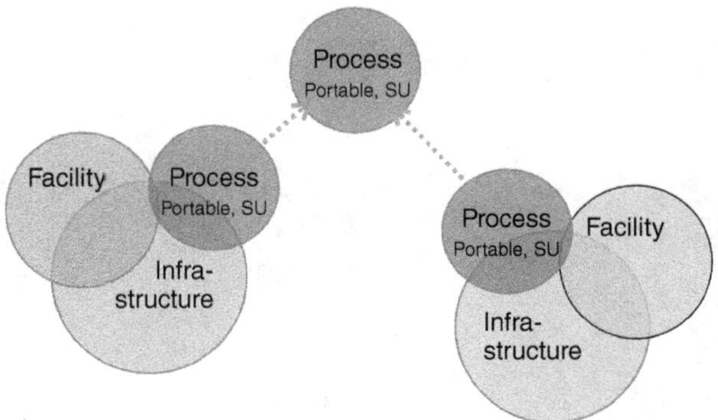

Figure 11.4 Using the new technologies, the facility elements are decoupled allowing them to be easily moved or modified to allow relocation, replication, or scale-out of manufacturing processes.

essentially separated from the facility because the unit operation skids can be moved anywhere within the facility or even moved to a different facility without impacting the performance of the process.

The process can also be easily cloned to increase capacity by replication (scale-out) within an existing facility, or duplicated and moved to a new facility quickly and efficiently. The manufacturing procedures and resulting batch records would also be less connected to the facility and more a part of the process because they are specific to the process as implemented in the SU skids not the specific facility. The process is therefore almost completely separated from the location in which it is run.

Once the process is separated from the facility, a wide variety of less expensive, simpler, and more flexible options becomes available for designing and constructing the manufacturing facility. One option is modular clean rooms that can be assembled faster and cheaper using standard components. Such technologies provide for more innovative and flexible clean room spaces as shown in Fig. 11.5.

For understanding, the enabling technologies can be incrementally incorporated with the manufacturing enterprise to evolve the facility of the future. However, depending on specific process requirements, the enabling technologies can be applied at any intermediate stage to accomplish specific process requirements or deal with unique process or product risks.

11.4 Evolution of the Facility of the Future

To better illustrate the impact from SU implementation, let us look at a basic block flow diagrams (BFDs) of a "traditional" SS-based stirred tank

Figure 11.5 Modular clean room POD. [Courtesy of G-CON.]

biomanufacturing platform and see how SU implementation and the other enabling technologies impact key facility elements.

Figure 11.6 depicts a BFD of a process like the one shown in Fig. 11.1. The upstream, downstream, and support areas are defined along with icons representing some of the key infrastructure elements required to operate the manufacturing process. Note the large number of interdependencies as reflected by the high number of solid arrows. The arrows reflect the complexity and interconnectedness of the enterprise's three elements. The connections can only be accomplished by the simultaneous design and construction of all the elements. Once built, the design provides little flexibility for adaptation or modifications required for process changes.

In Fig. 11.7, we begin the transition to SUSs like those shown in Fig. 11.2 where the SS-fixed vessels are replaced with SU components where possible. The number of solid arrows is reduced significantly by the elimination of the CIP systems and decreased utility requirements. Further reductions can be achieved by combining the media and buffer preparation areas using disposable material handling components.

With the introduction of SU components and the outsourcing of raw material preparation, we now see the reduction in utility systems as shown in Fig. 11.8. We begin to see the elimination of the solid arrows resulting in a flexible facility with fewer interdependencies required to support the process element.

With SU technology, the supply of raw materials, and possibly even waste materials, can be handled independently from the process, thus providing additional functional separation of the process from the facility.

We now have created what could be referred to as *localized working environments* where SU components reduce or even eliminate the need for classified clean room spaces and the high operating cost HVAC systems. Here, operators

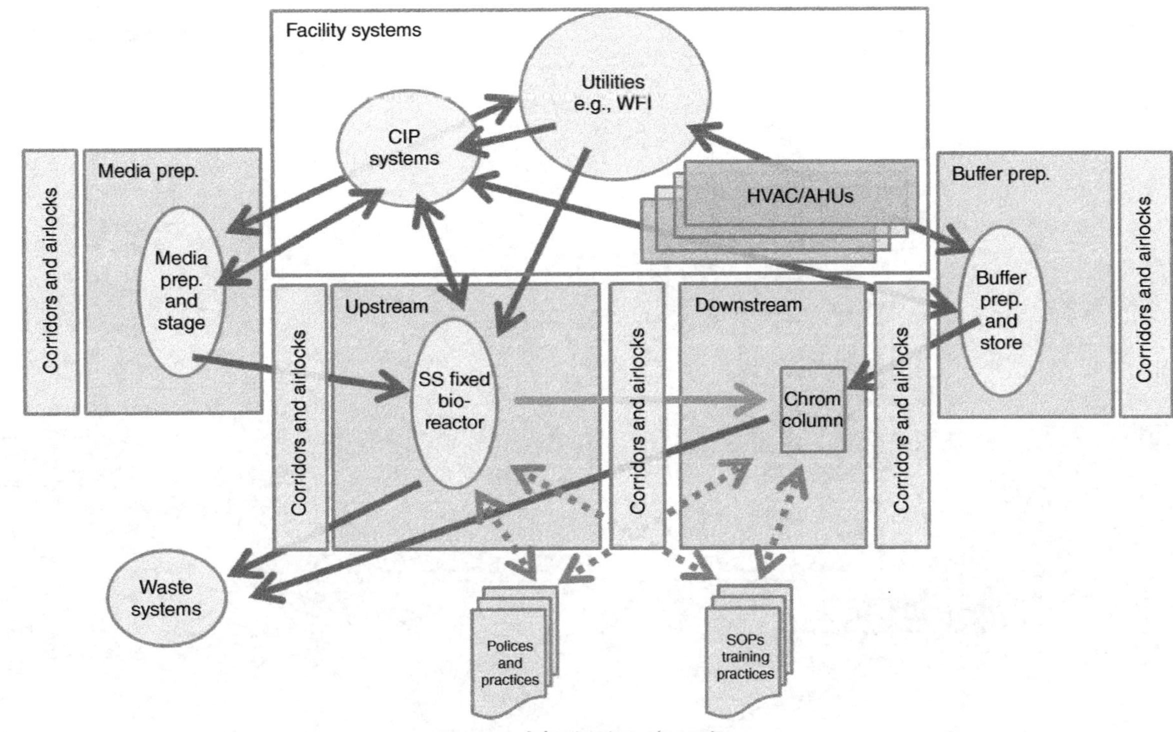

Figure 11.6 Traditional biomanufacturing facility BFD. The solid arrows represent hard piped SS connections between the manufacturing and support systems.

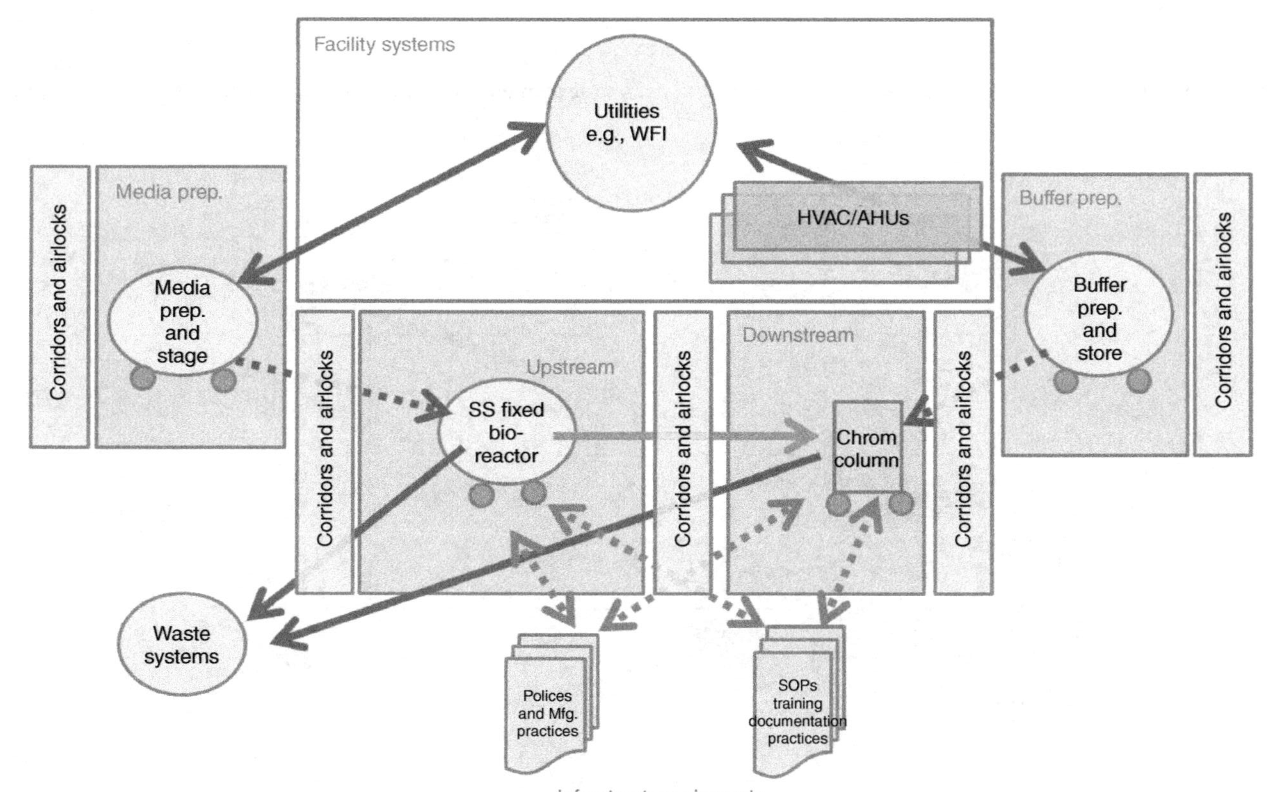

Figure 11.7 SUS replace fixed stainless steel systems.

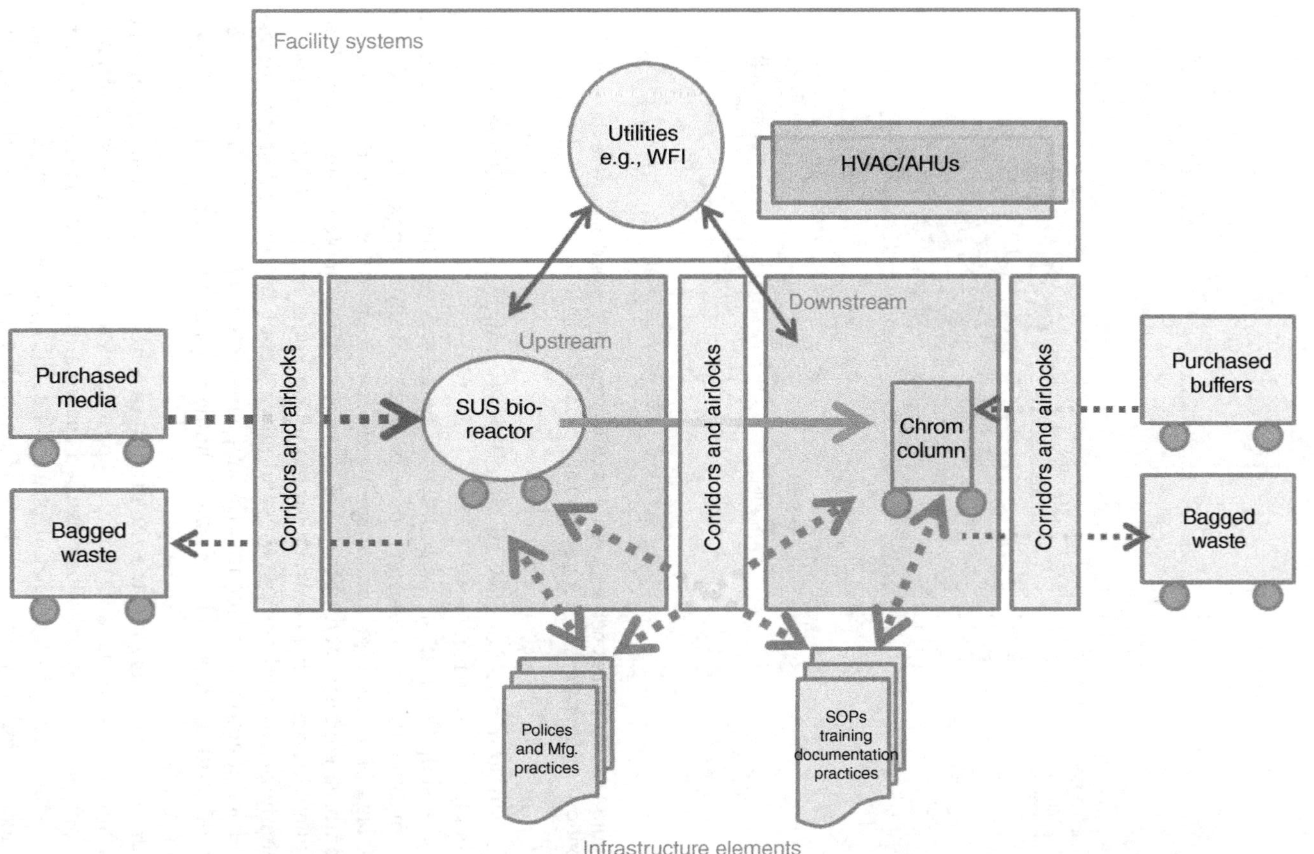

Figure 11.8 Reduced utility requirements.

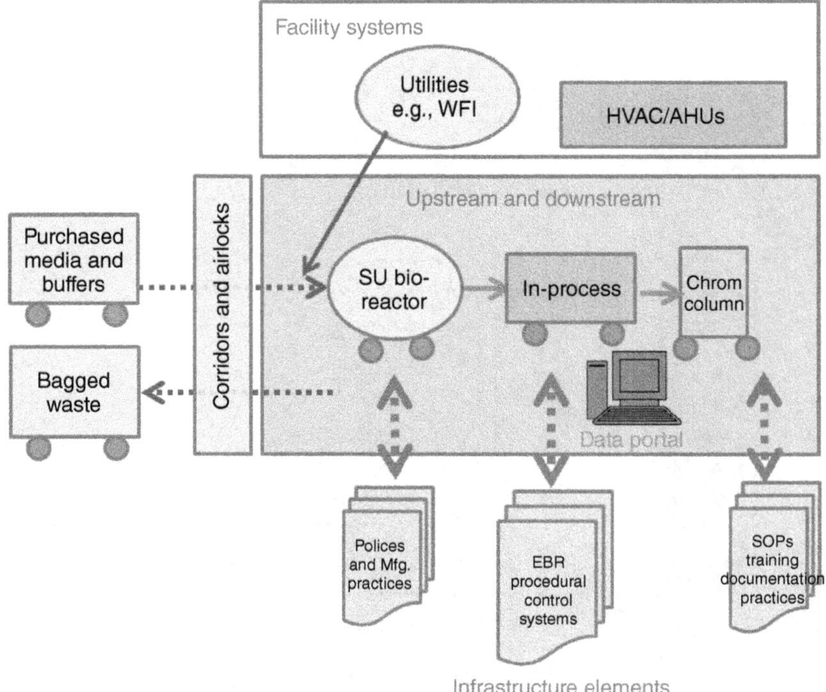

Figure 11.9 SU, single operating area, and EBR-based facility solution provide complete control of all operations assuring and proving product quality.

work in "gray space" or CNC space and move easily between unit operations in less segregated space.

The simplification of the facility can be taken even further by combining all manufacturing operations for a product within a single space as shown in Fig. 11.9. Because commingled processes require enhanced procedural controls, sophisticated procedural controls can be implemented using EBR along with positional control afforded by bar codes or RFID (radio frequency identification) methods assuring and proving complete control of the process and its operational sequence. The implementation of an EBR allows access to all critical documents through human machine interface (HMI) data portals.

Another advantage of the integration of automation technologies around SUS, closed systems, and modular approaches allows for an easing of environmental requirements for classified spaces and a more flexible approach to segregation. A shift into manufacturing in CNC space is an area receiving significant attention from both the manufacturing and regulatory organizations within the industry.

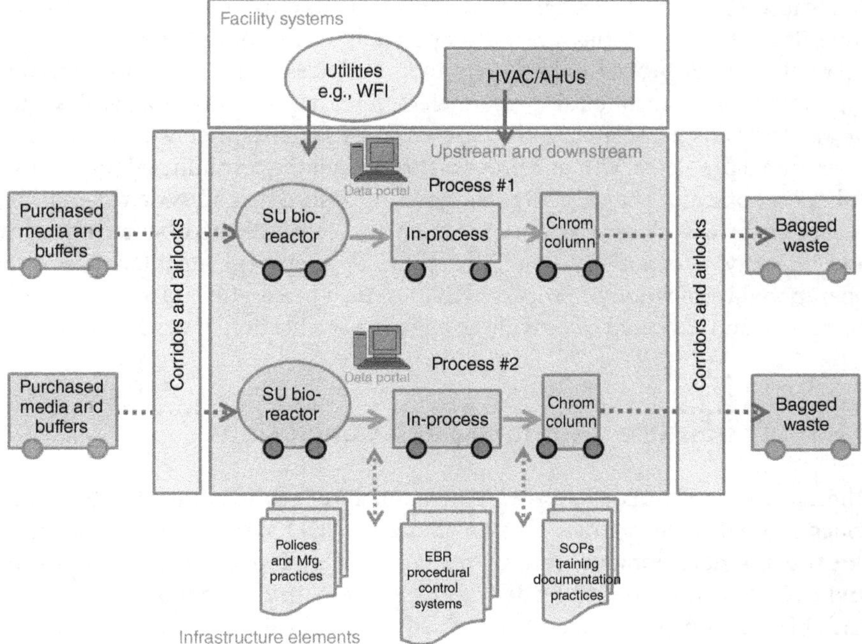

Figure 11.10 Multiproduct commingled processes in common space. Unidirectional flow used to aid in procedural control of incoming materials and waste flows.

The next and final progression is to design multiproduct facilities as shown in Fig. 11.10. Multiproduct facility with commingled processes provides a highly flexible facility option. However, to maintain complete control and assurance along with proof of control, an extensive Infrastructure is required. Comprehensive training, material control, and documentation are required to meet all control requirements necessary for complete product quality.

This simplified layout can even be developed around a ballroom concept where physical segregation of the manufacturing unit operations is eliminated by the use of SUS and closed system manufacturing (Fig. 11.10). For CNC implementation, there will also be a number of significant considerations that will impact the facility design:

- High degree of assurance that process equipment is closed.
- Defining risk-based and effective design and operating principles.
- Facility organization to manage the interface between manufacturing and external environments.
- Managing planned and unplanned system breaches.
- Assurance of environmental control.

Because of the high material flow requirements for complex SUS process time-lines, these facilities are best operated using unidirectional material flows along with procedural and temporal practices to maintain appropriate separation and control of all personnel, component, raw material, and waste flows. The risks associated with these types of facilities increase because all operational problems with one process threaten all the commingled processes. All operations must be carefully sequenced to assure that all systems are kept separated by the closed systems in which they are operated. System design and integrity, personnel training, procedure development, maintenance, and operational verification of process activities must be carefully planned, choreographed, and executed to provide continuous verification of product quality.

11.5 The Future—Summary and Conclusions

The enabling technologies provide a number of opportunities for building much more flexible facilities capable of lowering COG, more rapidly supporting process development, clinical manufacturing, and through to approval and commercial supply of product to patients. With the help of process and enterprise automation systems described above, facilities of the type shown in Fig. 11.10 are capable of providing multiproduct manufacturing for a wide variety of products through all the stages of manufacturing from preclinical to commercial manufacturing. Once the product is launched, the process can be moved to a long-term commercial manufacturing facility using either a scale-up or a multisite scale-out strategy as appropriate for supplying the product to the worldwide markets.

Because the SUSs within a simplified facility are more efficient, they also have the benefit of being environmentally friendly. These include sustainability and compatibility with a wide variety of manufacturing goals. In terms of sustainability, the modern faculties provided for the reduction in carbon footprint associated with SU implementation can be more than 50% as shown in Fig. 11.11. As shown in Fig. 11.11, significant reductions are realized in SIP, CIP, and WFI (water for injection) usage. At high production rates, these saving can be very significant.

A facility that can be quickly and easily reconfigured and then reused for many products is significantly more environmental friendly because it has the ability to support new products quickly and efficiently. The benefits of combining these enabling technologies (high productive process, SUS, and process automation) in manufacturing are numerous. They include the following:

- *Lower Operating Costs.* Reduced operating overhead through easy operation from simpler gowning and personnel flow transitions, reduced energy consumption, and simpler environmental monitoring.

Source	Avg USA Mix of Coal, Gas, and Other		
	DISP	SS	delta
SIP	—	338.5	(338.5)
CIP	19.3	688.4	(669.1)
Transporting plastic	148.5	—	148.5
Pumping water and wastewater	2.4	18.2	(15.8)
Steel fabrication amortized per batch	2,970.7	7,723.8	(4,753.1)
Polymerizing plastic per batch	505.8	—	505.8
Extruding plastic	316.1	—	316.1
WFI still	7,308.7	29,828.8	(22,520.1)
Cleanroom energy	83.6	129.2	(45.5)
Incinerating plastic	6,029.3	—	6,029.3
Workers driving CO_2 footprint	72,232.4	81,465.1	(9,232.7)
Total CO_2 LBs per batch	89,616.7	120,191.9	(30,575.2)
Total CO_2 LBs per batch without workers dri	17,384.3	38,726.8	(21,342.5)

Figure 11.11 Carbon footprint reduction in pounds/batch. The use of disposable (DISP) systems is compared to stainless steel (SS) systems.

- *Lower Capital and Construction Costs.* Fewer personnel and material airlocks, smaller AHU (air handling unit) and ducting systems for the HVAC systems, faster construction and qualification, and reuse of existing flexible facilities.
- *More Efficient Product Development.* Rapid construction and/or redeployment of existing facilities to provide manufacturing capacity from preclinical through commercial manufacturing.

These advantages will only come from a complete understanding of the product and process risks and their mitigation using the enabling technologies described above.

References

1 Odum J. Managing the design of a single-use facility for biomanufacturing. *Pharm Eng* 2015;**25**(3):25–32.
2 Witcher MF, Odum J. Biopharmaceutical manufacturing in the twenty-first century—the next generation manufacturing facility. *Pharm Eng* 2012;**32**(2):1–8.

Glossary

Active Product Ingredients (API) The ingredient in a pharmaceutical drug or biological medicine that is biologically active.

Aeration The process by which air is circulated through, mixed with or dissolved in a liquid or substance.

Airlock An architectural feature that allows for the passage of people, materials, and/or equipment into different segregated spaces within a facility. These are normally controlled via air-pressure differentials.

Amino Acids Any of a large number of compounds found in living cells that contain carbon, oxygen, hydrogen, and nitrogen, and join together to form proteins. Amino acids contain a basic amino group (NH_2) and an acidic carboxyl group (COOH), both attached to the same carbon atom.

Aseptic Free from contamination caused by harmful bacteria, viruses, or other microorganisms.

Ballroom Facility Design A facility concept for manufacturing where physical segregation between different production operations is eliminated due to the use of closed system design, resulting in an "open" space approach to equipment layout.

Bioburden Contamination Microbial contamination of nonsterile biopharmaceutical products that can occur by contact with the facility air, contaminated water, be introduced by operators, or process contaminants (nutrients) that promote microbial growth. Bioburden is typically reported as microbial colony forming units (CFU) per volume of liquid or surface area of a solid material. A closed process significantly reduces the possibility for bioburden contamination.

Biosafety Level (BSL) A set of biocontainment precautions required to isolate dangerous biological agents in an enclosed laboratory facility.

Biohazard Biological substances that pose a threat to the health of living organisms, primarily that of humans. This can include medical waste or samples of a microorganism, virus, or toxin (from a biological source) that can affect human health. It can also include substances harmful to other animals.

Process Architecture in Biomanufacturing Facility Design, edited by Jeffery Odum and Michael C. Flickinger.

Biorisk Combination of the probability of occurrence of harm and the severity of that harm where the source of harm is a biological agent or toxin. The source of harm may be an unintentional exposure, accidental release or loss, theft, misuse, diversion, unauthorized access, or intentional unauthorized releases.

Biosafety Cabinet (BSC) An enclosed, ventilated laboratory workspace for safely working with materials contaminated with (or potentially contaminated with) pathogens requiring a defined biosafety level.

Biosafety in Microbiological and Biomedical Laboratories (BMBL) Guidance document in the United States that defines biosafety practice and policy.

Biosimilar A biologic medical product that is essentially clinically equivalent to an original product that is manufactured by a different company (designated as the innovator company).

Building Management System (BMS) A computer-based control system installed in buildings that controls and monitors the building's mechanical and electrical equipment such as ventilation, lighting, power systems, fire systems, and security systems.

Clean-in-Place (CIP) A method of cleaning the interior surfaces of pipes, vessels, process equipment, filters, and associated fittings, without disassembly.

Closed Process A process step or system that utilizes equipment design techniques to ensure that product is not exposed to the surrounding room environment.

Controlled Non-classified (CNC) A cGMP-manufacturing area designed to produce a consistent and controlled environment, but not necessarily monitored to a given environmental classification.

Compendial Monograph (pharmacopeial) tests that are standardized methods and specification testing for raw materials and finished products in the pharmaceutical industry.

Cost of Goods (COG) The direct costs attributable to the production of the goods sold by a company. This amount includes the cost of the materials used in creating the good along with the direct labor costs used to produce the good.

Critical Process Parameter (CPP) Key variables affecting the production process.

Critical Quality Attributes (CQA) Chemical, physical, biological, and microbiological attributes that can be defined, measured, and continually monitored to ensure final product outputs remain within acceptable quality limits.

Cytokines Any of a number of substances, such as interferon, interleukin, and growth factors, that are secreted by certain cells of the immune system and have an effect on other cells.

DNA A molecule that carries the genetic instructions used in the growth, development, functioning, and reproduction of all known living organisms and many viruses.

Efficacy The maximum response achievable from an applied or dosed agent.

Electronic Batch Records (EBR) A computer-based document collection system that demonstrates accountability by providing proof of proper handling of every step in the production of each batch of a drug product. EBR documents must comply with 21 CFR Part 211.

Endotoxin A toxin that is present inside a bacterial cell and is released when the cell disintegrates. It is sometimes responsible for the characteristic symptoms of a disease (e.g., in botulism), or is pyrogenic—inducing fever in mammals (e.g., a pyrogenic lipopolysaccharide, LPS).

Facility Management System (FMS) A professional management discipline focused upon the efficient and effective delivery of support services for the organizations that it serves. It encompasses multiple disciplines to ensure functionality of the built environment by integrating people, systems, place, process, and technology.

Formulation The putting together of components in appropriate relationships or structures, according to a formula. Usually refers to combining the API (drug or biologic substance) with stabilizing agents to generate a stable drug product.

Full-Time Employee (FTE) A person who works a minimum number of hours defined as such by his/her employer.

Genetically Modified Organisms (GMO) Any organism whose genetic material has been altered using genetic engineering techniques.

Good (anything) Practice (GxP) A general term for good (anything) practice quality guidelines and regulations.

Good Laboratory Practice (GLP) A quality system of management controls for research laboratories and organizations to ensure the uniformity, consistency, reliability, reproducibility, quality, and integrity of chemical (including pharmaceuticals) nonclinical safety tests, from physiochemical properties through acute to chronic toxicity tests.

Good Manufacturing Practice (GMP) The practices required in order to conform to the guidelines recommended by agencies that control authorization and licensing for manufacture and sale of food, drug products, and active pharmaceutical products.

Gowning The process of wearing special garments in order to control particulate contamination.

Hazard Source, situation, or act with a potential for causing harm (adapted from OHSAS 18001:2007).

Human–Machine–Interface (HMI) A software application that presents information to an operator or user about the state of a process, and to accept and implement the operator's control instructions. Typically information is displayed in a graphic format (graphical user interface or GUI).

HVAC Acronym for heating, ventilation, and air-conditioning. The system is used to provide heating and cooling services to buildings.

Injectables Drugs capable of being injected into the patient. Parenteral human medicines.

International Organization for Standardization (ISO) ISO is an independent, nongovernmental international organization with a membership of 163 national standards bodies. The focus is to develop global standards for various industry activities.

In Vitro Studies that are performed in the laboratory with microorganisms, cells, or biological molecules outside their normal biological context.

In Vivo Studies that are tested on whole, living organisms, usually animals, including humans, and plants as opposed to a partial or dead organism.

Ionic Strength Total ion concentration in a solution.

Isolators Equipment technology that provides a comprehensive separation of the processing environment from the external facility/room environment. Isolators are normally associated with sterile drug product manufacturing.

Leadership in Energy and Environmental Design (LEED) A rating system devised by the US Green Building Council (USGBC) to evaluate the environmental performance of a building and encourage market transformation toward sustainable design.

Lyophilization A controlled freezing and sublimation/dehydration process typically used to preserve a perishable material or make the material more convenient for transport.

Monoclonal Antibody (mAb) Antibodies that are made by identical immune cells that are clones of a unique parent cell.

Multiproduct Comprising, manufacturing, or selling several products.

National Institute of Health (NIH) The primary agency of the US government responsible for biomedical and health-related research.

Operational Flow Diagram (OFD) A picture of the separate steps of a process in sequential order.

Organizational Change Management (OCM) A framework for managing the effect of new business processes, changes in organizational structure, or cultural changes within an enterprise.

Personal Protection Equipment (PPE) Protective clothing, helmets, goggles, or other garments or equipment designed to protect the wearer's body from injury or infection.

Polarity Having two opposite tendencies or opposite electrical charges.

Polymers Substances that have a molecular structure consisting chiefly or entirely of a large number of similar units bonded together, for example, many synthetic organic materials used as plastics and resins.

Post Translational Modifications (PTM) Covalent and generally enzymatic modification of proteins during or after protein biosynthesis.
Potent Compounds Pharmacologically active ingredients or intermediates with biological activity at 150 μg/kg of body weight or below in adults.
Process Analytical Technology (PAT) A mechanism to design, analyze, and control pharmaceutical manufacturing processes through the measurement of critical process parameters (CPP) that affect critical quality attributes (CQA).
Quality Control (QC) A procedure or set of procedures intended to ensure that a manufactured product or performed service adheres to a defined set of quality criteria or meets the requirements of the client or customer.
Quality Target Product Profile (QTPP) A prospective summary of the quality characteristics of a drug product that ideally will be achieved to ensure the desired quality, taking into account safety and efficacy.
Radio Frequency Identification (RFID) A technology that incorporates the use of electromagnetic or electrostatic coupling in the radio frequency (RF) portion of the electromagnetic spectrum to uniquely identify an object, animal, or person.
Real-Time Release Testing (RTRT) The ability to quickly evaluate and ensure the quality of in-process and/or final product based on process data.
Research and Development (R&D) Work directed toward the innovation, introduction, and improvement of products and processes.
Restricted Access Barriers (RABs) Equipment technology that provides sub-isolation protection of materials/product in a laminar flow environment.
Ribonucleic Acid (RNA) A nucleic acid present in all living cells. Its principal role is to act as a messenger carrying instructions from DNA for controlling the synthesis of proteins. Although in some viruses RNA rather than DNA carries the genetic information.
Safety, Health, and Environmental (SHE) An umbrella term for the laws, rules, guidance, and processes designed to help protect employees, the public, and the environment from harm.
Single-Use (SU) Process Equipment Equipment/components used in the manufacture of drug substance and product that are disposed of after one use. These components are generally manufactured using chemical composite materials such as plastics.
Solubility The amount of a substance that will dissolve in a given amount of another substance.
Total Project Cost (TPC) All costs specific to a project incurred through startup of a facility, but prior to the operation of the facility.
Toxicity The degree to which a substance can damage an organism.
Typology Study of or analysis or classification based on types or categories.
Unidirectional Moving or operating in a single direction.

Unit Operations Basic step in a process. Unit operations involve a physical change or chemical transformation such as separation, crystallization, evaporation, filtration, polymerization, isomerization, and other reactions.

Vapor-Phase Hydrogen Peroxide (VHP) Vapor form of hydrogen peroxide (H_2O_2) with applications as a low-temperature antimicrobial vapor used to decontaminate enclosed and sealed areas such as laboratory workstations, isolation, and pass-through rooms.

Viable Particulates Airborne particles that can affect the sterility of pharmaceutical product and generally range from ~0.2 to ~30 μm in size.

Water for Injection (WFI) Water that is intended for use in the manufacture of medicines for parenteral administration whose solvent is water. WFI has strict requirements that limit acceptable levels of bioburden and LPS contaminants for patient safety. WFI systems must be carefully maintained in order to consistently provide high-quality water.

World Health Organization (WHO) A specialized agency of the United Nations that is concerned with international public health.

Index

Process Architecture in Biomanufacturing Facility Design, edited by Jeffery Odum and Michael C. Flickinger.

c

d

g

h

i

l

r

t

u

www.ingramcontent.com/pod-product-compliance
Lightning Source LLC
Jackson TN
JSHW011918310125
78069JS00001B/81

* 9 7 8 1 1 1 8 8 3 3 6 7 4 *